Analytical Network and System
Administration

Analytical Network and System Administration

Managing Human–Computer Networks

Mark Burgess

Oslo University College, Norway

John Wiley & Sons, Ltd

Other Wiley Editorial Offices

John Wiley & Sons Inc., 111 River Street, Hoboken, NJ 07030, USA

Jossey-Bass, 989 Market Street, San Francisco, CA 94103-1741, USA

Wiley-VCH Verlag GmbH, Boschstr. 12, D-69469 Weinheim, Germany

John Wiley & Sons Australia Ltd, 33 Park Road, Milton, Queensland 4064, Australia

John Wiley & Sons (Asia) Pte Ltd, 2 Clementi Loop #02-01, Jin Xing Distripark, Singapore 129809

John Wiley & Sons Canada Ltd, 22 Worcester Road, Etobicoke, Ontario, Canada M9W 1L1

Wiley also publishes its books in a variety of electronic formats. Some content that appears
in print may not be available in electronic books.

British Library Cataloguing in Publication Data

A catalogue record for this book is available from the British Library

ISBN 0-470-86100-2

Typeset in 10/12pt Times by Laserwords Private Limited, Chennai, India
Printed and bound in Great Britain by Antony Rowe Ltd, Chippenham, Wiltshire
This book is printed on acid-free paper responsibly manufactured from sustainable forestry
in which at least two trees are planted for each one used for paper production.

Contents

Foreword   xi

Preface   xiv

1   Introduction   1
1.1   What is system administration? 1
1.2   What is a system? 2
1.3   What is administration? 2
1.4   Studying systems 3
1.5   What's in a theory? 6
1.6   How to use the text 10
1.7   Some notation used 10

2   Science and its methods   13
2.1   The aim of science 13
2.2   Causality, superposition and dependency 16
2.3   Controversies and philosophies of science 17
2.4   Technology 20
2.5   Hypotheses 20
2.6   The science of technology 21
2.7   Evaluating a system—dependencies 22
2.8   Abuses of science 22

3   Experiment and observation   25
3.1   Data plots and time series 26
3.2   Constancy of environment during measurement 27
3.3   Experimental design 28
3.4   Stochastic (random) variables 29
3.5   Actual values or characteristic values 30
3.6   Observational errors 30
3.7   The mean and standard deviation 31
3.8   Probability distributions and measurement 32
     3.8.1   Scatter and jitter 35
     3.8.2   The 'normal' distribution 35

	3.8.3	Standard error of the mean	36
	3.8.4	Other distributions	37
3.9		Uncertainty in general formulae	38
3.10		Fourier analysis and periodic behaviour	39
3.11		Local averaging procedures	41
3.12		Reminder	43

4 Simple systems — **45**

4.1		The concept of a system	45
4.2		Data structures and processes	46
4.3		Representation of variables	47
4.4		The simplest dynamical systems	48
4.5		More complex systems	49
4.6		Freedoms and constraints	50
4.7		Symmetries	51
4.8		Algorithms, protocols and standard 'methods'	52
4.9		Currencies and value systems	53
	4.9.1	Energy and power	53
	4.9.2	Money	54
	4.9.3	Social currency and the notion of responsibility	54
4.10		Open and closed systems: the environment	56
4.11		Reliable and unreliable systems	58

5 Sets, states and logic — **59**

5.1		Sets	59
5.2		A system as a set of sets	61
5.3		Addresses and mappings	61
5.4		Chains and states	62
5.5		Configurations and macrostates	64
5.6		Continuum approximation	65
5.7		Theory of computation and machine language	65
	5.7.1	Automata or State Machines	66
	5.7.2	Operators and operands	68
	5.7.3	Pattern matching and operational grammars	69
	5.7.4	Pathway analysis and distributed algorithms	70
5.8		A policy-defined state	71

6 Diagrammatical representations — **73**

6.1	Diagrams as systems	73
6.2	The concept of a graph	74
6.3	Connectivity	77
6.4	Centrality: maxima and minima in graphs	77
6.5	Ranking in directed graphs	80
6.6	Applied diagrammatical methods	84

7 System variables   **91**
    7.1   Information systems 91
    7.2   Addresses, labels, keys and other resource locators 92
    7.3   Continuous relationships 94
    7.4   Digital comparison 94

8 Change in systems   **97**
    8.1   Renditions of change 97
    8.2   Determinism and predictability 98
    8.3   Oscillations and fluctuations 99
    8.4   Rate of change 102
    8.5   Applications of the continuum approximation 103
    8.6   Uncertainty in the continuum approximation 105

9 Information   **109**
    9.1   What is information? 109
    9.2   Transmission 110
    9.3   Information and control 111
    9.4   Classification and resolution 111
    9.5   Statistical uncertainty and entropy 114
    9.6   Properties of the entropy 118
    9.7   Uncertainty in communication 119
    9.8   A geometrical interpretation of information 123
    9.9   Compressibility and size of information 127
    9.10   Information and state 128
    9.11   Maximum entropy principle 129
    9.12   Fluctuation spectra 133

10 Stability   **135**
    10.1   Basic notions 135
    10.2   Types of stability 135
    10.3   Constancy 136
    10.4   Convergence of behaviour 137
    10.5   Maxima and minima 138
    10.6   Regions of stability in a graph 139
    10.7   Graph stability under random node removal 141
    10.8   Dynamical equilibria: compromise 142
    10.9   Statistical stability 143
    10.10 Scaling stability 145
    10.11 Maximum entropy distributions 148
    10.12 Eigenstates 148
    10.13 Fixed points of maps 151
    10.14 Metastable alternatives and adaptability 155
    10.15 Final remarks 156

11 Resource networks   159
11.1   What is a system resource? 159
11.2   Representation of resources 160
11.3   Resource currency relationships 161
11.4   Resource allocation, consumption and conservation 162
11.5   Where to attach resources? 163
11.6   Access to resources 165
11.7   Methods of resource allocation 167
     11.7.1   Logical regions of systems 167
     11.7.2   Using centrality to identify resource bottlenecks 168
11.8   Directed resources: flow asymmetries 170

12 Task management and services   173
12.1   Task list scheduling 173
12.2   Deterministic and non-deterministic schedules 174
12.3   Human–computer scheduling 176
12.4   Service provision and policy 176
12.5   Queue processing 177
12.6   Models 178
12.7   The prototype queue M/M/1 179
12.8   Queue relationships or basic 'laws' 181
12.9   Expediting tasks with multiple servers M/M/k 186
12.10   Maximum entropy input events in periodic systems 188
12.11   Miscellaneous issues in scheduling 189

13 System architectures   191
13.1   Policy for organization 191
13.2   Informative and procedural flows 192
13.3   Structured systems and *ad hoc* systems 193
13.4   Dependence policy 193
13.5   System design strategy 195
13.6   Event-driven systems and functional systems 200
13.7   The organization of human resources 201
13.8   Principle of minimal dependency 202
13.9   Decision-making within a system 202
     13.9.1   Layered systems: Managers and workers 202
     13.9.2   Efficiency 203
13.10   Prediction, verification and their limitations 204
13.11   Graphical methods 205

14 System normalization   207
14.1   Dependency 207
14.2   The database model 209
14.3   Normalized forms 210

15 System integrity   215
15.1   System administration as communication? 215
15.2   Extensive or strategic instruction 219
15.3   Stochastic semi-groups and martingales 223
15.4   Characterizing probable or average error 224
15.5   Correcting errors of propagation 226
15.6   Gaussian continuum approximation formula 228

16 Policy and maintenance   231
16.1   What is maintenance? 231
16.2   Average changes in configuration 231
16.3   The reason for random fluctuations 234
16.4   Huge fluctuations 235
16.5   Equivalent configurations and policy 236
16.6   Policy 237
16.7   Convergent maintenance 237
16.8   The maintenance theorem 240
16.9   Theory of back-up and error correction 241

17 Knowledge, learning and training   249
17.1   Information and knowledge 250
17.2   Knowledge as classification 250
17.3   Bayes' theorem 252
17.4   Belief versus truth 254
17.5   Decisions based on expert knowledge 255
17.6   Knowledge out of date 259
17.7   Convergence of the learning process 260

18 Policy transgressions and fault modelling   263
18.1   Faults and failures 263
18.2   Deterministic system approximation 265
18.3   Stochastic system models 269
18.4   Approximate information flow reliability 273
18.5   Fault correction by monitoring and instruction 275
18.6   Policy maintenance architectures 279
18.7   Diagnostic cause trees 286
18.8   Probabilistic fault trees 290
       18.8.1 Faults 290
       18.8.2 Conditions and set logic 291
       18.8.3 Construction 293

19 Decision and strategy   295
19.1   Causal analysis 295
19.2   Decision-making 296
19.3   Game theory 297

19.4   The strategic form of a game 301
19.5   The extensive form of a game 302
19.6   Solving zero-sum games 303
19.7   Dominated strategies 304
19.8   Nash equilibria 305
19.9   A security game 309
         19.9.1 Zero-sum approximation 310
         19.9.2 Non-zero sum approximation 313
19.10   The garbage collection game 315
19.11   A social engineering game 321
19.12   Human elements of policy decision 328
19.13   Coda: extensive versus strategic configuration management 328

20 Conclusions   **331**

A Some Boolean formulae   **335**
A.1   Conditional probability 335
A.2   Boolean algebra and logic 336

B Statistical and scaling properties of time-series data   **339**
B.1   Local averaging procedure 339
B.2   Scaling and self-similarity 343
B.3   Scaling of continuous functions 344

C Percolation conditions   **347**
C.1   Random graph condition 347
C.2   Bi-partite form 350
C.3   Small-graph corrections 351

Bibliography   **353**

Index   **359**

Foreword

It is my great honor to introduce a landmark book in the field of network and system administration. For the first time, in one place, one can study the components of network and system administration as an evolving and emerging discipline and science, rather than as a set of recipes, practices or principles. This book represents the step from 'mastery of the practice' and 'scientific understanding', a step very similar to that between historical alchemy and chemistry.

As recently as ten years ago, many people considered 'network and system administration' to comprise remembering and following complex recipes for building and maintaining systems and networks. The complexity of many of these recipes—and the difficulty of explaining them to non-practitioners in simple and understandable terms—encouraged practitioners to treat system administration as an 'art' or 'guild craft' into which practitioners are initiated through apprenticeship.

Current master practitioners of network and system administration are perhaps best compared with historical master alchemists at the dawn of chemistry as a science. In contrast to the distorted popular image of alchemy as seeking riches through transmutation of base metals, historical research portrays alchemists as master practitioners of the subtle art of combining chemicals towards particular results or ends. Practitioners of alchemy often possessed both precise technique and highly developed observational skills. Likewise, current master practitioners of network and system administration craft highly reliable networks from a mix of precise practice, observational skills and the intuition that comes from careful observation of network behaviour over long time periods. But both alchemists and master practitioners lack the common language that makes it easy to exchange valuable information with others: the language of science.

Alas, the alchemy by which we have so far managed our networks is no longer sufficient. When networks were simple in structure, it was possible to maintain them through the use of relatively straightforward recipes, procedures and practices. In the post-Internet world, the administrator is now faced with managing and controlling networks that can dynamically adapt to changing conditions and requirements quickly and, perhaps, even unpredictably. These adaptive networks can exhibit 'emergent properties' that are not predictable in advance. In concert with adapting networks to serve human needs, future administrators must adapt themselves to the task of management by developing an ongoing, perpetually evolving, and shared understanding.

In the past, it was reasonable to consider a computer network as a collection of cooperating machines functioning in isolation. Adaptive networks cannot be analysed in this fashion; their human components must also be considered. Modern networks are not communities of machines, but rather communities of humans inextricably linked by machines; what the author calls 'cooperating ecologies' of users and machines. The behaviour of humans must be considered along with the behaviour of the network for making conclusions about network performance and suitability.

These pressures force me to an inescapable conclusion. System administrators cannot continue to be alchemist-practitioners. They must instead develop the language of science and evolve from members of a profession to researchers within a shared scientific discipline. This book shows the way.

Though we live thousands of miles apart, the author and I are 'kindred spirits'—forged by many of the same experiences, challenges and insights. In the late 1980s and early 1990s, both of us were faculty, managing our own computer networks for teaching and research. Neither of us had access to the contemporary guilds of system administration (or each other), and had to learn how to administer networks the hard way—by reading the documentation and creating our own recipes for success. Both of us realized (completely independently) that there were simple concepts behind the recipes that, once discovered, make the recipes easy to remember, reconstruct and understand. Concurrently and independently, both of us set out to create software tools that would avoid repeated manual configuration.

Although we were trained in radically differing academic traditions (the author from physics and myself from mathematics and computer science), our administrative tools, developed completely in isolation from one another, had very similar capabilities and even accomplished tasks using the same methods. The most striking similarity was that both tools were based upon the same 'principles'. For the first time, it very much looked like we had found an invariant principle in the art of system and network administration: the 'principle of convergence'. As people would say in the North Carolina backwoods near where I grew up, 'if it ain't broke, don't fix it'.

The road from alchemy to discipline has many steps. In the author's previous book, *Principles of Network and System Administration*, he takes the first step from practice ('what to do') to principles ('why to do it'). Recipes are not created equal; some are better than others. Many times the difference between good and poor recipes can be expressed in terms of easily understood principles. Good recipes can then be constructed top–down, starting at the principles. Practitioners have approached the same problem bottom-up, working to turn their tested and proven recipes into sets of 'best practices' that are guaranteed to work well for a particular site or application. Recently, many practitioners have begun to outline the 'principles' underlying their practices. There is remarkable similarity between the results of these two seemingly opposing processes, and the author's 'principles', and the practitioners' 'best practices' are now quickly meeting on a common middle ground of principles.

In this book, for the first time, the author identifies principles of scientific practice and observation that anyone can use to become proficient 'analysts' of network and system administration practices. This will not make one a better practitioner, but rather will allow one to discuss and evaluate the practice with others in a clear and concise manner. The reader will not find any recipes in this book. The reader will not find principles of practice. Rather, the book explains the principles behind the science and chemistry of cooking, so that one can efficiently derive one's own efficient and effective recipes for future networks. Proficient system administrators have always been capable of this kind of alchemy, but have found it challenging to teach the skill to others. This book unlocks the full power of the scientific method to allow sharing of analyses, so that future administrators can look beyond recipe, to shared understanding and discipline. In this way, now-isolated practitioners can form a shared scientific community and discipline whose knowledge is greater than the sum of its parts.

Looking at the table of contents, one will be very surprised to note that the traditional disciplines of 'computer science' and 'computer engineering'—long considered the inseparable partners of system administration—are not the basis of the new science. Rather, experimental physics has proven to be the Rosetta Stone that unlocks the mysteries of complex systems. To understand why, we must examine the fundamental differences in economics between the disciplines of computer science and engineering and the disciplines of network and system administration.

Traditional computer science and engineering (and, particularly, the sciences involved in building the systems that system administrators manage) are based upon either an operational or axiomatic semantic model of computing. Both models express 'what a program does' in an ideal computing environment. Software developers build complex systems in layers, where each subsequent layer presumes the correct function of layers upon which it is built. Program correctness at a given layer is a mathematical property based upon axioms that describe the behaviour of underlying layers. Fully understanding a very complex system requires understanding of each layer and its interdependencies and assumptions in dealing with other layers.

System administrators have a differing view of the systems they manage compared to that of the developers who designed the systems. It is not economically feasible to teach the deep knowledge and mathematical understanding necessary to craft and debug software and systems to large populations of human system administrators. System administrators must instead base their actions upon a high-level set of initial experimental hypotheses called the 'system documentation'. The documentation consists of hypotheses to be tested, not axioms to be trusted. As administrators learn how to manage a system, they refine their understanding top-down, by direct observation and ongoing evaluation of hypotheses.

Turning system and network administration into a discipline requires one to learn some skills, previously considered far removed from the practice. Evaluating hypotheses requires a rudimentary knowledge of statistics and the experimental method. These hypotheses are built not upon operational or axiomatic semantic models of computing, but upon specialized high-level mathematical models that describe behaviour of a complex system. With this machinery in hand, several advanced methods of analysis—prevalent in experimental physics and other scientific disciplines—are applied to the problem of understanding management of complex systems.

Proficient system administrators are already skilled experimental scientists; they just do not acknowledge this fact and cannot effectively communicate their findings to others. This book takes a major step towards understanding the profession of system and network administration as a science rather than as an art. While this step is difficult to take, it is both rewarding and necessary for those pioneers who will manage the next generation of networks and services. Please read on, and seek to understand the true nature of networking—as a process that involves connecting humans, not just computers.

Alva Couch
Tufts University, USA

Preface

This is a research document and a textbook for graduate students and researchers in the field of networking and system administration. It offers a theoretical perspective on human–computer systems and their administration. The book assumes a basic competence in mathematical methods, common to undergraduate courses. Readers looking for a less theoretical introduction to the subject may wish to consult (Burgess (2000b)).

I have striven to write a short book, treating topics briefly rather than succumbing to the temptation to write an encyclopædia that few will read or be able to lift. I have not attempted to survey the literature or provide any historical context to the development of these ideas (see Anderson et al. (2001)). I hope this makes the book accessible to the intelligent lay reader who does not possess an extensive literacy in the field and would be confused by such distractions. The more advanced reader should find sufficient threads to follow to add depth to the material. In my experience, too much attention to detail merely results in one forgetting why one is studying something at all. In this case, we are trying to formulate a descriptive language for systems.

A theoretical synthesis of system administration plays two roles: it provides a descriptive framework for systems that should be available to other areas of computer science and proffers an analytical framework for dealing with the complexities of interacting components. The field of system administration meets an unusual challenge in computer science: that of approximation. Modern computing systems are too complicated to be understood in exact terms.

In the flagship theory of physics, quantum electrodynamics, one builds everything out of two simple principles:

1. Different things can exist at different places and times.

2. For every effect, there must be a cause.

The beauty of this construction is its lack of assumptions and the richness of the results. In this text, I have tried to synthesize something like this for human–computer systems. In order to finish the book, and keep it short and readable, I have had to compromise on many things. I hope that the result nevertheless contributes in some way to a broader scientific understanding of the field and will inspire students to further serious study of this important subject.

Some of this work is based on research performed with my collaborators Geoff Canright, Frode Sandnes and Trond Reitan. I have benefited greatly from discussions with them and others. I am especially grateful for the interest and support of other researchers, most notably Alva Couch for understanding my own contributions when no one else did. Finally, I would like to thank several for reading the draft versions of the manuscript and commenting: Paul Anderson, Lars Kristiansen, Tore Jonassen, Anil Somayaji and Jan Bergstra.

Mark Burgess

1

Introduction

Technology: the science of the mechanical and industrial arts.
[Gk. *tekhne* art and *logos* speech].

—Odhams dictionary of the English language

1.1 What is system administration?

System administration is about the design, running and maintenance of human–computer systems. Human–computer systems are 'communities' of people and machines that collaborate actively to execute a common task. Examples of human–computer systems include business enterprises, service institutions and any extensive machinery that is operated by, or interacts with human beings. The human players in a human–computer system are often called the *users* and the machines are referred to as *hosts*, but this suggests an asymmetry of roles, which is not always the case.

System administration is primarily about the technological side of a system: the architecture, construction and optimization of the collaborating parts, but it also occasionally touches on softer factors such as user assistance (help desks), ethical considerations in deploying a system, and the larger implications of its design for others who come into contact with it. System administration deals first and foremost with the system as a whole, treating the individual components as black boxes, to be opened only when it is possible or practical to do so. It does not conventionally consider the design of user-tools such as third-party computer programs, nor does it attempt to design enhancements to the available software, though it does often discuss meta tools and improvised software systems that can be used to monitor, adjust or even govern the system. This omission is mainly because user-software is acquired beyond the control of a system administrator; it is written by third parties, and is not open to local modification. Thus, users' tools and software are treated as 'given quantities' or 'boundary conditions'.

Analytical Network and System Administration. Managing Human–Computer Networks Mark Burgess
© 2004 John Wiley & Sons, Ltd ISBN 0-470-86100-2

For historical reasons, the study of system administration has fallen into two camps: those who speak of *network management* and discuss its problems in terms of software design for the management of black box devices by humans (e.g. using SNMP), and those who speak of *system administration* and concern themselves with practical strategies of machine and software configuration at all levels, including automation, human–computer issues and ethical considerations. These two viewpoints are complementary, but too often ignore one another. This book considers human–computer systems in general, and refers to specific technologies only by example. It is therefore as much about purely human administrative systems as it is about computers.

1.2   What is a system?

A system is most often an organized effort to fulfil a goal, or at least carry out some predictable behaviour. The concept is of the broadest possible generality. A system could be a mechanical device, a computer, an office of workers, a network of humans and machines, a series of forms and procedures (a bureaucracy) etc. Systems involve themes, such as *collaboration* and *communication* between different actors, the use of *structure* to represent information or to promote efficiency, and the laws of *cause and effect*. Within any mechanism, *specialization* of the parts is required to build significant innovation; it is only through strategy of divide and conquer that significant problems can be solved. This implies that each division requires a special solution.

A computer system is usually understood to mean a system composed primarily of computers, using computers or supporting computers. A human–computer system includes the role of humans, such as in a business enterprise where computers are widely used. The principles and theories concerning systems come from a wide range of fields of study. They are synthesized here in a form and language that is suitable for scholars of science and engineering.

1.3   What is administration?

The word *administration* covers a variety of meanings in common parlance. The American Administration is the government of the United States, that is, a political leadership. A university administration is a bureaucracy and economic resource department that works on behalf of a board of governors to implement the university's policy and to manage its resources. The administrative department of a company is generally the part that handles economic procedures and payment transactions. In human–computer system administration, the definition is broadened to include all of the organizational aspects and also engineering issues, such as system fault diagnosis. In this regard, it is like the medical profession, which combines checking, management and repair of bodily functions. The main issues are the following:

- System design and rationalization

- Resource management

- Fault finding.

In order to achieve these goals, it requires

- Procedure

- Team work

- Ethical practices

- Appreciation of security.

Administration comprises two aspects: *technical solutions* and *arbitrary policies*. A technical solution is required to achieve goals and sub-goals, so that a problem can be broken down into manageable pieces. Policy is required to make the system, as far as possible, *predictable*: it pre-decides the answers to questions on issues that cannot be derived from within the system itself. Policy is therefore an arbitrary choice, perhaps guided by a goal or a principle.

The arbitrary aspect of policy cannot be disregarded from the administration of a system, since it sets the boundary conditions under which the system will operate, and supplies answers to questions that cannot be determined purely on the grounds of efficiency. This is especially important where humans are involved: human welfare, permissions, responsibilities and ethical issues are all parts of policy. Modelling these intangible qualities formally presents some challenges and requires the creative use of abstraction.

The administration of a system is an administration of temporal and resource development. The administration of a network of localized systems (a so-called *distributed system*) contains all of the above, and, additionally, the administration of the location of and communication between the system's parts. Administration is thus a flow of activity, information about resources, policy making, record keeping, diagnosis and repair.

1.4 Studying systems

There are many issues to be studied in system administration. Some issues are of a technical nature, while others are of a human nature. System administration confronts the human–machine interaction as few other branches of computer science do. Here are some examples:

- *System design* (e.g. how to get humans and machines to do a particular job as efficiently as possible. What works? What does not work? How does one know?)

- *Reliability studies* (e.g. failure rate of hardware/software, evaluation of policies and strategies)

- *Determining and evaluating methods for ensuring system integrity* (e.g. automation, cooperation between humans, formalization of policy, contingency planning etc.)

- *Observations that reveal aspects of system behaviour that are difficult to predict* (e.g. strange phenomena, periodic cycles)

- *Issues of strategy and planning.*

Usually, system administrators do not decide the purpose of a system; they are regarded as supporting personnel. As we shall see, this view is, however, somewhat flawed from the viewpoint of system design. It does not always make sense to separate the human and computer components in a system; as we move farther into the information age, the fates of both become more deeply intertwined.

To date, little theory has been applied to the problems of system administration. In a subject that is complex, like system administration, it is easy to fall back on *qualitative* claims. This is dangerous, however, since one is easily fooled by qualitative descriptions. Analysis proceeds as a dialogue between theory and experiment. We need theory to interpret results of observations and we need observations to back up theory. Any conclusions must be a consistent mixture of the two. At the same time, one must not believe that it is sensible to demand hard-nosed Popper-like falsification of claims in such a complex environment. Any numbers that we can measure, and any models we can make must be considered valuable, provided they actually have a sensible interpretation.

Human–computer interaction

The established field of human–computer interaction (HCI) has grown, in computer science, around the need for reliable interfaces in critical software scenarios (see for instance Sheridan (1996); Zadeh (1973)). For example, in the military, real danger could come of an ill-designed user interface on a nuclear submarine; or in a power plant, a poorly designed system could set off an explosion or result in blackouts.

One can extend the notion of the HCI to think less as a programmer and more as a physicist. The task of physics is to understand and describe what happens when different parts of nature interact. The interaction between fickle humans and rigid machinery leads to many unexpected phenomena, some of which might be predicted by a more detailed functional understanding of this interaction. This does not merely involve human attitudes and habits; it is a problem of systemic complexity—something that physics has its own methods to describe. Many of the problems surrounding computer security enter into the equation through the HCI. Of all the parts of a system, humans bend most easily: they are often both the weakest link and the most adaptable tools in a solution, but there is more to the HCI than psychology and button pushing. The issue reaches out to the very principles of science: what are the relevant timescales for the interactions and for the effects to manifest? What are the sources of predictability and unpredictability? Where is the system immune to this interaction, and where is the interaction very strong? These are not questions that a computer science analysis alone can answer; there are physics questions behind these issues. Thus, in reading this book, you should not be misled into thinking that physics is merely about electrons, heat and motion: it is a broad methodology for 'understanding phenomena', no matter where they occur, or how they are described. What computer science lacks from its attachment to technology, it must regain by appealing to the physics of systems.

Policy

The idea policy plays a central role in the administration of systems, whether they are dominated by human or technological concerns.

Definition 1 (Policy—heuristic) *A policy is a description of what is intended and desirable about a system. It includes a set of ad hoc choices, goals, compromises, schedules, definitions and limitations about the system. Where humans are involved, compromises often include psychological considerations, and welfare issues.*

A policy provides a frame of reference in which a system is understood to operate. It injects a relativistic aspect into the science of systems: we cannot expect to find absolute answers, when different systems play by different rules and have different expectations. A theory of systems must therefore take into account policy as a basic axiom. Much effort is expended in the chapters that follow to find a tenable definition of policy.

Stability and instability

It is in the nature of almost all systems to change with time. The human and machine parts of a system change, both in response to one another, and in response to a larger environment. The system is usually a predictable, known quantity; the environment is, by definition, an unknown quantity. Such changes tend to move the system in one or two directions: either the system falls into disarray or it stagnates. The meaning of these provocative terms is different for the human and the machine parts:

- Systems will fall into a stable repetition of behaviour (a limit cycle) or reach some equilibrium at which point further change cannot occur without external intervention.

- Systems will eventually invalidate their assumptions and fail to fulfil their purpose.

Ideally, a machine will perform, repetitively, the same job over and over again, because that is the function of mechanisms: stagnation is good for machines. For humans, on the other hand, this is usually regarded as a bad thing, since humans are valued for their creativity and adaptability. For a system mechanism to fall into disarray is a bad thing.

The relationship between a system and its environment is often crucial in determining which of the above is the case. The inclusion of human behaviour in systems must be modelled carefully, since humans are not deterministic in the same way that machines (automata) can be. Humans must therefore be considered as being part system and part environment. Finally, policy itself must be our guide as to what is desirable change.

Security

Security is a property of systems that has come to the forefront of our attention in recent times. How shall we include it in a theory of system administration?

Definition 2 (Security) *Security concerns the possible ways in which a system's integrity might be compromised, causing it to fail in its intended purpose. In other words, a breach of security is a failure of a system to meet its specifications.*

Security refers to 'intended purpose', so it is immediately clear that it relates directly to *policy* and that it is a property of the entire system in general. Note also that, while we associate security with 'attacks' or 'criminal activity', natural disasters or other occurrences are equally to be blamed for the external perturbations that break systems.

A loss of integrity can come from a variety of sources, for example, an internal fault, an accident or a malicious attack on the system. Security is a property that requires the analysis of assumptions that underpin the system, since it is these areas that one tends to disregard and that can be exploited by attackers, or fail for diverse reasons. The system depends on its components in order to function. Security is thus about an analysis of *dependencies*. We can sum this up in a second definition:

Definition 3 (Secure system) *A secure system is one in which every possible threat has been analysed and where all the risks have been assessed and accepted as a matter of policy.*

1.5   What's in a theory?

This book is not a finished theory, like the theory of relativity, or the theory of genetic replication. It is not the end of a story, but a beginning. System administration is at the start of its scientific journey, not at its end.

Dramatis personae

The players in system administration are the following:

- The computer

- The network

- The user

- The policy

- The system administrator.

We seek a clear and flexible language (rooted in mathematics) in which to write their script. It will deal with basic themes of

- time (when events occur or should occur),

- location (where resources should be located),

- value (how much the parts of a system contribute or are worth),

- randomness and predictability (our ability to control or specify).

It must answer questions that are of interest to the management of systems. We can use two strategies:

- Type I (pure science) models that describe the behaviour of a system without attempting to interpret its value or usefulness. These are 'vignettes' that describe what we can observe and explain in impartial terms. They provide a basic understanding of phenomena that leads to expertise about the system.

- Type II (applied science) models add interpretations of value and correctness (policy) to the description. They help us in making decisions by impressing a rational framework on the subjectivities of policy.

A snapshot of reality

The system administrator rises and heads for the computer, grabs coffee or cola and proceeds to catch up on e-mail. There are questions, bug reports, automatic replies from scripted programs, spam and lengthy discussions from mailing lists.

The day proceeds to planning, fault finding, installing software, modifying system parameters to implement (often ad hoc) policy that enables the system to solve a problem for a user, or which makes the running smoother (more predictable)—see fig. 1.1. On top of all of this, the administrator must be thinking about what users are doing. After all, they are the ones who need the system and the ones who most often break it. How does 'the system' cope with them and their activities as they feed off it and feed back on it? They are, in every sense, a part of the system. How can their habits and skills be changed to make it all work more smoothly? This will require an appreciation of the social interactions of the system and how they, in turn, affect the structures of the logical networks and demands placed on the machines.

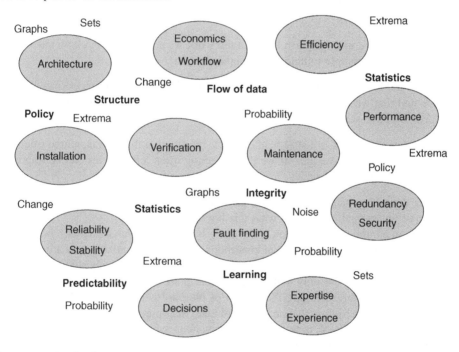

Figure 1.1: The floating islands of system administration move around on a daily basis and touch each other in different ways. In what framework shall we place these? How can we break them down into simpler problems that can be 'solved'? In `courier` font, we find some primitive concepts that help to describe the broader ideas. These will be our starting points.

There are decisions to be made, but many of them seem too uncertain to be able to make a reliable judgement on the available evidence. Experimentation is required, and searching for advice from others. Unfortunately, you never know how reliable others' opinions and assertions will be. It would be cool if there were a method for turning the creative energy into the optimal answer. There is ample opportunity and a wealth of tools to collect information, but how should that information be organized and interpreted? What is lacking is not software, but theoretical tools.

What view or philosophy could unify the different facets of system administration: design, economics, efficiency, verification, fault-finding, maintenance, security and so on? Each of these issues is based on something more primitive or fundamental. Our task is therefore to use the power of abstraction to break down the familiar problems into simpler units that we can master and then reassemble into an approximation of reality. There is no unique point of view here (see next chapter).

Theory might lead to better tools and also to better procedures. If it is to be of any use, it must have predictive power as well as descriptive power. We have to end up with formulae and procedures that make criticism and re-evaluation easier and more effective. We must be able to summarize simple 'laws' about system management (thumb-rules) that are not based only on vague experience, but have a theoretical explanation based on reasonable cause and effect.

How could such a thing be done? For instance, how might we measure how much work will be involved in a task?

- We would have to distinguish between the work we actually do and how much work is needed in principle (efficiency and optimization).

- We would look for a mathematical idea with the characteristics or properties of work. We find that we can map work into the idea of 'information' content in some cases (now we have something concrete to study).

- Information or work is a statistical concept: information that is transmitted often can be compressed on average—if we do something often, efficiencies can be improved through economies of scale.

By starting down the road of analysis, we gain many small insights that can be assembled into a deeper understanding. That is what this book attempts to do.

The system administrator wonders if he or she will ever become redundant, but there is no sign of that happening. The external conditions and requirements of users are changing too quickly for a system to adapt automatically, and policy has to be adjusted to new goals and crises. Humans are the only technology on the planet that can address that problem for the foreseeable future. Besides, the pursuit of pleasure is a human condition, and part of the enjoyment of the job is that creative and analytical pursuit.

The purpose of this book is to offer a framework in which to analyse and understand the phenomenon of human–computer management. It is only with the help of theoretical models that we can truly obtain a deeper understanding of system behaviour.

Studies

The forthcoming chapters describe a variety of languages for discussing systems, and present some methods and issues that are the basis of the author's own work. Analysis is the scientific method in action, so this book is about analysis. It has many themes:

1. *Observe*—we must establish a factual basis for discussing systems.

2. *Deduce cause*—we establish probable causes of observed phenomena.

3. *Establish goals*—what do we want from this information?

4. *Diagnose 'faults'*—what is a fault? It implies a value judgement, based on policy.

5. *Correct faults*—devise and apply strategies.

Again, these concepts are intimately connected with 'policy', that is, a specification of right and wrong. In some sense, we need to know the 'distance' between what we would like to see and what we actually see.

This is all very abstract. In the day-to-day running of systems, few administrators think in such generalized, abstract terms—yet this is what this book asks you to do.

Example 1 (A backup method) *A basic duty of system administrators is to perform a backup of data and procedures: to ensure the integrity of the system under natural or unnatural threats. How shall we abstract this and turn it into a scientific enquiry?*

We might begin by examining how data can be copied from one place to another. This adds a chain of questions: (i) how can the copying be made efficient? (ii) what does efficient mean? (iii) how often do the data change, and in what way? What is the best strategy for making a copy: immediately after every change, once per day, once per hour? We can introduce a model for the change, for example, a mass of data that is more or less constant, with small random fluctuating changes to some files, driven by random user activity. This gives us something to test against reality. Now we need to know how users behave, and what they are likely to do. We then ask: what do these fluctuations look like over time? Can they be characterized, so that we can tune a copying algorithm to fit them? What is the best strategy for copying the files?

The chain of questions never stops: analysis is a process, not an answer.

Example 2 (Resource management) *Planning a system's resources, and deploying them so that the system functions optimally is another task for a system administrator. How can we measure, or even discuss, the operation of a system to see how it is operating? Can important (centrally important) places be identified in the system, where extra resources are needed, or the system might be vulnerable to failure? How shall we model demand and load? Is the arrival of load (traffic) predictable or stochastic? How does this affect our ability to handle it? If one part of the system depends on another, what does this mean for the efficiency or reliability? How do we even start asking these questions analytically?*

Example 3 (Pattern detection) *Patterns of activity manifest themselves over time in systems. How do we measure the change, and what is the uncertainty in our measurement? What are their causes? How can they be described and modelled? If a system changes its pattern of behaviour, what does this mean? Is it a fault or a feature?*

In computer security, intrusion detection systems often make use of this kind of idea, but how can the idea be described, quantified and generalized, hence evaluated?

Example 4 (Configuration management) *The initial construction and implementation of a system, in terms of its basic building blocks, is referred to as its configuration. It is a measure of the system's state or condition. How should we measure this state? Is it a fixed pattern, or a statistical phenomenon? How quickly should it change? What might cause it to change unexpectedly? How big a change can occur before the system is damaged? Is it possible to guarantee that every configuration will be stable, perform its intended function, and be implementable according to the constraints of a policy?*

In each of the examples above, an apparently straightforward issue generates a stream of questions that we would like to answer. Asking these questions is what science is about; answering them involves the language of mathematics and logic in concert with a scientific inquiry: science is about extracting the essential features from complex observable phenomena and modelling them in order to make predictions. It is based on observation and approximate verification. There is no 'exact science' as we sometimes hear about in connection with physics or chemistry; it is always about suitably idealized approximations to the truth, or 'uncertainty management'. Mathematics, on the other hand, is not to be confused with science—it is about rewriting assumptions in different ways: that is, if one begins with a statement that is assumed true (an axiom) and manipulates it according to the rules of mathematics, the resulting statement is also true by the same axiom. It contains no more information than the assumptions on which it rests. Clearly, mathematics is an important language for expressing science.

1.6 How to use the text

Readers should not expect to understand or appreciate everything in this book in the short term. Many subtle and deep-lying connections are sewn in these pages that will take even the most experienced reader some time to unravel. It is my hope that there are issues sketched out here that will provide fodder for research for at least a decade, probably several. Many ideas about the administration of systems are general and have been discussed many times in different contexts, but not in the manner or context of system administration.

The text can be read in several ways. To gain a software-engineering perspective, one can replace 'the system' with 'the software'. To gain a business management perspective, replace 'the system' with 'the business', or 'the organization'. For human–computer administration, read 'the system' as 'the network of computers and its users'.

The first part of the book is about observing and recording observations about systems, since we aim to take a scientific approach to systems. Part 2 concerns abstracting and naming the concepts of a system's operation and administration in order to place them into a formal framework. In the final part of the book, we discuss the physics of information systems, that is, the problem of how to model the time-development of all the resources in order to determine the effect of policy. This reflects the cycle of development of a system:

- Observation

- Design (change)

- Analysis.

1.7 Some notation used

A few generic symbols and notations are used frequently in this book and might be unfamiliar.

The function $q(t)$ is always used to represent a 'signal' or quality that is varying in the system, that is, a scalar function describing any value that changes in time. I have found it more useful to call all such quantities by the same symbol, since they all have the same status.

$q(x, t)$ is a function of time and a label x that normally represents a spatial position, such as a memory location. In structured memory, composed of multiple objects with finite size, the addresses are multi-dimensional and we write $q(\vec{x}, t)$, where $\vec{x} = (x_1, \ldots, x_\ell)$ is an ℓ-dimensional vector that specifies location within a structured system, for example, (6,3,8) meaning perhaps bit 6 of component 3 in object 8.

In describing averages, the notation $\langle \ldots \rangle$ is used for mean and expectation values, for example, $\langle X \rangle$ would mean an average over values of X. In statistics literature, this is often written as $E(X)$.

In a local averaging procedure, a large set X is reduced to a smaller set x of compounded objects; thus, it does not result in a scalar value but a smaller set whose elements are identified by a new label. For example, suppose we start with a set of 10 values, X. We could find the mean of all values $\langle X \rangle_{10}$ giving a single value. Group them into five groups of two. Now we average each pair and end up with five averaged values: $\langle X(x) \rangle_2$. This still has a label x, since it is a set of values, where $x = 1 \ldots 5$.

Applications and Further Study 1

- *Use these broad topics as a set of themes for categorizing the detailed treatments in forthcoming chapters.*

2

Science and its methods

Science is culture,
Technology is art.

—Author's slogan.

A central theme of this book is the application of scientific methodologies to the design, understanding and maintenance of human–computer systems. Ironically, 'Computer Science' has often lacked classical scientific thinking in favour of reasoned assertion, since it has primarily been an agent for technology and mathematics. The art of observation has concerned mainly those who work with performance analysis.

While mathematics is about reasoning (it seeks to determine logical relationships between assumed truths), the main purpose of science is to interpret the world as we see it, by looking for suitably idealized descriptions of observed phenomena and quantifying their uncertainty. Science is best expressed with mathematics, but the two are independent. There are many philosophies about the meaning of science, but in this book we shall be pragmatical rather than encyclopedic in discussing these.

2.1   The aim of science

Let us define science in a form that motivates its discussion in relation to human–computer systems.

Principle 1 (Aim of science) *The principal aim of science is to uncover the most likely explanation for observable phenomena.*

Science is a procedure for making sure that we know what we are talking about when discussing phenomena that occur around us. It is about managing our uncertainty. Science does not necessarily tell us what the correct explanation for a phenomenon is, but it provides

Analytical Network and System Administration. Managing Human–Computer Networks   Mark Burgess
© 2004 John Wiley & Sons, Ltd   ISBN 0-470-86100-2

us with tools for evaluating the likelihood that a given explanation is true, given certain experimental conditions. Thus, central to science is the act of observation.

Observation is useless without interpretation, so experiments need theories and models to support them. Moreover, there are many strategies for understanding observable phenomena: it is not necessary to have seen a phenomenon to be able to explain it, since we can often predict phenomena just by guesswork, or imagination[1]. The supposed explanation can then be applied and tested once the phenomenon has actually been observed.

The day-to-day routine of science involves the following themes, in approximately this order:

Observation of phenomena

Normally, we want to learn something about a system, for example, find a pattern of behaviour so that we might predict how it will behave in the future, or evaluate a property so that we can make a choice or a value judgement about it. This might be as simple as measuring a value, or it might involve plotting a set of values in a graph against a parameter such as time or memory.

Example 5 *Performance analysts measure the rate at which a system can perform its task. They do this with the larger aim of making things faster or more efficient. Computer anomaly detectors, on the other hand, look for familiar patterns of behaviour so that unusual occurrences can be identified and examined more closely for their significance.*

Estimation of experimental error

In observing the world, we must be cautious about the possibility of error in procedure and interpretation: if we intend to base decisions on observations, we need to know how certain we are of our basis. Poor data can mislead (garbage in; garbage out). Any method of observation admits the possibility of error in relation to one's assumptions and methods.

- We make a mistake in measurement (either at random or repeatedly).

- The measuring apparatus might be unreliable.

- The assumptions of the experiment are violated (e.g. inconstant environmental conditions).

Although it is normal to refer to this as 'experimental error', a better phrase is *experimental uncertainty*. We must quantify the uncertainty in the experimental process itself, because this contributes an estimation of how correct our speculations about the results are. Uncertainties are usually plotted as 'error bars' (see fig. 2.1).

Identification of relationships

Once we know the main patterns of behaviour, we try to quantify them by writing down mathematical relationships. This leads to empirical relationships between variables, that is,

---

[1] This is how black holes were 'discovered' in astrophysics. It is now believed that there is unambiguous evidence for black holes.

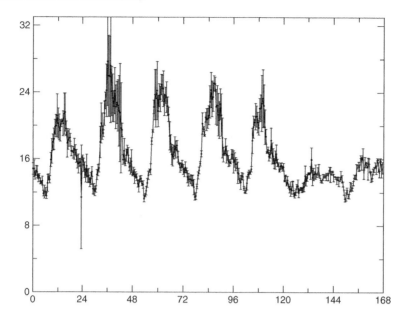

Figure 2.1: A pattern of process behaviour. The solid curve is the measured expectation value of the behaviour for that time of week. The error bars indicate the standard deviation, which also has a periodic variation that follows the same pattern as the expectation value; that is, both moments of the probability distribution of fluctuations has a daily and a weekly period.

it tells us how many of the variables we are able to identify are *independent*, and how many are *determined*.

Example 6 *It is known that the number of processes running on a college web server is approximately a periodic function (see fig. 2.1). Using these observations, we could try to write down a mathematical relationship to describe this. For example,*

$$f(t) = A + Be^{-\gamma(t-t_0)} \sin(\omega t), \tag{2.1}$$

where t is time along the horizontal axis, and $f(t)$ is the value on the vertical axis, for constants $A, B, \omega, \gamma, t_0$.

In the example above, there are far too many parameters to make a meaningful fit. It is always possible to fit a curve to data with enough parameters ('enough parameters to fit an elephant' is a common phrase used to ridicule students); the question is how many are justified before an alternative explanation is warranted?

Speculation about mechanisms

Expressing observations in algebraic form gives us a clue about how many parameters are likely to lie behind the explanation of a phenomenon. Next, we speculate about the plausible explanations that lead to the phenomena, and formulate a theory to explain the relationships. If our theory can predict the relationships and the data we have provided, it is reasonable to call the speculation a *theory*.

Confirmation of speculations

One must test a theory as fully as possible by comparing it to existing observations, and by pushing both theory and observation to try to predict something that we do not already know.

Quantification of uncertainty

In comparing theory and observation, there is much uncertainty. There is a basic uncertainty in the data we have collected; then there is a question of how accurately we expect a theory to reproduce those data.

Example 7 *Suppose the formula above for fig. 2.1, in eqn. (2.1) can be made to reproduce the data to within 20% of the value on either side, that is, the approximate form of the curve is right, but not perfect. Is this an acceptable description of the data? How close do we have to be to say that we are close enough? This 'distance from truth' is our uncertainty.*

In a clear sense, science is about uncertainty management. Nearly all systems of interest (and every system involving humans) are very complex and it is impossible to describe them fully. Science's principal strategy is therefore to simplify things to the point where it is possible to make some concrete characterizations about observations. We can only do this with a certain measure of uncertainty. To do the best job possible, we need to control those uncertainties. This is the subject of the next chapter.

2.2   Causality, superposition and dependency

In any dynamical system in which several processes can coexist, there are two possible extremes:

- Every process is independent of every other. System resources change additively (linearly) in response to new processes.

- The addition of each new process affects the behaviour of the others in a non-additive (non-linear) fashion.

The first case is called *superposition*, that is, that two processes can coexist without interfering. This is not true or possible in general, but it can be a useful viewpoint for approximating some system regimes. The latter case is more general and often occurs when a system reaches some limitation, or constraint on its behaviour, such as when there is contention over which process has the use of critical resources.

The principle of causality governs all systems at a fundamental level. It is simply stated as follows:

Principle 2 (Causality) *Every change or effect happens in response to a cause, which precedes it.*

This principle sounds intuitive and even manifestly obvious, but the way in which cause and effect are related in a dynamical system is not always as clear as one might imagine. We would often like to be able to establish a causal connection between a change of a specific parameter and the resulting change in the system. This is a central skill in fault finding, for instance; however, such causal links are very difficult to determine in complex systems. This is one of the reasons why the administration of systems is hard.

2.3   Controversies and philosophies of science

Science and philosophy have long been related. Indeed, what we now call science was once 'natural philosophy', or pondering about the natural world. Those who practice science today tend to think little about its larger meaning, or even its methodology. Science has become an 'industry'—the high ideals that were afforded to it in the seventeenth century have since been submerged in the practicalities of applying it to real problems.

Here are some assertions that have been made of science by philosophers (Horgan (1996)):

- 'Science cannot determine the truth of an explanation, only its likelihood'.

- 'Science can only determine the falsity of a theory, not whether it is true'.

- 'We must distinguish between truth, which is objective and absolute, and certainty which is subjective'.

To the casual technologist, such assertions are likely to draw only scepticism as to the value of philosophy. However, those willing to reflect more deeply on the whole investigative enterprise will find many ideas in the philosophy of science that are both interesting and of practical importance. The difficulty in presenting the labours of careful thought in such a brief and summarized form is that it is easy to misrepresent the philosophers' detailed arguments[2]. No doubt they would be horrified by this summary if they were alive to read it.

One of the first modern philosophers of science was Sir Francis Bacon, of the sixteenth century. Bacon (who died of pneumonia after stuffing a chicken with ice to see if it would preserve its flesh—thus anticipating the deep freeze) maintained that the task of science is to uncover a thing's character, by noting the presence or the absence of telltale qualities. Thus, to understand heat, for instance, we must examine a list of hot and cold things and discern what features are relevant and irrelevant to the production of heat; for example, exposure to sunlight is relevant, but the width of an object is not. Next, we would examine instances in which a phenomenon is present in varying degrees, noting what circumstances also vary. For example, to understand heat, we must observe things at different temperatures and note what circumstances are present in varying degrees. Bacon recognized that we cannot examine an endless number of instances: at some point we must stop and survey the instances so far.

Especially in the seventeenth century, philosophy became intertwined with mathematics, or analytical thinking. The philosopher Descartes used geometry for his inspiration as to how best to conduct an impartial inquiry. John Locke, an understudy of Isaac Newton, hoped to draw inspiration from the phenomenal success of Newton's laws of motion and the calculus, and derive an analytical way of addressing a 'method of inquiry'—what, today, we would call a 'scientific method'. His philosophy, now called *empiricism*, implies a reliance on experience as the source of ideas and knowledge.

Newton was a significant source of inspiration to philosophers because, for the first time, his work had made it possible to calculate the outcome of a hypothetical situation that no one had ever observed before, that is, predict the future for idealized physical systems.

---

[2] At this point, it would be natural to give a reference to a book in which a nice summary was presented. Alas, I have yet to find a clear exposition of the philosophy of science printed in English.

During the Enlightenment, philosophers even came to believe that scientific inquiry could yield truths about human nature and thus that ethical principles might be best derived from such truths; this would therefore be a basis for a new order of society.

In the eighteenth century, others began to realize that this vision was flawed. David Hume discovered an important twist, namely that predictions about events that are not observed cannot be *proven* to be true or false, not even to be probable, since observation alone cannot see into the future, and cannot attempt to assess the *cause* of a phenomenon. He asserted that there are two sources of knowledge: analytical knowledge that is certain (provable assertions) but which cannot directly represent reality, and empirical knowledge or observations that are uncertain but which apply to the real world.

The empirical observation that releasing a stone causes it to fall to the ground is insufficient to prove, beyond doubt, that every stone will always fall to the ground in the future. This is a good example of how our limited experience shapes our view of the world. Before humans went into space, the assertion was always true; however, away from gravity, in the weightlessness of space, the observation becomes meaningless. Hume's point is that we do not know what we don't know, so we should not make unwarranted assumptions.

Although Hume's ideas had an impact on philosophy, they were not generally accepted in science. Immanuel Kant and John Stuart Mill made attempts to solve some of Hume's problems. Kant claimed to solve some of them by assuming that certain facts were to be regarded as axioms, that is, articles of faith that were beyond doubt; that is, that one should always set the stage by stating the conditions under which conclusions should be deemed 'true'.

Kant supposed, moreover, that our perception of the world is important to how we understand it. In what sense are things real? How do we know that we are not imagining everything? Thus, how do we know that there are not many equally good explanations for everything we see? His central thesis was that the possibility of human knowledge presupposes the participation of the human mind. Instead of trying, by reason or experience, to make our concepts match the nature of objects, Kant held that we must allow the structure of our concepts shape our experience of objects.

Mill took a more pragmatic line of inquiry and argued that the truth of science is not absolute, but that its goals were noble; that is, science is a self- correcting enterprise that does not need axiomatic foundations per se. If experience reveals a flaw in its generalities, it can be accommodated by a critical revision of theory. It would eventually deal with its own faults by a process of refinement.

Epistemology is a branch of philosophy that investigates the origins and nature, and the extent of human knowledge. Although the effort to develop an adequate theory of knowledge is at least as old as Plato, epistemology has dominated Western philosophy only since the era of Descartes and Locke, largely as an extended dispute between *rationalism* and *empiricism*. Rationalism believes that some ideas or concepts are independent of experience and that some truth is known by reason alone (e.g. parallel lines never meet). Empiricism believes truth must be established by reference to experience alone.

Logical positivism is a twentieth-century philosophical movement that used a strict principle of *verifiability* to reject non-empirical statements of metaphysics, theology and ethics. Under the influence of Hume and others, the logical positivists believed that the only meaningful statements were those reporting empirical observations; the tautologies of logic and mathematics could not add to these, but merely re-express them. It was thus a mixture of rationalism and empiricism.

The *verifiability principle* is the claim that the meaning of a proposition is no more than the set of observations that would determine its truth, that is, that an empirical proposition is meaningful only if it either actually has been verified or could at least in principle be verified. Analytic statements (including mathematics) are non-empirical; their truth or falsity requires no verification. Verificationism was an important element in the philosophical program of logical positivism.

One of the most influential philosophers of science is Karl Popper. He is sometimes referred to as the most important philosopher of science since Francis Bacon. Karl Popper's ideas have proven to be widely influential for their pragmatism and their belief in the rational. Popper rejected that knowledge is a social phenomenon—it is absolute. He supposed that we cannot be certain of what we see, but if we are sufficiently critical, we *can* determine whether or not we are wrong, by deductive *falsification* or a process of conjecture and refutation (see fig. 2.2).

Figure 2.2: A pastiche of Rene Magritte's famous painting 'Ceci n'est pas une pipe'. The artist's original paintings and drawings are pictures of a pipe, on which is written the sentence 'this is not a pipe'. The image flirts with paradox and illustrates how uncritical we humans are in our interpretation of things. Clearly the picture is not a pipe—it is a picture that represents a pipe. However, this kind of pedantic distinction is often important when engaging in investigative or analytical thought.

Popper believed that theories direct our observations. They are a part of our innate desire to impose order and organization on the world, that is, to systematize the phenomena we see, but we are easily fooled and therefore we need to constantly criticize and retest every assumption to see if we can *falsify* them. Hume said we can never prove them right, but Popper says that we can at least try to see if they are wrong.

Paul Feyerabend later argued that there is no such thing as an objective scientific method. He argued that what makes a theory true or false is entirely a property of the world view of which that assertion is a part. This is *relativism*, that is, objectivity is a myth. We are intrinsically locked into our own world view, perceiving everything through a particular filter, like a pair of sunglasses that only lets us see particular things.

We need only one flaw in an explanation to discount it; but we might need to confirm hundreds of facts and details to be sure about its validity, that is, 'truth'. In the context of this book, science itself is a system that we shall use to examine others. We summarize with a pragmatic view of science:

Principle 3 (Controlled environment) *Science provides an impartial method for investigating and describing phenomena within an idealized environment, under controlled conditions.*

2.4   Technology

Science, we claim, is an investigative enterprise, whose aim is to characterize what is already there. Technology, on the other hand, is a creative enterprise: it is about tool-building.

The relationship between science and technology is often presented as being problematical by technologists, but it is actually quite clear. If we do not truly understand how things work and behave, we cannot use those things to design tools and methods. In technology, we immediately hit upon an important application of science, namely its role in making *value judgements*. A value judgement is a subjective judgement, for example, one tool can be better than another, one system or method can be better than another—but how are such judgements made? Science cannot answer these questions, but it can assist in evaluating them, if the subjectivity can be defined clearly.

The situation is somewhat analogous to that faced by the seventeenth century philosophers who believed that ethics could be derived from scientific principles. Science cannot tell us whether a tool or a system is 'good' or 'bad', because 'good' and 'bad' have no objective definitions. Science craves a discipline in making assertions about technology, and perhaps even guides us in making improvements in the tools we make, by helping us to clarify our own thoughts by quantification of technologies.

2.5   Hypotheses

Although science sometimes springs from serendipitous discovery, its systematic content comes from testing existing ideas or theories and assertions. Scientific knowledge advances by undertaking a series of studies, in order to either verify or falsify a hypothesis. Sometimes these studies are theoretical, sometimes they are empirical and frequently they are a mixture of the two. *Statistical reproducibility* is an important criterion for any result, otherwise it is worthless, because it is *uncertain*. We might be able to get the same answer twice by accident, but only repeated verification can be trusted.

In system administration, software tools and human methods form the technologies that are used. Progress in understanding is made with the assistance of the tools only if investigation leads to a greater predictive power or a more efficient solution to a problem.

- Scientific progress is the gradual refinement of the conceptual model that describes the phenomenon we are studying. In some cases, we are interested in modelling tools. Thus, technology is closely related to science.

- Technological progress is the gradual creative refinement of the tools and methods referred to by the technology. In some cases, the goal is the technology itself; in other situations, the technology is only an implement for assisting the investigation.

All problems are pieces of a larger puzzle. A complete scientific study begins with a *motivation*, followed by an *appraisal* of the problems, the construction of a *theoretical model* for understanding or solving the problems and finally an *evaluation* or *verification* of the *approach used* and the *results obtained*. Recently, much discussion has been directed towards finding suitable methods for evaluating technological innovations in computer science as well as to encouraging researchers to use them. Nowadays, many computing systems are of comparable complexity to phenomena found in the natural world and our understanding of them is not always complete, in spite of the fact that they were designed to fulfil a specific task. In short, technology might not be completely predictable, hence there is a need for experimental verification.

2.6 The science of technology

In technology, the act of observation has two goals: (i) to gather information about a problem in order to motivate the design and construction of a technology which solves it, and (ii) to determine whether or not the resulting technology fulfils its design goals. If the latter is not fulfilled in a technological context, the system may be described as faulty, whereas in natural science there is no right or wrong. In between these two empirical bookmarks lies a theoretical model that hopefully connects the two.

System administration is a mixture of science, technology and sociology. The users of computer systems are constantly changing the conditions for observations. If the conditions under which observations are made are not constant, then the data lose their meaning: the message we are trying to extract from the data is supplemented by several other messages that are difficult to separate from one another. Let us call the message we are trying to extract *signal* and the other messages that we are not interested in *noise*. Complex systems are often characterized by very noisy environments.

In most disciplines, one would attempt to reduce or eliminate the noise in order to isolate the signal. However, in system administration, it would be no good to eliminate the users from an experiment, since it is they who cause most of the problems that one is trying to solve. In principle, this kind of noise in data could be eliminated by statistical sampling over very long periods of time, but in the case of real computer systems, this might not be possible since seasonal variations in patterns of use often lead to several qualitatively different types of behaviour that should not be mixed. The collection of reliable data might therefore take many years, even if one can agree on what constitutes a reasonable experiment. This is often impractical, given the pace of technological change in the field.

2.7   Evaluating a system—dependencies

Evaluating a model of system administration is a little bit like evaluating the concept of a bridge. Clearly, a bridge is a structure with many components, each of which contributes to the whole. The bridge either fulfils its purpose in carrying traffic past obstacles or it does not. In evaluating the bridge, should one then consider the performance of each brick and wire individually? Should one consider the aesthetic qualities of the bridge? There might be many different designs, each with slightly different goals. Can one bridge be deemed better than another on the basis of objective measurement? Perhaps only the bridge's maintainer is in a position to gain a feeling for which bridge is the most successful, but the success criterion might be rather vague: a collection of small differences that make the perceptible performance of the bridge optimal, but with no measurably significant data to support the conclusion. These are the dilemmas of evaluating a complex technology.

The options we have for performing experimental studies are as follows:

- Measurements

- Simulations

- User surveys.

with all of the incumbent difficulties which these entail.

Simplicity

Conceptual and practical simplicity are often deemed to be positive attributes of systems and procedures. This is because simple systems are easy to understand and their behaviours are easy to predict. We prefer that systems that perform a function do so predictably.

Evaluation of individual mechanisms

For individual pieces of a system, it is sometimes possible to evaluate the efficiency and correctness of the components. Efficiency is a relative concept and, if used, it must be placed in a context. For example, efficiency of low-level algorithms is conceptually irrelevant to the higher levels of a program, but it might be practically relevant, that is, one must say what is meant by efficiency before quoting results. The correctness of the results yielded by a mechanism/algorithm can be measured in relation to its design specifications. Without a clear mapping of input/output, the correctness of any result produced by a mechanism is a heuristic quality. Heuristics can only be evaluated by experienced users expressing their informed opinions.

2.8   Abuses of science

Science is about constantly asking questions and verifying hypotheses to see if one's world view holds up to scrutiny. However, the authority that science has won is not always been wielded in a benign way. History is replete with illegitimate ideas that have tried to hide behind the reputation of science, by embracing its terminology without embracing its forms.

Marketeers are constantly playing this game with us, inventing scientific names for bells and whistles on their products, or claiming that they are 'scientifically proven' (an oxymoron). By quoting numbers or talking about 'ologies', there are many uncritical forces in the world who manipulate our *beliefs*, assuming that most individuals will not be able to verify them or discount them[3]. In teaching a scientific method, we must be constantly aware of abuses of science.

Applications and Further Study 2 The observation and analysis of systems involves these themes:

- Variables or measurables

- Determinism or causality

- Indeterministic, random or stochastic influences

- Systems and their environments

- Accounting and conservation.

---

[3] Eugenics is one classic example where the words and concepts discovered by science were usurped for illegitimate means to claim that certain individuals were genetically superior to others. This was a classic misunderstanding of a scientific concept that was embraced without proper testing or understanding.

3

Experiment and observation

Trust, but verify!

—Russian Proverb

Collecting data to support an idea or a hypothesis is central to the scientific method. We insist on the existence of evidence that can be examined and related analytically (by mathematics or other reasoning) to the phenomenon under consideration, because our trust in random observation or hearsay is only limited. The paraphrased proverb, 'Trust but verify' is often cited in connection with system security, but it is equally pertinent here. In a sense, the scientific method is the security or quality assurance system for 'truth'.

To study human–computer systems, we draw on analytical methods from the diverse branches of science, but our conclusions must be based on observed fact. Reliable observational evidence is most easily obtained where one can perform experiments to gather numerical data, then derive relationships and conclusions. Descriptive sciences do not always have this luxury and are forced to use a form of data collection that involves visual observation, classification or even interview. This is less focused and therefore harder to use to support specific conclusions.

Example 8 *A zoologist might find no problem in measuring the weight of animals, but might find it difficult to classify the colours of animals in order to relate this to their behaviour. When is red really brown? Fuzzy classifiers from day-to-day experience lead to difficulties for science—qualitative descriptions are prone to subjective interpretation.*

Example 9 *In human–computer systems, it is easy to measure numerical quantities such as rate of change of data, but qualitative features such as 'lawfulness' of users seem too vague to quantify.*

Difficulties with qualitative characterizations can sometimes be eliminated by going to a lower level, or to a smaller scale of the system: for example, the classification of animals

Analytical Network and System Administration. Managing Human–Computer Networks   Mark Burgess
© 2004 John Wiley & Sons, Ltd   ISBN 0-470-86100-2

might be done more precisely by looking at their DNA, and the lawfulness of a user might be measured by examining the policy conformance of each file and the changes made by the user.

3.1 Data plots and time series

In the observation of real systems, measurements are made and data are collected. If the data are collected at regular intervals, they are usually represented either as *time-series*, that is, plots of measured values versus the time at which the measurements were made, or as *histograms* that count the numbers of measurements that fall into certain domains (called classes) in the data values. Both types of diagram play important roles in understanding systems. In addition, various kinds of graphical representations are used to elucidate relationships between variables, such as plots of one variable against another, or log–log plots of the same variables that indicate power-law relationships.

Figure 3.1 shows a typical series of measurements made from a computer system over the course of several weeks. By plotting all of the data against a timescale of a week, one sees not only a clear pattern in the data, but also a scatter in the values measured at each

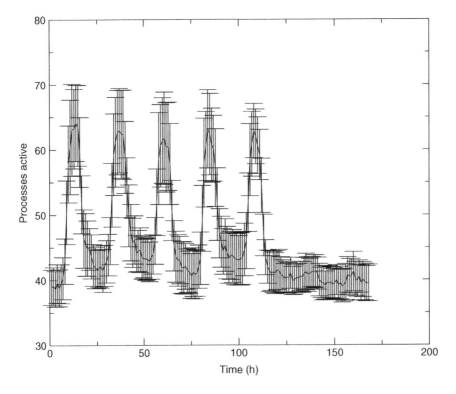

Figure 3.1: A time series of measurements taken over repeated weeks. Notice that repeated measurements at the same time of week are averaged over and 'error bars' are used to represent the width of the scatter. The result is a plot of the mean value $\langle q \rangle \pm \sigma$.

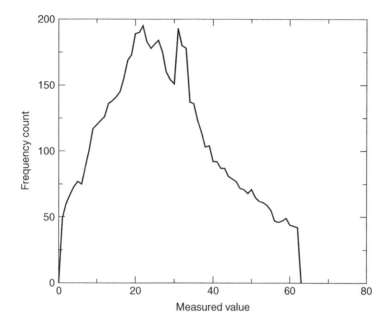

Figure 3.2: A frequency plot of the numbers of measurements of a given value.

time. Error bars are drawn at each point where there are repeated measurements. These show the width of the standard deviation $\pm\sigma$ centered about the mean value. It is important to plot the scatter in data as a visual guide to the uncertainty in the claimed result (see section 3.4).

Since, at each time, in fig. 3.1 there is a distribution of values, we can plot that distribution on a frequency plot like that in fig. 3.2. This is a kind of histogram in which the columns have been joined into an approximate curve. If the area under a frequency plot like this is normalized to unity, it represents a probability distribution for the measured values $P(q)$. The probability distribution for the measured values is important in gauging the stability of a system as well as in characterizing its fluctuation spectrum, as we shall see in chapter 8.

3.2   Constancy of environment during measurement

In science, our aim is to take small steps, by stripping away everything down to single cause–effect relationships, and then gradually putting things back together. Einstein is famous for having said that everything should be made as simple as possible, but no simpler. By this, he meant that we should neither overcomplicate nor oversimplify an explanation.

Most phenomena are governed by a number of parameters; for example, suppose the rate of a computer is affected by three parameters:

$$R = R(c, m, s) \tag{3.1}$$

where c is the CPU rate, m is the amount of memory and s is the speed of memory. If we want to discover just how R depends on each of these, we must test each parameter individually, holding the others constant, else we might mix up the dependence on each parameter. Science ends up with neat formulae relating measurables, only because this isolation is possible. Such formulae describe the real world, but they do not really describe the 'real environment' because the environment is messy. Science therefore strives to ensure idealized environmental conditions for investigating phenomena, in order to take one thing at a time.

In the real world of human–computer systems, there are many variables and influences that affect a system, so we must strive to maintain constant conditions in all variables but the one we would like to test. This is rarely possible, and thus there is an inevitable *uncertainty* or *experimental error* in any experiment. An important task of science is to quantify this uncertainty.

Principle 4 *Scientific observation strives to isolate single cause–effect relationships, by striving to keep environmental conditions constant during measurement. The impossibility of completely constant external conditions makes it necessary to quantify the uncertainty in each measurement.*

Note that by isolating 'single' cause–effect relationships, we do not mean to imply that there is always a single variable that controls a process, only that each independent change can be identified with an independent parameter.

The way we do this for simple measurable values is relatively easy and is described in this chapter. However, not all situations are so easily quantifiable. Qualitative experiments, such as those of biology (e.g. classifying types of behaviour) also occur in the study of human–computer systems. If we do not actually begin with hard numbers, the estimate of uncertainty has to be made by finding a numerical scale, typically through a creative use of *classification* statistics; for example, how many animals have exhibited behaviour A and how many behaviour B? Or how far is behaviour A from behaviour B on some arbitrary scale, used only for comparison?

All scales are arbitrary in science (that is why we have many different units for weight, height, frequency etc.), what is important is how we relate these scales to observables.

3.3   Experimental design

The cleverness of an experiment's design can be crucial to its success in providing the right information. Our aim is to isolate a single channel of cause–effect at a time. We must ensure that the experimental observation does not interfere with the system we are measuring. Often, an experiment yields unexpected obstacles that must be overcome. There can be a lot of work involved in answering even a simple question. (For examples from computer performance analysis, see Jain (1991).)

Example 10 *Suppose we wish to compare the behaviour of two programs for mirroring (copying) files, for backup. We notice that one program seems to complete its task very quickly, presenting a high load to the source and destination machines. The other takes much longer but presents almost no load. How shall we determine the reason?*

We might begin by finding some data to copy. Data is composed of files of different sizes. Size might be important, so we shall be interested in how size affects the rate of copying, if at all. The first time we copy the files, every file must be transferred in full. On subsequent updates, only changes need to be copied. One program claims to copy only those bytes that are different; the other has to copy a whole file, even if only one byte has changed, so file size again becomes important.

We could investigate how the total time for copying is related to the total amount of data (i) of all files, (ii) of files that are copied. We might also be interested in what dependencies the programs have: do they use the Internet Protocol with TCP or UDP, IPv4 or IPv6? Does the computer kernel or operating system affect the performance of the two programs?

The stream of questions never ceases; we must decide when to stop. Which questions are we interested in, and when have they been sufficiently answered? This is a value judgement that requires experience and inquisitiveness from the investigator.

3.4 Stochastic (random) variables

Our inability to control, or even follow every variable in a system's environment means that some of the changes appearing in the system seem random, or inexplicable.

> **Definition 4 (Random process)** *A random process is one in which there are too many unknowns to be able to trace the channels of cause and effect.*

A *stochastic* or *random* variable is a variable whose value depends on the outcome of some underlying random process. The range of values of the variable is not at issue, but which particular value the variable has at a given moment is random. We say that a stochastic variable X will have a certain value x with a probability $P(x)$.

Usually, in an experiment, a variable can be said to have a certain random component (sometimes called its *error* from the historical prejudice that science is deterministic and the only source of randomness is the errors incurred by the experimental procedure) and an average stable value. We write this as

$$x = \langle x \rangle + \Delta x, \tag{3.2}$$

where x is the actual value measured, $\langle x \rangle$ is the *mean* or *expectation value* of all measurements (often written $E(x)$ in statistical literature), and Δx is the deviation from the mean. The mean value changes much more slowly than Δx. For example:

- Choices made by large numbers of users are not predictable, except on average.

- Measurements collected over long periods of time are subject to a variety of fluctuating conditions.

Measurements can often appear to give random results, because we do not know all of the underlying mechanisms in a system. We say that such systems are *non-deterministic* or that there are *hidden variables* that prevent us from knowing all the details. If a variable has a fixed value, and we measure it often enough and for long enough, the random components will often fall into a *stable distribution*, by virtue of the *central limit theorem* (see, for instance Grimmett and Stirzaker (2001)). The best-known example of a stable distribution is the Gaussian type of distribution.

3.5   Actual values or characteristic values

There is a subtle distinction in measurement between an observable that has an actual 'true' value and one that can only be characterized by a typical value.

For example, it is generally assumed that the rest mass of an electron has a 'true' value that never changes. Yet when we measure it, we get many different answers. The conclusion must be that the different values result from *errors* in the measurement procedure. In a different example, we can measure the size of a whale and we get many different answers. Here, there is no 'true' or 'standard' whale and the best we can do is to measure a typical or *expected* value of the size.

In human–computer systems, there are few, if any, measurements of the first type, because almost all values are affected by some kind of variation. For example, room temperature can alter the maximum transmission rate of a cable. We must therefore be careful about what we claim to be constant, and what is the reason for the experimental variation in the results. Part of the art in science is in the interpretation of results, within the constraints of cause and effect.

3.6   Observational errors

All measurements involve certain errors. One might be tempted to believe that, where computers are involved, there would be no error in collecting data, but this is false. Errors are not only a human failing; they occur because of unpredictability in the measurement process, and we have already established throughout this book that computer systems are nothing but unpredictable. We are thus forced to make estimates of the extent to which our measurements can be in error. This is a difficult matter, but approximate statistical methods are well known in the natural sciences, methods that become increasingly accurate with the amount of data in an experimental sample.

The ability to estimate and treat errors should not be viewed as an excuse for constructing a poor experiment. Errors can only be minimized by design. There are several distinct types of error in the process of observation.

The simplest type of error is called *random error*. Random errors are usually small deviations from the 'true value' of a measurement that occur by accident, by unforeseen jitter in the system, or by some other influence. By their nature, we are usually ignorant of the cause of random errors, otherwise it might be possible to eliminate them. The important point about random errors is that they are distributed evenly about the mean value of the observation. Indeed, it is usually assumed that they are distributed with an approximately *normal* or *Gaussian* profile about the mean. This means that there are as many positive as negative deviations and thus random errors can be averaged out by taking the mean of the observations.

It is tempting to believe that computers would not be susceptible to random errors. After all, computers do not make mistakes. However, this is an erroneous belief. The measurer is not the only source of random errors. A better way of expressing this is to say that random errors are a measure of the unpredictability of the measuring process. Computer systems are also unpredictable, since they are constantly influenced by outside agents such as users and network requests.

The second type of error is a *personal error*. This is an error that a particular experimenter adds to the data unwittingly. There are many instances of this kind of error in the history of science. In a computer-controlled measurement process, this corresponds to any particular bias introduced through the use of specific software, or through the interpretation of the measurements.

The final and most insidious type of error is the *systematic error*. This is an error that runs throughout all of the data. It is a systematic shift in the true value of the data, in one direction, and thus it cannot be eliminated by averaging. A systematic error leads also to an error in the mean value of the measurement. The sources of systematic error are often difficult to find, since they are often a result of misunderstandings, or of the specific behaviour of the measuring apparatus.

Principle 5 *In a system with finite resources, the act of measurement itself leads to a change in the value of the quantity one is measuring.*

Example 11 *In order to measure the pressure of a bicycle tyre, we have to release some of the pressure. If we continue to measure the pressure, the tyre will eventually be flat.*

In order to measure the CPU usage of a computer system, for instance, we have to start a new program that collects that information, but that program inevitably uses the CPU also and therefore changes the conditions of the measurement. These issues are well known in the physical sciences and are captured in principles such as Heisenberg's Uncertainty Principle, Schrödinger's cat and the use of infinite idealized heat baths in thermodynamics. We can formulate our own verbal expression of this for computer systems:

Principle 6 (Uncertainty) *The act of measuring a given quantity in a system with finite resources, always changes the conditions under which the measurement is made, that is, the act of measurement changes the system.*

For instance, in order to measure the pressure in a tyre, you have to let some of the air out, which reduces the pressure slightly. This is not noticeable in a car tyre, but it can be noticeable in a bicycle tyre. The larger the available resources of the system, compared to the resources required to make the measurement, the smaller the effect on the measurement will be.

3.7 The mean and standard deviation

In the theory of errors, we use the ideas above to define two quantities for a set of data: the mean and the standard deviation. Contrary to what one sometimes reads, these quantities are not necessarily tied to the normal distribution: they are just expressions of scale that can be used to characterize data sets. They are also called the first and second *moments* of the data.

The situation is now as follows: we have made a number N of observations of values v_1, $v_2, v_3 \ldots, v_N$, which have a certain randomness and we are trying to find out a characteristic value v for the measurement. Assuming that there are no systematic errors, that is, assuming that all of the deviations have independent random causes, we define the value $\langle v \rangle$ to be the arithmetic mean of the data:

$$\langle v \rangle = \frac{v_1 + v_2 \ldots v_N}{N} = \frac{1}{N} \sum_{i=1}^{N} v_i. \tag{3.3}$$

Next, we treat the deviations of the actual measurements as our guesses for the error in the measurements:

$$\Delta g_1 = \langle v \rangle - v_1$$
$$\Delta g_2 = \langle v \rangle - v_2$$
$$\vdots$$
$$\Delta g_N = \langle v \rangle - v_N$$

and define the *standard deviation* of the data by

$$\sigma = \sqrt{\frac{1}{N} \sum_{i=0}^{N} \Delta g_i^2}. \qquad (3.4)$$

This is clearly a measure of the scatter in the data due to random influences. σ is the root mean square (RMS) of the assumed errors. These definitions are a way of interpreting measurements, from the assumption that one really is measuring the true value, affected by random interference.

Definition 5 (Gaussian signal power) *A random signal that is distributed according to a Gaussian distribution has a characteristic amplitude σ, and thus a squared amplitude of σ^2. Since the squared amplitude of a signal is associated with the* power *(in Watts) of a physical signal, the variance is often assumed to measure power.*

An example of the use of standard deviation can be seen in the error bars of the figures in this chapter. Whenever one quotes an average value, the number of data and the standard deviation should also be quoted in order to give meaning to the value. In system administration, one is interested in the average values of any system metric that fluctuates with time.

3.8   Probability distributions and measurement

Whenever we repeat a measurement and obtain different results, a distribution of different answers is formed. The spread of results needs to be interpreted. There are two possible explanations for a range of values:

- The quantity being measured does not have a fixed value.

- The measurement procedure is imperfect and incurs a range of values due to error or uncertainty.

Often, both of these are the case. In order to give any meaning to a measurement, we have to repeat the measurement a number of times and show that we obtain approximately the same answer each time. In any complex system, in which there are many things going on which are beyond our control (read: just about anywhere in the real world), we will never obtain exactly the same answer twice. Instead, we will get a variety of answers that we can plot as a graph: on the x-axis, we plot the actual measured value and on the y-axis we plot the number of times we obtained that measurement divided by a normalizing factor, such as the total number of measurements. By drawing a curve through the points, we obtain an idealized picture that shows the probability of measuring the different values.

Over time, measurements often develop stable average behaviour, so that a time series $x = \{x_1, x_2, x_3, \ldots\}$ has an average that tends towards a stable value. This is written in a variety of notations in the literature:

$$\bar{x} = E(x) = \langle x \rangle \equiv \frac{1}{N} \sum_{i=1}^{N} x_i \to \mu, \tag{3.5}$$

where N is the number of data. Few, if any, of the actual measurements will actually be equal to μ; rather, they are scattered around the average value in some pattern, called their distribution $P(x)$. The normalization factor is usually chosen so that the area under the curve is unity, giving a probabilistic interpretation.

Definition 6 (Probability) *The probability $P(x)$ of measuring a value x in original data set is defined to be the fraction of values that fell into the range $x \pm \Delta x/2$, for some class width Δx.*

$$P(x) = \frac{N(x - \Delta x/2, x + \Delta x/2)}{N_{\text{total}}}. \tag{3.6}$$

Here, $N(x, y)$ is the number of observations between x and y.

This probability distribution is the histogram shown in fig. 3.3.

There are two extremes of distribution: complete certainty (fig. 3.4) and complete uncertainty (fig. 3.5). If a measurement always gives precisely the same answer, then we say that there is no error. This is never the case in real measurements. Then the curve is just a sharp spike at the particular measured value. If we obtain a different answer each time we measure a quantity, then there is a spread of results. Normally, that spread of results will be concentrated around some more or less stable value (fig. 3.6). This indicates that the probability of measuring that value is biased, or tends to lead to a particular range of

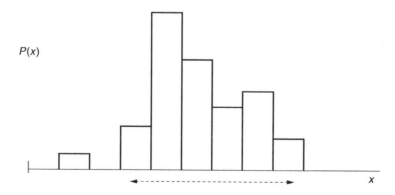

Figure 3.3: The scatter is an estimate of the width of the populated regions of the probability distribution.

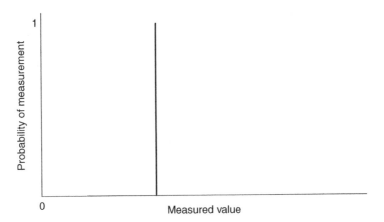

Figure 3.4: The delta distribution represents complete certainty. The distribution has a value of 1 at the measured value.

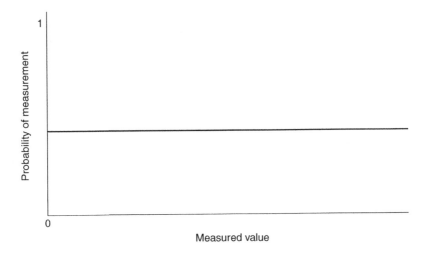

Figure 3.5: The flat distribution is a horizontal line indicating that all measured values, within the shown interval, occur with equal probability.

values. The smaller the range of values, the closer we approach fig. 3.4. But the converse might also happen: in a completely random system, there might be no fixed value of the quantity we are measuring. In that case, the measured value is completely uncertain, as in fig. 3.5. To summarize, a flat distribution is unbiased, or completely random. A non-flat distribution is biased, or has an expectation value, or probable outcome. In the limit of complete certainty, the distribution becomes a spike, called the *delta distribution*.

We are interested in determining the shape of the distribution of values on repeated measurement for the following reason. If the variation of the values is symmetrical about some preferred value, that is, if the distribution peaks close to its mean value, then we can

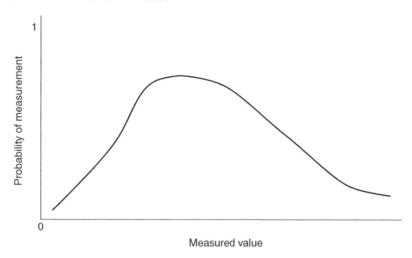

Figure 3.6: Most distributions peak at some value, indicating that there is an expected value (expectation value) that is more probable than all the others.

probably infer that the value of the peak or of the mean is the true value of the measurement and that the variation we measured was due to random external influences. If, on the other hand, we find that the distribution is very asymmetrical, some other explanation is required and we are most likely observing some actual physical phenomenon that requires explanation.

3.8.1 Scatter and jitter

The term *scatter* is often used to express the amount of variation in the measurements about the mean. It is estimated as the 'width' of the histogram $P(x)$. The term *jitter* is often used when describing the scatter of arrival times between measurements in the time series. Decades of artificial courses on statistics have convinced many scientists that the distribution of points about the mean must follow a Gaussian 'normal' distribution in the limit of large numbers of measurements. This is not true, however; there are ample cases in which the scatter is asymmetric or less uniform than the 'normal distribution'.

3.8.2 The 'normal' distribution

It has been stated that 'Everyone believes in the exponential law of errors; the experimenters because they think it can be proved by mathematics; and the mathematicians because they believe it has been established by observation' (Whittaker and Robinson (1929)). Some observational data in science closely satisfy the normal law of error, but this is by no means universally true. The main purpose of the normal error law is to provide an adequate idealization of error treatment that applies to measurements with a 'true value' (see section 3.5), which is simple to deal with, and which becomes increasingly accurate with the size of the data sample.

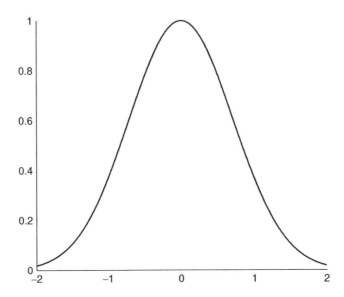

Figure 3.7: The Gaussian normal distribution, or bell curve, peaks at the arithmetic mean. Its width characterizes the standard deviation. It is therefore the generic model for all measurement distributions.

The normal distribution was first derived by DeMoivre in 1733 while dealing with problems involving the tossing of coins; the law of errors was deduced theoretically in 1783 by Laplace. He started with the assumption that the total error in an observation was the sum of a large number of independent deviations, which could be either positive or negative with equal probability, and could therefore be added according to the rule explained in the previous sections. Subsequently, Gauss gave a proof of the error law based on the postulate that the most probable value of any number of equally good observations is their arithmetic mean. The distribution is thus sometimes called the Gaussian distribution, or the bell curve.

The Gaussian normal distribution is a smooth curve that is used to model the distribution of discrete points distributed around a mean. The probability density function $P(x)$ tells us with what probability we would expect measurements to be distributed about the mean value $\bar{x}$ (see fig. 3.7).

$$P(x_i) = \frac{1}{(2\pi\sigma^2)^{1/2}} \exp\left(-\frac{(x_i - \bar{x})^2}{2\sigma^2}\right).$$

It is based on the idealized limit of an infinite number of points.

3.8.3 Standard error of the mean

No experiments have an infinite number of points, so we need to fit a finite number of points to a normal distribution as well as we can. It can be shown that the most probable choice is to take the mean of the finite set to be our estimate of the mean of the ideal set. Of course, if we select at random a sample of N values from the idealized infinite set, it is not clear that they will have the same mean as the full set of data. If the number in the sample N is large, the two will not differ greatly, but if N is small, they might. In fact, it can be shown that if we take many random samples of the ideal set, each of size N, they will have mean values that are themselves normally distributed, with a standard deviation equal to $\sigma/\sqrt{N}$. The quantity

$$\alpha = \frac{\sigma}{\sqrt{N}},$$

where σ is the standard deviation, is therefore called the *standard error of the mean*. This is clearly a measure of the accuracy with which we can claim that our finite sample mean agrees with the actual mean. In quoting a measured value *which we believe has a unique or correct value* (e.g. the height of the Eiffel Tower), it is therefore normal to write the mean value, plus or minus the standard error of the mean:

$$\text{Result} = \bar{x} \pm \sigma/\sqrt{N} \text{ (for } N \text{ observations)}, \tag{3.7}$$

where N is the number of measurements. Otherwise, if we believe that the measured value should have a distribution of values (e.g. the height of a river on the first of January of each year), one uses the standard deviation as a measure of the error. Many transactional operations in a computer system do not have a fixed value (see next section).

The law of errors is not universally applicable, without some modification, but it is still almost universally applied, for it serves as a convenient fiction that is mathematically simple[1].

3.8.4 Other distributions

Another distribution that appears in the periodic rhythms of system behaviour is the exponential form. There are many exponential distributions, and they are commonly described in textbooks. Exponential distributions are used to model component failures in systems over time, that is, most components fail quickly or live for a long time.

The Planck distribution is one example that can be derived theoretically as the most likely distribution to arise from an assembly of fluctuations in equilibrium with a large source (see Burgess et al. (2001)). The precise reason for its appearance in computer systems is subtle, but has to do with the periodicity imposed by users' behaviours, as well as the interpretation of transactions as fluctuations. The distribution has the form

$$D(\lambda) = \frac{\lambda^{-m}}{e^{1/\lambda T} - 1},$$

where T is a scale, and m is usually an integer greater than 2. When $m = 3$, a single degree of freedom is represented. The shape of the graph is shown in fig. 3.8.

Internet network traffic analysis studies (see Paxson and Floyd (1995); Willinger et al. (1996)) show that the arrival times of data packets within a stream has a long-tailed

---

[1] The applicability of the normal distribution can, in principle, be tested with a χ^2 test, but this is seldom used in physical sciences, since the number of observations is usually so small as to make it meaningless.

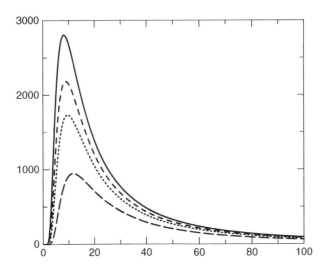

Figure 3.8: The Planck distribution for several temperatures. This distribution is the shape generated by random fluctuations from a source that is unchanged by the fluctuations. Here, a fluctuation is a computing transaction, a service request or new process.

distribution, often modelled as a Pareto distribution (a power law) in the asymptotic limit, for constants α and β:

$$f(\omega) = \beta\, a^\beta\, \omega^{-\beta-1}. \tag{3.8}$$

This can be contrasted with the Poissonian arrival times of telephonic data traffic. It is an important consideration to designers of routers and switching hardware. It implies that a fundamental change in the nature of network traffic has taken place. A partial explanation for this behaviour is that packet arrival times consist not only of Poisson random processes for session arrivals but also of internal correlations within a session. Thus, it is important to distinguish between measurements of packet traffic and measurements of numbers of sockets (or TCP sessions). The power law behaviour exhibited by Pareto tails is often indicative of clustered behaviour. If one event arrives, several tend to arrive in a cluster or a *burst*.

3.9   Uncertainty in general formulae

Suppose we measure the values of N variables that feed into a mathematical expression for something:

$$S = S(x, y, z, \ldots) \tag{3.9}$$

Assuming that errors are small, we can estimate the effect of an error in one of the parameters on the calculated expression by calculating the gradient (rate of change) of the function at the approximate value of the parameter and by multiplying this by our estimate of the error in the parameter. This tells us the expected error in S, given an estimate of the error in x. We use the first-order Taylor expansion for each variable and then treat each contribution as an orthogonal perturbation and use Pythagoras formula to express the combined error. Knowing the errors Δx, Δy, Δz, etc, we may evaluate the error in S:

$$\Delta S \equiv \sqrt{\left(\frac{\partial S}{\partial x}\right)^2 \Delta x^2 + \left(\frac{\partial S}{\partial y}\right)^2 \Delta y^2 + \ldots} \qquad (3.10)$$

Example 12 *The average rate of user transactions per second in a database is given by*

$$R = N/T, \qquad (3.11)$$

where N is the total number of transactions recorded, and T is the interval of time over which the measurement was made. We assume that the uncertainty in N is ΔN (caused by the fact that we cannot exactly separate every user transaction from administrative transactions), and that the uncertainty in T is ΔT, (caused by not being able to tell the exact moment when the measurements started and stopped, owing to context switching). Using the formula above, we find that

$$\frac{\partial S}{\partial N} = \frac{1}{T}$$
$$\frac{\partial S}{\partial T} = -\frac{N}{T^2}, \qquad (3.12)$$

so that

$$\Delta S = \sqrt{\left(\frac{1}{T}\right)^2 \Delta N^2 + \left(-\frac{N}{T^2}\right)^2 \Delta T^2}. \qquad (3.13)$$

Thus, if the total number of transactions was 1046, with approximately 20% ($0.2 \times 1046 \sim 200$) being administrative transactions, and the time for measurement was 200 seconds, give or take a few milliseconds, then:

$$\Delta S = \sqrt{(1/200)^2 \times 200^2 + (1046/4000)^2 \times 0.001^2},$$
$$\simeq \frac{200}{200},$$
$$\simeq 1. \qquad (3.14)$$

Thus, we quote the value for S to be

$$S = 1046/200 \pm 1 = 523 \pm 1. \qquad (3.15)$$

Note that this is an estimate based on a continuum approximation, since N and T are both discrete, non-differentiable quantities. As we are only estimating, this is acceptable.

3.10   Fourier analysis and periodic behaviour

Many aspects of computer system behaviour have a strong periodic quality, driven by the human perturbations introduced by users' daily rhythms. Other natural periods follow from the largest influences on the system from outside. For instance, hourly updates, or automated backups. The source might not even be known: for instance, a potential network intruder attempting a stealthy port scan might have programmed a script to test the ports periodically over a length of time. Analysis of system behaviour can sometimes benefit from knowing these periods. For example, if one is trying to determine a causal relationship between one part of a system and another, it is sometimes possible to observe the signature of a process that is periodic and thus obtain direct evidence for its effect on another part of the system.

Periods in data are the realm of Fourier analysis. What a Fourier analysis does is to assume that a data set is built up from the superposition of many periodic processes. Any curve can be represented as a sum of sinusoidal-waves with different frequencies and amplitudes. This is the complex Fourier theorem:

$$f(t) = \int d\omega \, f(\omega) e^{-i\omega t},$$

where $f(\omega)$ is a series of coefficients. For strictly periodic functions, we can represent this as an infinite sum:

$$f(t) = \sum_{n=0}^{\infty} c_n e^{-2\pi i \, nt/T},$$

where T is some timescale over which the function $f(t)$ is measured. What we are interested in determining is the function $f(\omega)$, or equivalently the set of coefficients c_n that represent the function. These tell us how much of which frequencies are present in the signal $f(t)$, or its *spectrum*. It is a kind of data prism, or spectral analyser, like the graphical displays one finds on some music players. In other words, if we feed in a measured sequence of data and Fourier-analyse it, the spectral function show the frequency content of the data that we have measured.

The whys and wherefores of Fourier analysis are beyond the scope of this book; there are standard programs and techniques for determining the series of coefficients. What

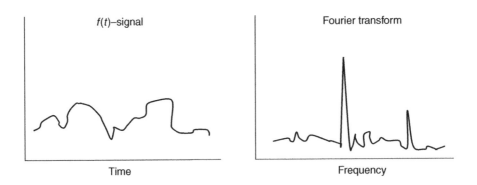

Figure 3.9: Fourier analysis is like a prism, showing us the separate frequencies of which is signal is composed. The sharp peaks in this figure illustrate how we can identify periodic behaviour that might otherwise be difficult to identify. The two peaks show that the input source conceals two periodic signals.

is more important is to appreciate its utility. If we are looking for periodic behaviour in system characteristics, we can use Fourier analysis to find it. If we analyse a signal and find a spectrum such as the one in fig. 3.9, then the peaks in the spectrum show the strong periodic content of the signal. To discover these smaller signals, it will be necessary to remove the louder ones (it is difficult to hear a pin drop when a bomb explodes nearby).

3.11   Local averaging procedures

One of the most important techniques for analysing data in time series is that of coarse graining, or local averaging. This is a smoothing procedure in which we collect together a sequence of measurements from a short interval of time Δt and replace them with a single average value for that interval. It is a way of smoothing out random fluctuations in data and extracting the trends. It also used as a way of characterizing the pattern of change in a time series.

Computer systems and human systems have often quite different patterns of behaviour. When they are combined, the result is often complex and hence local averaging is a straightforward approach to extracting or suppressing detail about the signal.

Let us define a local averaging procedure using fig. 3.10. See also Appendix B for more details.

The local averaging procedure re-averages data, moving from a detailed view to a less detailed view, by grouping neighbouring data together. In practice, one always deals with data that are sampled at discrete time intervals, but the continuous time case is also important for studying the continuum approximation to systems.

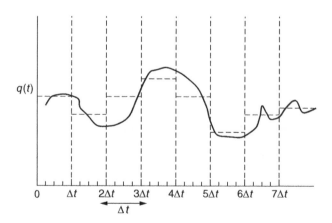

Figure 3.10: A coarse-graining, or local averaging procedure involves averaging over intervals larger than the basic resolution of the data. The flat horizontal lines represent the coarse-grained histogrammatic representation of the function. The scaling hypothesis say that if one 'zooms out' far enough, and views the fundamental and coarse-grained representations from a sufficiently high level ($\delta t \gg \Delta t$), then they are indistinguishable for all calculational purposes.

Discrete time data

Consider the function $q(t)$ shown in figs. 3.10 and 3.11. Let the small ticks on the horizontal axis represent the true sampling of the data, and label these by $i = 0, 1, 2, 3, \ldots, I$. These have unit spacing. Now let the large ticks, which are more coarsely spread out, be labelled by $k = 1, 2, 3, \ldots, K$. These have spacing $\Delta t = m$, where m is some fixed number of the smaller ticks. The relationship between the small and the larger ticks is thus

$$i = (k - 1)\Delta t = (k - 1)m. \tag{3.16}$$

In other words, there are $\Delta t = m$ small ticks for each large one. To perform a coarse-graining, we replace the function $q(t)$ over the whole kth cell with an average value, for each non-overlapping interval Δt. We define this average by

$$\langle q(k) \rangle_{\mathrm{m}} \equiv \frac{1}{\Delta t} \sum_{i=(k-1)\Delta t+1}^{k\Delta t} q(i). \tag{3.17}$$

We have started with an abstract function $q(t)$, sampled it at discrete intervals, giving $q(i)$, and then coarse-grained the data into larger contiguous samples $\langle q(k) \rangle_{\mathrm{m}}$:

$$q(t) \rightarrow q(i) \rightarrow \langle q(k) \rangle_{\mathrm{m}}. \tag{3.18}$$

Continuous time data

We can now perform the same procedure using continuous time. This idealization will allow us to make models using continuous functions and functional methods, such as functional

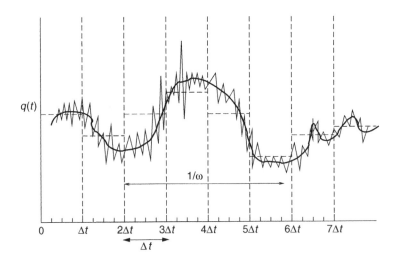

Figure 3.11: A jagged signal can be separated into local fluctuations plus a slowly varying local average, only if the variance is always finite.

integrals. Referring once again to the figure, we define a local averaging procedure by

$$\langle q(\bar{t}) \rangle_{\Delta t} = \frac{1}{\Delta t} \int_{\bar{t}-\Delta t/2}^{\bar{t}+\Delta t/2} q(\tilde{t}') \, d\tilde{t}'. \tag{3.19}$$

The coarse-grained variable $\bar{t}$ is now the more slowly varying one. It is convenient to define the parameterization

$$\tilde{t} = (t - t') \tag{3.20a}$$

$$\bar{t} = \frac{1}{2}(t + t'), \tag{3.20b}$$

on any interval between points t and t'. The latter is the mid-point of such a cell, and the former is the offset from that origin.

3.12 Reminder

Although much of the remainder of the book explores mathematical ways of describing and understanding information from human–computer systems, assuming that observations have been made, one should not lose sight of the importance of measurement. Science demands measurement. Mathematics alone only re-describes what we feed into it. Thus, at every stage of investigation into human–computer systems, one should ask: how can I secure an empirical basis for these assertions?

Applications and Further Study 3

- *Developing critical and analytical thinking.*

- *Formulating and planning experiments to gather evidence about systems.*

- *Estimating the uncertainties inherent in observational knowledge.*

- *Diagnostic investigations into anomalous occurrences.*

4

Simple systems

This chapter relates an approach to describing systems and their behaviour in terms of their components, using the ideas of predictability and utility.

4.1 The concept of a system

The concept of a *system* is intuitively familiar to us. In our daily lives, we are surrounded by so many systems that we scarcely notice them or think about them, until they go wrong. From the simplest wristwatch, whose mechanical parts cooperate to provide a time service, to public transport systems, to the Byzantine convolutions of our taxation systems, which serve to distribute resources throughout a larger social collective, systems pervade society at every level.

A modern computer system is a collection of hardware and software components that cooperate to achieve a goal for users. When users employ computers to carry out a task, the users themselves become a part of the system, both working on behalf of the machine when it prompts them, and instructing the machine on the direction it should take next. If users have access to several computers, which cooperate, then the system is said to be *distributed* in location. A single computer program can itself be regarded as a system; computer programs often consist of multiple functions and procedures that work together in order to evaluate some larger algorithm. Computer systems can be described using various kinds of diagrams and languages that show where information flows from component to component, and how it changes; a whole field of study has built up around this, and we shall draw upon that field here, since it is a formal framework, which admits analysis.

Any ordinary workplace also has the elements of a system. The concept of a system applies to whole organizations, or indeed any subset of an organization that can function independently. This might be a company, a branch office, a computer, a network of computers or even a single-celled organism on the keyboard of a computer.

The principles of system design and improvement are quite general, and need not be tied to any one of these examples, but it is useful to adopt the language of computer systems

---

Analytical Network and System Administration. Managing Human–Computer Networks   Mark Burgess (information systems) in what follows (see fig. 4.1). This is both our primary area of interest and a more rigorous language than the corresponding terms of the social sciences, and it ties the discussion immediately to one of its most important applications.

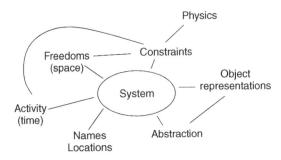

Figure 4.1: An informal diagram of associations within a system, which shows the main aspects for consideration.

4.2   Data structures and processes

There are many ways to classify systems. At the most basic level, one may speak of two kinds of system: those that are *dynamic* and those that are *static*.

A *static* system is often referred to as a *data structure*, rather than a system, that is, it is a systematic organization of its parts or resources (a form of data), which does not change. An archive is an example of this, as is a building, as is the book that you are reading.

Definition 7 (Data structure) *A data structure is an ordered (systematic) collection of resources or records that are interrelated.*

Example 13 *A building is a static data structure: it is a regular arrangement of parts that contribute towards a function. A library, or archive, is another example in which individual records are organized in an orderly pattern, with a scheme for relocating the information. The book you are reading is a third example. The functional components in these examples are somewhat diverse in their variety, but all the examples share a common feature: they are organized patterns designed to serve a purpose.*

For a system to exceed the archival character of a museum, there has to be some activity. A *dynamical system* is a system that evolves in time with a *rate of change*; it usually produces something and experiences a number of different operational states, such as *running*, *stopped*, *broken*, *waiting*, and so on. Dynamical systems are a more interesting class of systems, because they open up literally a whole new dimension (time) for organization. Human–computer communities belong to this group.

In order to describe the activity within a system, it is useful to define the notion of a *process*. A system may comprise several independent processes.

Definition 8 (Process) *A process is a unit of activity within a system. In order for something to happen, there needs to be something with a freedom to change. A process therefore comprises a set of resources, together with an algorithm (a sequence of operations) for manipulating them. A process is unleashed by the freedoms of the system and restricted by its constraints.*

A process can operate on a data structure and alter it. The sum of a process and a data structure is thus a dynamical system.

Definition 9 (Dynamical system) *A dynamical system is a set of processes that act on a data structure.*

A computer program is a process in which an algorithm changes the data in the computer. A maintenance procedure, such as cleaning a building, is a process that changes the state of organization of the building's resources.

Example 14 Active processes: *a process is a combination of resources associated with an executed task. It includes a code text, which contains instructions and algorithms for data manipulation, and it comprises data and resources associated with the task. In a multi-process environment, each process is an independent object with its own progress documentation (stack and index markers in computers), so that, if the process should be interrupted, its current state can be saved and later resumed, without loss of integrity.*

Example 15 Passive data: *the purpose of a process is to manipulate some data. Such data are often arranged in some non-trivial structure, which makes them easily accessible to the algorithms of the process. A file system is an example of passive data, as is a database.*

4.3   Representation of variables

To describe a system in definite terms, one needs to identify properties of its resources that can change or vary. Without any such variation, a system would be truly uninteresting. There are two kinds of information in a system:

Definition 10 (Resource information) *The information that is used and produced by the system as part of its functioning. This is the fuel and produce of the system. This information might be sent back and forth between different functional elements in the system, between* clients *and external* service providers, *or even between the system and its storage.*

Definition 11 (Algorithmic information) *The information on how to achieve the task, or generate the produce of the system is contained in a detailed programme of steps and procedures. This includes control information on which way to branch as a result of a question, and information about the initial state of the system.*

Describing the actual characters of a system's resources requires abstraction. One must use functions that vary with the basic parameters of the system (e.g. time) to *represent* those properties. The basic properties of a dynamical system are usually labelled q_i, for $i = 1, 2, 3, \ldots$. A function is a mapping from the parameter space into a *range of values*. The range is the set of possible values that the variable can change into.

What might $q(t)$ represent?

1. A number of objects

2. The value of an object

3. An average value

4. The size of an object

5. The shape of an object.

There are no rules about what can and cannot be represented by a variable. The challenge is only to find a sensible and sufficient representation of a phenomenon, to some appropriate level of approximation.

Example 16 *In a university or a college, students and lecturers follow a system in which they are allocated rooms to be in, at different times. We could choose to analyse the behaviour of this system in a number of ways. One could, for instance, measure the number of students in each lecture room, as a function of time. One could measure the average number of students to get an idea of attendance. One can characterize the rate of work, or how fast the contents of the blackboard change. One could attempt to measure the rate of learning (how would we measure this?). One could measure room temperature, or air quality and relate it to student attentiveness, and rate of learning. In each case, one must decide what it is that one is interested in finding out and characterize that mathematically in terms of measurable quantities.*

The lesson one learns from natural sciences is that models of systems are *suitably idealized representations*, not exact and unique facsimiles of reality.

4.4   The simplest dynamical systems

Complex systems can often be built up by combining simpler ones. An understanding of the behaviour of simple systems is therefore a reasonable place to begin, in order to fathom the greater complexities of more realistic systems. There is something to be learned from even the simplest mechanical devices.

The property that characterizes dynamical systems is that their resources change in time; but how? What kind of change can be expected? The theory of change is the study of differential and difference equations. These predict three basic kinds of change:

- Monotonic change (growth, decay)

- An oscillatory change

- Chaotic change.

These behaviours can be found in well-known places.

Example 17 *A simple pendulum is a system that provides a time service. It can go wrong if one of its components fails. The pendulum is a collaboration between a force, a pivot, a string and a mass. A pendulum has an oscillatory motion that gradually dies away. It is a combination of oscillation and decay.*

It is important to capture the essence of systems in a way that is conducive to analysis. For that reason, we begin with the simplest of mathematical considerations. A mathematical definition of a system is the following:

- A set of variables $\{q_i(t)\}$, where $i = 0, 1, 2, \ldots$ (information describing resources) whose values can change in time.

- A set of rules $\{\chi_i\}$ that describes how the variables change with time.

- A definition of the rates $\{\frac{dq_i}{dt}\}$ at which the different variables change in time, for deterministic systems, or a probable rate with which transitions occur non-deterministically.

These three things are the basic requirements that are necessary and sufficient to comprise any system that changes in time. This extends from the simplest of mechanical devices to the most complex chaotic combinations of elements. To make predictions about a system, one also needs to know what state it was in at some known or *initial time*. Such information is known as a boundary condition.

Example 18 *In a computer system, one has variables that characterize state; for example, the amount of data on a disk, the rate of processing, the number of users, and so on. A set of physical rules governs the hardware at the level of electronics, software provides a set of rules for program execution and policy provides an even higher-level set of rules about how the hardware and software should be used. The system is not static, so we find a rate of change of data on a disk, a rate of change of number of users, and even a rate of change of 'rate of processing', as jobs are turned on and off.*

Any larger definition of a system that we concoct must contain the basic definition above. In system administration, we are concerned with far more complex systems than can be described with the aid of a few simple variables. One is forced to deal with approximation, as one does in the natural sciences. This is an unusual remedy for computer science, which is more at home with logical propositions and exact theorems. Nonetheless, the lesson of the natural sciences is that complexity abhors precision, and forces one to embrace approximation as a tool for making headway.

4.5 More complex systems

In general, we need to describe a complete organization, with interconnecting data structures and inter-communicating sub-processes. Note that the word 'organization' is ambiguous in this context, in that it describes both an attribute of a system (how well it is organized) and a name for a system (a company or an other institution). We shall henceforth limit the word 'organization' to the first of these meanings and refer to the second by the term *enterprise*. Adding these notions, we have the following:

Definition 12 (System) *A system is a set of resources (variables and processes) and consumers, together with descriptions of how those resources are organized and how they develop in time. This total description defines the arena in which the system develops. It prescribes the possible freedoms one has for change and the constraints imposed externally and internally for activity within the system.*

At a superficial level, we can identify the key elements in common systems.

Example 19 *A public transport system has a set of resources (buses and trains) $q(t)$ that are constrained to move on roads or on rails, and which run on expended fuel. The details of change in the system are partly determined by policy (the schedule), partly by environmental considerations and critically by the natural laws that govern the physical processes driving the system (all these are in χ). The rate of flow of transport $\frac{dq}{dt}$ is related to the overall change in the system.*

Example 20 *A web server is like a query-handling system or a help desk. These are all systems in which the number of incoming, unanswered requests $q(t)$ is changing with time. A protocol for handling the requests constrains their expedition by various algorithms that manipulate resources. The rate at which the system expedites requests is $\frac{dq}{dt}$.*

4.6   Freedoms and constraints

Any system that has some kind of predictable or regular behaviour is a balance between the freedom to develop and change and a number of constraints that limit the possibilities for change to predictable avenues (see table 4.1).

> **Definition 13 (Degree of freedom I)** *A degree of freedom is a potential for change within a system. Freedom to change is usually represented by a parameter that can be varied over a range of values.*

For example, in a service-based system, the freedom to accept new clients permits the system to expand.

Example 21 *A program might run equally well on an array of 1 or 20 computers. The number of computers is thus a freedom. Similarly, a system might have the freedom to increase its use of memory; that is then a freedom of that system.*

At the machine level, one has the freedom to choose the software and the hardware platform that is used to carry out a job.

> **Definition 14 (Constraint)** *A constraint is a limitation on the possible changes that can occur on variables or parameters in a system. This often takes the form of a rule, or a parameter inequality.*

Example 22 *In the client-service system above, a constraint could take the form of limited resources for handling client queries (a maximum number that can be expedited per unit time); similarly, it could represent a policy constraint that denies the service to certain clients, or limits their availability to the service. Another constraint is a productivity goal.*

At the machine level, it might be that a program runs equally well only on computers with a particular operating system; in that case, the choice of operating system becomes a constraint. Similarly, the amount of memory available to a program is normally limited by

Table 4.1: Freedoms and constraints in organizations.

Freedoms	Constraints
Expansion	Available budget
Time	Deadline, or limited duration
Space	Specific location, limited space
Rate of service	Throughput of bottleneck

the total amount on that computer; the amount of memory is thus a constraint on program execution.

Note the words 'freedom' and 'constraint' are used in a strict sense. Their usage does not imply the advantages or disadvantages they confer on a system. Do not be tempted to bestow these terms with social connotations; for example, a smart human might creatively use a constraint to his or her advantage, but that is not the same as it being a freedom. For example, a geographical constraint in which an organization is limited to one building might be turned into a positive attribute by arguing that this lends cohesion to the organization. This does not mean that the constraint is really a freedom. The advantage here is made possible only by virtue of another freedom, namely the freedom to be creative in that context.

In addition to objects that are manipulated and changed, most systems have input and output channels, where information is exchanged with external actors—the 'environment'.

4.7   Symmetries

Symmetries are descriptions of what can be changed in a system without affecting the system's function. Determining what is *not* important to the functioning of a system is a way of identifying degrees of freedom that could be manipulated for strategic advantage. Knowing about these freedoms might be an advantage to someone managing the system. A change that does not affect a sub-system might nevertheless result in an advantage elsewhere. If, for instance, we come up with a system model in which results depend on a specific choice, where in fact that choice does not matter, then we know that model must be wrong.

Example 23 *If one moves every user of a desktop workstation to a different workstation, the organization will still function in exactly the same way, if the workstations are all alike. This freedom to reseat people might allow groups of workers to sit in close proximity, for verbal communication, or it might allow workers to be spread out to balance the load*

Example 24 *If one swaps all of the buses on route 20 with those on route 37, the service will not be affected, provided the buses are comparable. So buses can be rotated and checked for maintenance in parallel to those in service.*

Example 25 *If one relabels every file on the system, in all references, it will continue to work as before.*

4.8   Algorithms, protocols and standard 'methods'

Systems embody a cooperation of parts and are often surrounded by an environment of unpredictable occurrences. The addressing of the cooperation between parts of a system needs to be formalized by defining the mechanisms that contribute to it; to cope with the unpredictable, external events, mechanisms and procedures need to be introduced that offer predictability. These requirements are covered by the concepts of algorithm and protocol.

Definition 15 (Algorithm) *An algorithm is a recipe, or a sequence of steps and decisions that solve a particular problem within a system. Algorithms are sometimes referred to as* methods *in the parlance of programming. A formal definition of an algorithm can be provided in terms of Turing machines (see Lewis and Papadimitriou (1997)).*

An algorithm is a reasoned flow of logic, designed to efficiently perform an operation or a sequence of operations. An algorithm is what one finds if one opens up the black box of an operator and peers at its inner workings. Algorithms are not arbitrary, though there might be several algorithms that solve the same problem. In that situation, a policy decision is required to determine which algorithm is to be used.

Definition 16 (Protocol) *A protocol is a standard of behaviour, or a strict rule of conduct that ensures that one part of a system behaves in a predictable and comprehensible way. Protocols ensure that one part of a system is able to cooperate with another, and that the integrity of the process is maintained, that is, information is not lost or misunderstood. A protocol is formally a* constraint *on a process.*

Protocols are used to ensure consistency and to avoid error when executing a process. Typical examples of protocols are used when two communicating parties must understand one another.

Example 26 *When a computer sends data over a network, it does so by coding the data into a stream of bit pulses. The machine receiving the message would have no idea how to decode the message unless a pre-agreed standard of behaviour were established in advance: for example, the first 8 bits are a number, the next 24 bits are three 8-bit characters, and so on. Without such a protocol for interpreting the stream of bits, the meaning of each bit would be lost.*

Another example of a protocol is a response plan in case of emergency, for example, fire, security breach, war, and so on. The purpose of a strict code of behaviour here is to minimize the damage caused by the event as well as to mobilize countermeasures. This ensures that all parts of the system are activated and informed in the right order, thus avoiding confusion and conflict.

Protocols are a part of system policy. They are arbitrary, even though some properties of protocols can be analysed for efficiency and efficacy. They are strategic choices.

Example 27 *Computer security intrusion: (i) freeze the state of the system, (ii) gather standardized information and evidence (copy the current memory to a file), (iii) pull the plug to avoid setting off any logic traps, (iv) report the intrusion to law enforcement, (v) reboot the system with a CD-ROM or a trusted read-only medium to avoid logic traps set by an attacker.*

The goals of this protocol are as follows: protect the system, secure evidence, repair the problem and obey the law, in that order. If a different set of priorities were in force, the protocol might be changed. It is designed by a person with experience of intrusion, to take into account all of the interests above. An inexperienced person might not understand the protocol, but by following it, the goals of the system will be upheld. The technology of computer security is to a large extent a litany of protocols, designed to avoid unwanted behaviour.

The difference between an algorithm and a protocol is subtle. An algorithm is a sequence of instructions or steps; a protocol is only a specification of what kind of steps is allowed. One is a process specification, and the other is a constraint on a specification. Protocols do not make decisions; they are pre-agreed standards.

4.9 Currencies and value systems

While autonomous, mechanical and electronic systems can be described purely by simple physical principles, once humans are involved in a system, human values necessarily become part of the equation. This complicates a system, and many engineers find this interaction disturbing because society has an ingrained culture of treating human values as fundamentally different to physical measurements.

Some might even say that the idea of modelling human values, in the same way that one models physical processes, would be disrespectful. As we shall see, however, there is no basic impediment to writing formal rules for human values and concerns; these merely extend the complexity of systems by introducing additional constraints and boundary conditions. Indeed, to represent financial and economic aspects of a system, one already does precisely this.

Human emotion allows us to attach importance, and hence value, to almost anything. Not all values can necessarily be traded, as money or goods can. Happiness, for example, cannot normally be traded, say, for food, but it might be reasonable to say that happiness of a workforce could be traded for efficiency, in a system with a Draconian work ethic. The key to analysing the interactions between human values and physical resources is to assign to them arbitrary scales that can then be calibrated so that rules can be written down.

4.9.1 Energy and power

Nature's fundamental system bookkeeping currency is called *energy* in the physical sciences. Energy is simply an abstract representation of the level of activity in different parts of a system. It has an arbitrary value, which is calibrated against the real world by convention, that is, by using some system of units for measurements, and by adjusting certain constants in physical laws to 'make it right'.

When parts of a system interact, they are said to exchange energy. Each new interaction has its own equation, and requires a 'coupling constant' that calibrates the effect of the arbitrarily dimensioned energy transfer from one part, to the measured effect in the other part. Whatever new age writers might believe, there is really only one kind of energy, just as there is really only one kind of money. People have different conventions for referring to energy or money in different places, but the idea is the same.

Just as saved money can be traded for services, stored (potential) energy can be traded for activity. Physicists have long been dealing with this simple bookkeeping concept without questioning its validity, so it should not be a surprise that the same idea can be applied to any form of currency.

Example 28 *Basic energy requirements are at the heart of all systems, grounded in the physical world. Machines require power and humans require nutrition. A failure to provide these leads to a failure in the system.*

4.9.2 Money

In days of old, one used silver, gold and other riches to trade for goods. Sufficient gold is still kept in reserves around the world to be able to trade paper currency for its value in gold if the bearer demands it. Today, however, we use money to represent only the promise of something real. Our abstraction of wealth has reached such extremes that we buy and sell shares in the estimated value that a company *might have*, if its assets were sold. The value of something, in our modern world, is clearly not an intrinsic physical property, like its electric charge, or its mass. It is a fictitious quantity based as much on trends and feelings as it is on physical resources.

In short, money is worth what everyone *believes* it is worth; no more and no less. It is a sobering thought, therefore, that our contemporary society and all of its systems run on money. Money is used to measure the cost of building a system, the cost of running it and the value of what it produces. We speak of *assets* or *resources* to mean the things of value that are recognized by a potential buyer.

How should money be represented in formal (mathematical) models of systems? How does it relate to other measurable resources, like time, space, equipment, and so on? Clearly, 'time is money' because humans will not work for nothing: we require the promise of reward (money). Space and commodities 'are money' because we value anything that is not in infinite supply. But, if the value of these things depends on the fickle moods of the human actors in the system, how can it be used to represent these other things?

To say that a relationship exists between time t and money m, for instance, is to say that

$$t = f(m), \tag{4.1}$$

for some function f. In the simplest case, this would be a simple linear relationship:

$$t = km + c, \tag{4.2}$$

where k and c are arbitrary constants, which can be fixed by calibration. They represent what it costs to buy someone's time.

The value of objects is arbitrary (a matter of policy), so any relationship could change from system to system, or from time to time within a system. It is not always necessary to think in terms of money. If the issue is that of a cost to the system, then money takes the form of a constraint or a limitation.

4.9.3 Social currency and the notion of responsibility

Money is not the only form of abstract currency. Humans value other things too: peer status, privilege and responsibility are all used as measures of social value, every bit as real as money.

The meaning of responsibility is taken for granted in common speech, but it has several different meanings in the running of systems. If we are to analyse the concept formally, it is necessary to relate responsibility to more tangible concepts.

- **Cause and effect**: For a machine, one says that a component is responsible for an action if the action depends on it somehow. Responsibility is thus associated with an implied trust or *dependency*. It is simply the law of cause and effect, which is central to the analyses in this text. For example, the printer is responsible for writing information to paper; electricity is responsible for making a lamp work.

For humans, responsibility has two meanings, and these are somewhat subtle:

- **Responsibility for completing a task (policy)**: This refers to the assignment of a task to some agent within the system (a person or a department, for instance). It is often laced with connotations of a penalty for failure, or a reward for success. For example, 'You are responsible for getting this job done before March!' This is a combination of an arbitrary policy decision and a constraint. The correct way to model this is thus as an externally controlled condition, together with the relevant constraint on the system resources (in this case: time). Rewards and penalties can be modelled by introducing a fictitious currency (see below).

- **Responsibility for actions (blame)**: As in the case above, this is used to imply the direction of a reward or a punishment associated with the completion, or failure of something in the system. However, this kind of responsibility can be assumed or transferred from one object to another. A manager or a commanding officer will often be made (or held) responsible for a whole department or unit, for instance. This transfer of responsibility is not necessarily related to cause and effect, since the manager's actions do not necessarily have any direct influence on what transpires. It covers the situation where a leader trusts in the outcome of a hidden process. For example, an accountant trusts computer software to calculate a correct answer and holds the software producer responsible for the result, but his boss holds him responsible. This can be modelled as a combination of policy with a fictitious currency scheme.

In modelling human values, one deals with issues such as status, prestige and other emotional considerations. These are social currencies, analogous to material wealth or resource riches. Social status amongst humans can affect the details of how policy applies to them, and can act as an incentive or a deterrent to certain kinds of behaviour.

Example 29 *A person with sufficient privilege might be given access to parts of the system that are unavailable to others. Emotional reward can be a sufficient motivation to complete a task, or conversely, emotional pressure can be a hindrance or even a factor motivating sabotage. Disgruntled employees have been responsible for the theft of millions of dollars through their abuse of human–computer systems, particularly in the financial sector.*

One begins to appreciate the complexity of human–computer communities when one attempts to represent the exchanges that take place on the human side of systems. For computer administrators, this type of modelling is normally only done in connection with the security of the system, that is, in threat analysis. Software engineers who design software for critical systems need to think about such issues that might lead to human error. Businesses and people-run enterprises such as universities and the military depend critically on the actions of humans; thus, social currency tends to dominate these systems.

4.10   Open and closed systems: the environment

If we wish to describe the behaviour of a system from an analytical viewpoint, we need to be able to write down a number of variables that capture its behaviour. Ideally, this characterization would be numerical since quantitative descriptions are more reliable than qualitative ones, though this might not always be feasible. In order to properly characterize a system, we need a theoretical understanding of the system or sub-system that we intend to describe. These are a few important points to be clear about.

Dynamical systems fall into two categories, depending on how one is able to analyse them. These are called *open systems* (partial systems) and *closed systems* (complete, independent systems).

- **Open system:** This is a *sub-system* of some greater whole. An open system can be thought of as a black box that takes in input and generates output, that is, it communicates with its environment. The names *source* and *sink* are traditionally used for the input and output routes. What happens in the black box depends on the state of the environment around it. The system is open because input changes the state of the system's internal variables and output changes the state of the environment. Every piece of computer software is an open system. Even an isolated total computer system is an open system as long as any user is using it. If we wish to describe what happens inside the black box, then the source and the sink must be modelled by two variables that represent the essential behaviour of the environment. Since one cannot normally predict the exact behaviour of what goes on outside a black box (it might itself depend on many complicated variables), any study of an open system tends to be incomplete. The source and sink are essentially unknown quantities. Normally, one would choose to analyse such a system by choosing some special input and consider a number of special cases. An open system is internally *deterministic*, meaning that it follows strict rules and algorithms, but its behaviour is not necessarily determined, since the environment is an unknown.

- **Closed system:** This is a system which is complete, in the sense of being isolated from its environment. A closed system receives no input and normally produces no output. Computer systems can only be approximately closed for short periods of time. The essential point is that a closed system is neither affected by, nor affects its environment. In thermodynamics, a closed system always tends to a steady state. Over short periods, under controlled conditions, this might be a useful concept in analysing computer sub-systems, but only as an idealization. In order to speak of a closed system, we have to know the behaviour of all the variables that characterize the system. A closed system is said to be completely *determined*[1].

---

[1] This does not mean that it is exactly calculable. Non-linear, chaotic systems are deterministic but inevitably inexact over any length of time.

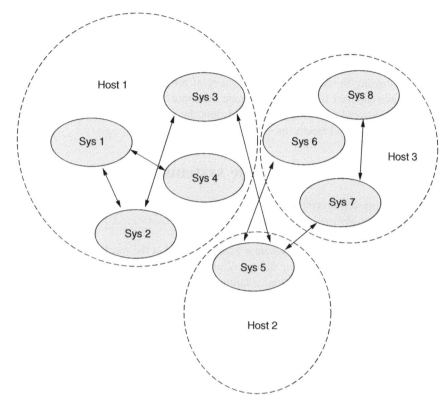

Figure 4.2: A complex system is a causal web or network of inter-communicating parts. It is only possible to truly isolate a sub-system if we can remove a piece of the network from the rest without cutting a connection. If we think of the total system as $S(x_1 \ldots x_n)$, and the individual sub-systems as $s_1(x_1 \ldots x_p)$, $s_2(x_p \ldots x_n)$, and so on, then one can analyse a sub-system as an open system if the sub-systems share any variables, or as a closed system if there are no shared variables.

Suppose we want to consider the behaviour of a small sub-system within the entirety of a much larger system (e.g. a computer on the Internet, or an animal in a complex ecology); first, we have to define what we mean by the sub-system that we are studying. This might be a straightforward conceptual partitioning of the total system, but conceptual decompositions do not necessarily preserve causal relationships (see fig. 4.2).

In fact, we might have to make special allowances for the fact that the sub-system might not be completely described by a closed set of variables. By treating a sub-system as though it were operating in isolation, we might be ignoring important links in the causal web. If we ignore some of the causal influences to the sub-system, its behaviour will seem confusing and unpredictable.

The principle of causality tells us that unpredictable behaviour means that we have an *incomplete description* of the sub-system. An important difference between an open system and a closed system is that an open system is not always in a steady state. New input changes the system. The internal variables in the open system are altered by external perturbations from the source, and the sum state of all the internal variables (which can be called the system's *macrostate*) reflect the history of changes that have occurred from outside. For example, suppose that we are analysing a word processor. This is clearly an open system: it receives input and its output is simply a window on its data to the user. The buffer containing the text reflects the history of all that was inputted by the user and the output causes the user to think and change the input again. If we were to characterize the behaviour of a word processor, we would describe it by its internal variables: the text buffer, any special control modes or switches, and so on.

4.11   Reliable and unreliable systems

Definition 17 (Unreliable system) *An unreliable system is one that attempts to carry out its function without verification or guarantee of success.*

Unreliable systems are used either where a trust is placed on the mechanisms of the system to perform their duty, or where a failure is unimportant. For instance, if an unreliable system repeats its actions fairly often, a failure might be corrected at a later time with a high probability. In some cases, it does not matter whether a process fails; we might use an unreliable system to approximately probe or test a random variable, without too much ado.

Definition 18 (Reliable system) *A reliable system is one that tries, verifies and repeats its actions until the result succeeds.*

Reliable systems can prove to be expensive, since resources have to be applied to monitor and correct any errors or failures that occur.

Example 30 *The Internet Protocol (IP) has two control layers: the User Datagram Protocol (UDP) and the Transmission Control Protocol (TCP) that are unreliable and reliable respectively. UDP is used for 'one shot' requests, such as name service look-ups and route-tracing probes, where a reply is not necessarily expected. TCP is used for more formal behaviour of network communication protocols where certainty is demanded.*

Applications and Further Study 4

- *Understanding fundamental issues in system analysis.*

- *Examining and classifying systems using the concepts described here.*

- *Isolating the basic behavioural traits of a system.*

- *Identifying the freedoms and constraints to better understand how a system might be modified or improved.*

5

Sets, states and logic

The concept of a *state* or condition of a system will be central to several discussions. In order to discuss states, we need to have variable quantities that can take on a set of values. The state of any object is its value at a given place and time. Usually, variables cannot take on just any value: they take values from a specified *set* of values.

We need a language that is general enough to be applied to a wide range of situations, but which is general enough to make clear and verifiable statements about systems. The language of sets and mathematics allows us to state things precisely, and will be of great utility in describing Networks and System Administration.

5.1 Sets

A set is the mathematical notion of a collection of related things. The idea is general and is not limited simply to numbers. Sets describe all manners of useful collections. We must take care not to confuse the term 'group' with 'set', as these words both have special meanings in mathematics. Sets are denoted by lists enclosed by curly braces $S = \{\ldots\}$.

Example 31 *Example sets include days of the week:*

$$D = \{\text{Sunday, Monday, Tuesday, Wednesday, Thursday, Friday, Saturday}\} \quad (5.1)$$

Types of operating system:

$$O = \{\text{Windows, Solaris, GNU/Linux, MacOS}\} \quad (5.2)$$

Files owned by user mark:

$$F = \{\texttt{INBOX, file.txt}, \ldots\} \quad (5.3)$$

Directories owned by mark:

$$D' = \{\texttt{Mail, Documents}, \ldots\} \quad (5.4)$$

Analytical Network and System Administration. Managing Human–Computer Networks   Mark Burgess
© 2004 John Wiley & Sons, Ltd   ISBN 0-470-86100-2

System load average values measured in an experiment:

$$L = \{0.34, 0.42, 0.45, \ldots\} \tag{5.5}$$

Example 32 *A set of rules is often called a* policy.

The union of two sets is the combination of all elements from both sets, for example, the catalogue set is the union of directories D' and files F:

$$C = F \cup D'. \tag{5.6}$$

The union does not contain more than one copy of an element, so $\{A, B\} \cup \{A\} = \{A, B\}$. Similarly, files F are a subset of the catalogue C:

$$F \subset C. \tag{5.7}$$

If we let S be the set of all secure policies, and P be the set of policies in use by an organization, then

$$P_S = S \cap P \tag{5.8}$$

is the intersection or overlap between these sets (see fig. 5.1) and represents the sets that are both secure and in use. If S is a subset of a set X, then the rest of X that is not in S is called the complement and is written $X - S$ or simply $-S$ (also written $\neg S$).

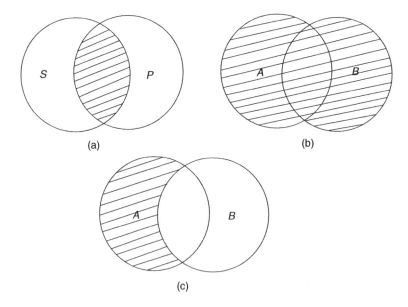

Figure 5.1: Operations on sets: (a) the intersection of sets $A \cap B$ also written A **AND** B; (b) the union of sets $A \cup B$ also written A **OR** B; and (c) the difference $A - B$ also written A **AND NOT** B.

Example 33 *Some common sets include the following:*

- *∅ is the empty set, containing no elements.*

- *R^1 The set of real numbers (often written simply **R** or with a calligraphic R).*

- *Z is the set of integers.*

- *$R^n = R^1 \times R^1 \ldots R^1$ is the n-dimensional Cartesian space of real numbers, e.g. R^3 in three-dimensional Euclidean space.*

Note that the notation ¬ is used interchangeably with **NOT** and the complement operator '−' for sets.

5.2   A system as a set of sets

To describe a system, we must have a number of objects that can be related to one another and change in some way. Sets enter this picture in two ways:

- There is a set of objects that comprises the components of the system.

- Each component object can change its value or properties, by taking a new value in the set of values that it can possibly take.

Thus, our abstract picture of a system is a set of variable objects, each of which can take on a value from another set of values. These objects can be related to one another, and the values they take can change according to rules that are determined by the system. The objects are typically areas of memory in a computer system, specific people in a team, or even sets of users who work together. In chapter 6, we shall see that this picture also describes a *graph*.

Example 34 *Strings of symbols or operations (or symbols that represent operations) can be used to formulate some systems. If one can express a system in these terms, the development of the system becomes akin to the problem of transmitting information over a communications channel. This is one of the themes of this book.*

5.3   Addresses and mappings

Since we need to refer to the elements in a set, we give each element in a system a *name* or an *address*, that is, a label that uniquely identifies it. There is thus a mapping from addresses to objects (see figs. 5.3 and 5.4).

Example 35 (Functional mappings and dependencies) *In fig. 5.2, we see a mapping from a set of five elements to another set of three elements. In this case, the mapping does not exist for all elements. A similar illustration could be used to create a mapping from the set of computer architectures to the set of operating systems that can run on them, for example:*

$$O = \{\text{Windows, Solaris, GNU/Linux, MacOS}\}$$
$$A = \{\text{Sparc, Intel, ARM}\} \tag{5.9}$$

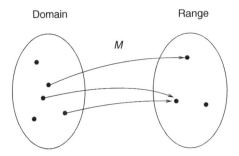

Figure 5.2: A function M is a mapping from one set, called its domain, to another set called its range. The domain and the range of the function do not have to be sets of the same size.

So that, if we create a mapping function f, one may write

$$o = f(a) \tag{5.10}$$

where $o \in O$ and $a \in A$, e.g. Windows $= f(\text{Intel})$.

The advantage of writing the function is that it can be written for any variable that takes values in the respective sets. This allows us to solve equations for valid mappings, for example, what are the solutions of

$$\text{Windows} = f(x)? \tag{5.11}$$

Depending on whether the map is *single–valued* or *multi–valued*, there might be zero, one or more solutions.

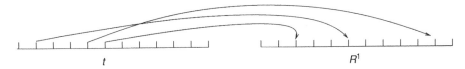

Figure 5.3: A function of time, $q(t)$, is a mapping from an arbitrary set of times into the set of real numbers R^1.

If we have a continuous set of values, that is, a set that takes on a different value for any real number parameter, the function can be described in terms of known differentiable functions. See the example in fig. 5.3.

5.4   Chains and states

Using the idea of sets, we can define states to be the current values of the objects in the system:

Definition 19 (State) *A state of the system (also called a microstate) is a value $q \in Q$ associated with a single addressable entity or location within the system.*

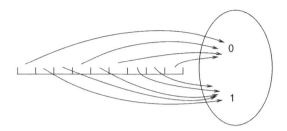

Figure 5.4: The memory of a computer is a mapping from bit–addresses to the binary values $\{0, 1\}$.

A chain is a sequence of events X_n $(n = 0, 1, \ldots, N)$, where each event X_n takes a value q_i $(i = 1, \ldots I)$ called a *state* at address n, belonging to a set of values Q. The integers n normally label the development of the process in discrete steps, and are interpreted as time intervals, but they could also be spatial locations, for example, a sequence of values on a hard disk.

The *transition matrix* T_{ij}, is written in a variety of notations in the literature, including the following:

$$T_{ji} = p_{ji} = |\langle q_j | q_i \rangle|^2$$
$$= P(X_{n+1} = q_j | X_n = q_i). \tag{5.12}$$

It represents the probability that the next event X_{n+1} will be in a state q_j, given that it is currently in the state q_i. By discussing the *probability* for transitions, we leave open the issue of whether such transitions are deterministic or stochastic. There are a number of possibilities. If $T_{ij} = 1$ at X_n, for some i, j, the process is *deterministic* and one may write the development of the chain as a rule

$$X_{n+1} = U_t(X_n, \ldots, X_0), \tag{5.13}$$

in the general case. If $T_{ij} < 1$ at X_n, for all i, j, n, the process is *stochastic*. If T depends only on the current state of the system,

$$P(X_{n+1} = q_i | X_0, X_1, \ldots, X_n) = P(X_{n+1} = q_i | X_n), \ \forall \, n \geq 1, \tag{5.14}$$

then the chain is said to be a *Markov chain*, or *memoryless*. Markov processes are also called *steady state*, or *equilibrium processes*. If T depends on the whole history of $\{X\}$, then it is called a non-equilibrium, non-Markov process. A state is called *persistent* if

$$P(X_{n+m} = q_i | X_m = q_i) = 1, \text{ for some } \ n \geq 1 \tag{5.15}$$

and *transient* if

$$P(X_{n+m} = q_i | X_m = q_i) < 1. \tag{5.16}$$

The terms periodic, aperiodic and ergodic also describe chains in which the processes return to the same state. Readers are referred to Grimmett and Stirzaker (2001) for more about this.

5.5   Configurations and macrostates

If we want to collectively talk about the state of all of the objects or locations in a system, at a given time, we use the term configuration. In large systems, a precise system configuration consists of too many independent values to describe in detail and we move to a statistical, averaged description.

Example 36 *One of the central issues in system administration is the management of device and software configurations.*

Definition 20 (Configuration) *A configuration of the system is the vector of values (microstates) of all its resource variables $q_i(t, x, \ldots)$, at a given time t, for all positions x and other parameters so on.*

Definition 21 (Macrostate) *A macrostate is an averaged, collective description of a configuration that captures its statistical behaviour in a simplified, high–level description. Any function of all the microstates that leads to a summarized value can be called the macrostate of the system. It can therefore be represented as a set of probability distributions for the probabilities that the ith resource variable is in a state s_i: $P_i(x = s_i)$.*

The concepts of microstates and macrostates are not necessarily unique; they depend on a particular viewpoint of a system, and thus they can be defined, in each case, to discuss a particular issue in a particular way. The main advantage of a description in terms of states is the clarity and definiteness that such a description brings to a discussion.

Example 37 *To discuss the security configuration of a computer, it is not necessary to go to the level of bits. We may take the microstates of the system as being the different permission combinations for the users, for example, $q_1 = (\texttt{read}, \texttt{write}; \texttt{mark})$, $q_2 = (\texttt{read}; \texttt{sally}), \ldots$. A macrostate of the system is one possibility for the collective permissions of all files and objects of the system, for example, $(\texttt{file1}, q_1), (\texttt{file2}, q_2), \ldots$.*

Example 38 *A signal sent from a computer to another, as a binary stream, is represented as a function $q(t)$. The values that $q(t)$ can take, at any time, lie in the set $Q = \{0, 1\}$, which is the state space for the data–stream.*

Example 39 *A user types at a computer terminal, onto a QWERTY keyboard. The data stream can be represented as a function $q(t)$, that maps a moment in time to a key typed by the user. The state space of the stream is the set $Q = \{q, w, e, r, t, \ldots\}$.*

Example 40 *Data stored on a computer disk are recorded as binary patterns on a series of concentric rings. Each bit is located physically at a location parametrized by its distance r, from the centre, and angle θ from some reference line on the disk. The changing binary pattern of data on the disk is thus a function $q(r, \theta, t)$ of position and time. At its lowest level of representation, the state space of the data is the binary set $Q = \{0, 1\}$.*

5.6   Continuum approximation

It is often helpful and even important to be able to describe systems in terms of smoothly varying variables. Just as one would not imagine describing the flow of water in terms of individual discrete atoms, one does not attempt to describe collective behaviour of many discrete sources in terms of discrete digital changes. The transition to continuous processes is straightforward. A discrete chain

$$X_0 = q_i, X_1 = q_j, \ldots X_n = q_k, \tag{5.17}$$

is replaced by a function q of a continuous parameter t, so that a time interval from an initial time t_i to a final time t_f maps into the state space Q:

$$q(t) : [t_i, t_f] \to Q. \tag{5.18}$$

The discrete event notation X_n is now redundant and we can now speak of the value of the state at time t as $X(t)$. X is the symbol used in most mathematical literature, but we prefer the symbol $q(t)$ here. A set of parallel chains, labelled by a parameter x, and time parameter t is thus written $q(x, t)$. The transition matrix is now a function of two times:

$$T(t, t') = |\langle q(t')|q(t)\rangle|^2 = T(\tilde{t}, \bar{t}) \tag{5.19}$$

where

$$\tilde{t} = t - t'$$
$$\bar{t} = \frac{1}{2}(t + t'). \tag{5.20}$$

If there is no dependence on the absolute time $\bar{t}$, the process is said to be *homogeneous* or *translationally invariant*, otherwise it is *inhomogeneous*. A process is said to be memoryless if it does not depend on $\bar{t}$, since then it is in a steady state with nothing to characterize how it got there.

5.7   Theory of computation and machine language

In what we might call Traditional Computer Science, computer systems are described in terms of logical propositions—as automata, working on a usually deterministic set of rules. This area of computer science includes algorithmic complexity theory and automated theorem proving, amongst other things (see for instance Lewis and Papadimitriou (1997)). Recently, the attempt to formalize simple processes has been applied to software engineering, with language-like constructions such as the Unified Modelling Language (UML) (see for instance Sommerville (2000)) that apply the forms of rigour to processes that include unpredictability. Although such theories rule a domain of computer science that deals with determinism, this book begins, in a sense, by questioning their broad validity as an approach to human–computer management. Rather than applying a machine model to human–computer interaction, the claim here is that one should apply the scientific, behaviourist approach, which has been developed to deal with real-world complexity and uncertainty in a systematic way.

Logic is a description of very elementary processes—too elementary to capture the essence of the interaction between networks of humans and computers in actual use; nonetheless, one cannot describe the world without understanding it 'atoms'. We need logic and reasoning to describe the elementary building blocks before adding the complicating interactions, such as patterns and algorithms. We also need descriptive theories of unpredictable behaviour (computer input is always unpredictable) and of multiple levels of complexity. But as layer upon layer of complexity are compounded, we also need broader scientific ideas that retain determinism in the form of *causation*, but which abandon the idea of having a complete and exact description about systems.

5.7.1 Automata or State Machines

A finite automaton, or finite state machine (FSM), is a theoretical system that responds to changes transmitted to it by a data stream, by signalling *state*. Many mechanisms work in this way. Automata can be represented or realized in different ways (e.g. cellular automata, push-down automata). The important feature of automata is that they remember state information, that is, they have a memory of what has happened to them in the recent past. The amount of memory determines the type of automaton.

A finite automaton receives a string of input and makes changes in its internal states (i.e. its memory) depending on the values that it reads at its input. There are physical systems that behave in this way: for instance, multi-state atoms, atoms moving around on metal surfaces etc. If the input is a finite set of photons with frequencies matching different transitions, then that is a simple FSM. Some input strings result in final states, which are defined to be 'acceptable' and all others are unacceptable. Thus, an FSM can accept or reject strings of digital information. A finite automaton is described by the following components:

1. Q—a complete but finite set of internal states (i.e. memory/registers).

2. Σ—a finite alphabet of symbols or digits to be read (e.g. 0, 1; $A, B, C, \ldots$).

3. $T = Q \times \Sigma$—a transition function ($T_{12} = \langle q_1 | q_2 \rangle$) of deterministic state mappings, or of transition probabilities.

4. A start state (boundary condition) $|q_s\rangle$.

5. A constraint on the allowed transitions.

Automata can do only one type of job: they can distinguish between legal and illegal input. If the system reads in a string, and ends up in an acceptable final state, then the string is said to be legal. If the accepted final state is not amongst the set of legal states, then the input is classified as illegal.

If an input string (a chain) satisfies well-defined rules of construction, it is said to satisfy a *grammar*. Not all grammars can be parsed by just any automaton. For instance, a simple FSM cannot parse any language that admits an arbitrary level of recursion. The classic example is to think of how you could arrange the states and the transitions to parse text containing parentheses.

Example 41 *An FSM can parse a parenthesis embedded string, with maximum nesting level n if it has sufficient internal states. How many states does it need?*

$$(\,.\,.\ \ (\ \,.\,.\ (\,.\,.)\ \ \ .\,.\,.\ \ \ (\,.\,.)\,)\,)$$

To tackle arbitrary recursion, one needs a stack *that can remember symbols. The automaton can place a value onto the stack, and retrieve values from it. The stack structure is used in every modern computer because it mimics recursion precisely. If we take the parenthesis example, for every left bracket we add a number to the stack, then for every right bracket we remove one from the stack. The level of the stack is an indicator of which nesting level we are at. At the end of a valid string, the stack should be empty, else the brackets are not balanced. Automata with stacks are called* push-down automata.

Example 42 *The World Wide Web's Hypertext transfer protocol, HTTP, is a stateless protocol in the sense that when you visit a web page, it does not remember that you have been there before. However, extensions to the basic protocol using cookies and server-side sessions allow the server to remember the state of transactions that have transpired in the past, for example, if you have typed in a password, the state changes from 'not authenticated' to 'authenticated'.*

The generalization of the FSM is the Turing machine and is the model for all modern digital computers. We shall not discuss Turing machines in this book.

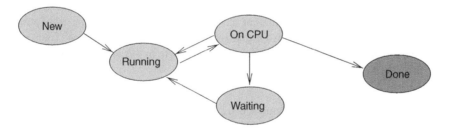

Figure 5.5: A finite state machine representation of a computer process scheduling history.

State machines are a common and useful way of describing certain kinds of systems in both a qualitative or quantitative manner. A state machine that is well known to computer science students is the transition diagram of an operating system process dispatcher (fig. 5.5). This describes transitions between certain states of the system for a given process. These states are not the most microscopic level of describing the system: we do not know the internal code instructions that take place in the operating system of the computer, but these states are characteristic of a 'black box' description of the process. The diagram encapsulates rules that the system uses to alter its behaviour.

A similar state machine is found in a rather different system: a service help desk, run by humans (fig. 5.6). The task being performed by these two systems is qualitatively similar to the dispatcher, so the similarity of states should not be a surprise. Of course, humans are not deterministic machines, so they cannot be modelled by exact rules, but the essence of their behaviour can be modelled statistically in any manner that is convenient to the

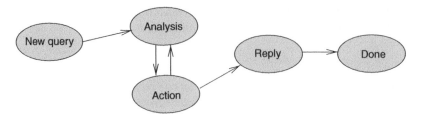

Figure 5.6: A finite state machine representation for a system administrative help desk.

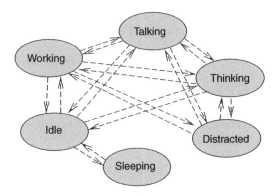

Figure 5.7: A finite state machine representation for a human being. This is clearly contrived, but this might be a way of modelling the main kinds of activity that humans undergo. A model of transitions between these states is unlikely to be describable by any realistic algorithm, but one could measure the time spent in each state and the likelihood of transitions from one state to another in order to improve the efficiency of the individual's work pattern.

design of a human–computer system (see fig. 5.7). The procedural aspect of the human system can be represented as an automaton, since it is predetermined and has a simple logical structure. Drawing the transition diagram for an FSM can be a very useful way of debugging problems in a system.

- Are there states that can never be reached?

- Is it possible for an FSM to get into a state from which it cannot emerge?

It might be necessary to change the rules of the system to disallow certain transitions, add more or even add extra states to distinguish between situations that a system cannot cope with. For systems that are probabilistic in nature, so-called hidden Markov models can be used. In a hidden Markov model, there are hidden variables that control how the transitions occur. The transition rules become themselves state-dependent.

5.7.2 Operators and operands

In order to discuss the active changes made within systems, using a formal framework, we introduce the notions of operators and operands. These terms are encountered routinely in mathematics and in computer science, and have a familiar and well-defined meaning.

Definition 22 (Operator) *An operator is a process that changes the value of a resource variable within the system, by performing an operation on it. It invokes a unit of change* Δq.

Definition 23 (Operand) *An operand is a resource q, within the system, which is operated on by an operator.*

A computer program is an operator that acts on the data within a computer. Human beings are operators when we perform operations. If we want to describe them formally, we need only find a suitable representation for an operator, acting on some operand that mimics the behaviour we need to model.

5.7.3 Pattern matching and operational grammars

Input to a system is, by definition, unpredictable. If it were not, it could be eliminated and the system could be replaced by a closed automaton. Systems interpret input, usually in the form of symbols, though 'analogue' systems often measure continuous input from sensors. To determine the meaning of the data at the input of a system, one must therefore interpret the stream of data and determine its intended meaning. This is a problem that is still a matter of considerable research. There are two main approaches that are often complementary: grammatical and statistical recognition methods.

If the behaviour of a system is deterministic, it can be described in terms of operations that are executed in some well-defined sequence. A description of this 'language' of operation can rightfully be called a machine language. The set of all legal sentences in this language is called the *syntax* of the language. It does not matter whether humans ever see this language of operation; it exists and is well-defined without ever having to be written down.

The syntax of any language can be modelled by a general theory of its structure, called a *grammar*. Grammatical methods assume that arriving data form a sequence of digital symbols (called an alphabet) and have a structure that describes an essentially hierarchical coding stream. The meaning of the data is understood by *parsing* this structure to determine what information is being conveyed. The leads us to the well-known Chomsky hierarchy of transformational grammars (see for instance Lewis and Papadimitriou (1997)).

Using statistical methods of recognition, patterns are digitized and learned, regardless of whether they began in digital form or not. One then gathers statistical evidence about the meaning of previously seen patterns of symbols and tries to use it to guess the meaning of new patterns. This method has been interestingly employed in bio-informatics in recent times (Durbin et al. (1998)) to interpret gene sequences. Even in this case, the idea of grammar is useful for classifying the patterns. A classification of a pattern is a way of assigning one-to-one meaning to it.

The complexity of patterns is related to the level of sophistication required to decipher their structures. Linguist Noam Chomsky defined a four-level hierarchy of languages called the Chomsky hierarchy of transformational grammars that corresponds precisely to four classes of automata capable of parsing them. Each level in the hierarchy incorporates the lower levels: that is, anything that can be computed by a machine at the lowest level can also be computed by a machine at the next highest level.

State machine	Language class
Finite Automata	Regular Languages
Push-down Automata	Context-free Languages
Non-deterministic Linear Bounded Automata	Context-sensitive Languages
Turing Machines	Recursively Enumerable Languages

State machines are therefore important ways of recognizing input, and thus play an essential part in human–computer systems.

Example 43 *Regular expressions are simple regular languages with their own grammar that are used for matching simple lexical patterns. They are widely used for searching computer systems with 'wildcards' for particular filename patterns. Regular expressions are also used for finding patterns of input in Network Intrusion Detection Systems.*

Because of their regularity and conduciveness to formalization, computer science has seized upon the idea of grammars and automata to describe processes (see section 13.10). We shall make some use of these idea, especially in chapter 15, but will not attempt to cover this huge subject in depth here. Symbolic logics are used to describe everything from computer programs and language (Logic (n.d.)) to biological mechanisms that describe processes like the vertebrate immune response (Jerne (1964)). Readers are referred to texts like (Lewis and Papadimitriou (1997); Watt (1991)) for authoritative discussions. For a cultural discussion of some depth (see Hofstadter (1979/1981)).

5.7.4   Pathway analysis and distributed algorithms

In networks or graphs, deterministic methods for locating features of the network are often algorithmically complex and many belong to the class NP of algorithms for which there is no known solution that will execute in polynomial time. Fault isolation is one such problem, as is finding the shortest path through a network. Approximations are often used to find a reasonable solution to these problems.

Distributed algorithms are often discussed in connection with management of networks (see for instance Steinder and Sethi (2002)). There is a powerful prejudice towards the use of deterministic algorithms, and in defaulting to heuristics when these fail to yield desired progress. Computer science has its roots more in logic than in statistical methods, though some of the latter enter in the field of artificial intelligence (see, for instance, Pearl (1988) and Pearl (2000)). However, increasingly there is a realization that an algorithmic approach is too elementary for describing systems at a higher level, and thus we spend little time on discussing such algorithms in this book. A nice overview of the traditional system management viewpoint is given in (Herzog (1994)), for instance. Other authors describe algorithmic tricks for elementary management processes, some of which will make a brief appearance later in the book (see Burgess (2004); Couch and Sun (2003); Couch et al. (2003)).

5.8 A policy-defined state

A theme that recurs in system administration is that of configuration management, that is, ensuring that devices, computers and organizations are primed with the correct state configurations in order so as to behave as we would like. We refer to a description of what we would like a system to do as a *policy*, and can imagine that there is an *ideal configuration* that represents this policy's wishes[1]. Can we make a system tend towards this state by itself?

In (Burgess (1998b)), this idea is called a computer immune system, and the desired state is thought of as being the 'healthy configuration' of the system. If a system will automatically tend to fall into its own desired configuration, this would require little maintenance (see fig. 5.8). The system would be self-healing. The task is therefore to design

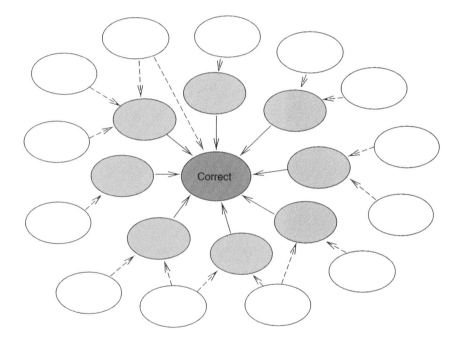

Figure 5.8: If a system has an 'ideal' configuration, we would like this state to be a basin of attraction for the transitions. The state transitions should therefore converge towards a fixed point that we consider to be 'correct' according to policy.

a system with a transition matrix that tends to lead it towards a fixed, singular configuration, indicated at the centre of the figure. We refer to such a transition matrix as being convergent[2].

---

[1] Later we shall advocate more strongly the idea that an ideal state can be associated with policy. For now we can say that we choose a particular desired state as a matter of policy.

[2] This is sometimes expressed by saying that, in a convergent policy, a repair never makes the system worse.

Applications and Further Study 5

- *Describing systems in terms of objects and their classifications.*

- *Classification of devices and configuration parameters.*

- *Classification of roles and cliques of components in a system.*

- *Descriptions of operations and instructions.*

- *Verification of designs and procedures.*

6

Diagrammatical representations

Diagrams are useful in planning, designing and understanding systems, because they either conceal details that are not relevant at the architectural level or amplify details that one normally suppresses.

Diagrams have been used to design and understand systems for thousands of years, in engineering and in medicine, for instance. Complex component diagrams have been in use ever since the first electronic circuits were built. Many themes in electronics have direct analogues in human–computer systems: for example, flow, amplification, diode (one-way flow), resistance, storage, input and output. From the experience of electronic engineering, we know that even fairly simple diagrams can be difficult to understand; huge diagrams are impossible for humans to digest.

Diagrams are helpful as maps, as a blueprint for construction, and even for diagnostics, but they become quickly unwieldy unless they are broken down into manageable pieces. One thus uses a strategy of divide and conquer, or modularity to make systems comprehensible. This results naturally in a hierarchical substructure to any large system in which high levels assume and *depend on* the correct functioning of lower levels.

It is not easy to formalize diagrams. A diagram is itself a data structure, in which the only resources one has are spatial extent, shapes and colours. Since the space on a page is limited, one usually runs into space limitations long before a formal scheme for organizing that space succeeds in showing anything like a realistic example, that is, before becoming unwieldy and impossible to grasp. Moreover, there is a limit to the number of ways of making distinctions in an intuitive fashion. For this reason, diagrams are mainly used as visual aids for more formal descriptions using algebra or some kind of functional pseudo-code.

There are top-down and bottom-up approaches to diagrams.

6.1 Diagrams as systems

Diagrams are systems in which one uses space as a resource to represent information about something else. An effective diagram, therefore, requires a proper allocation of the resource

Analytical Network and System Administration. Managing Human–Computer Networks   Mark Burgess
© 2004 John Wiley & Sons, Ltd   ISBN 0-470-86100-2

Table 6.1: Freedoms and constraints for printed diagrams

Freedom	Constraint
Position	Page size
Object size	Proximity
Shape	Difficulty of recognition
Colour	Colour perception
Direction	Only 2 dimensions

of space to the problem of representing information. The freedoms and constraints one has in a diagram are shown in table 6.1.

Given that a diagram is dependent on these factors, the task is one of how to allocate the space and shapes creatively in order to achieve the goal of the diagram. How is space used? Ideally, every diagram would indicate what its premises are. Some diagrams use space to indicate time, some use space to indicate extent. Some use distance to indicate some measurable property of the system.

Visual representations are, of course, not the only way of representing information. Underwater animals, such as dolphins, communicate mainly by sound, blind people mainly by touch, and so on. Gilfix and Couch (2000) applied a 'sound diagram' to the problem of representing a computer network. An audio representation presents different limitations such as ease of distraction from the environment. Electronic diagrams on computers can combine sight and sound and other senses into a single representation, but this goes beyond the limitations of a diagram on a printed page.

Example 44 *Electrical circuit diagrams: the flow of activity is carried by electricity, but the function of the circuit might be something as abstract as playing music (a radio). The actual function is rather hard to see from the internal algorithms, and yet those details are required at the level of flow. The mechanics of individual electrons is not required, because one trusts that the components behave in a predictable fashion.*

6.2   The concept of a graph

There are many situations in which we draw dots that are joined together by lines, perhaps with arrows on them, to denote some kind of information. The links between nodes often represent qualities such as

- *A* dominates *B* (directed),

- *A* depends on *B* (directed),

- *A* is associated with *B* (undirected),

and so on. Note that some of these relationships are one-way (directed) and others are multi-way (undirected).

D. König suggested the name *graph* for all such diagrams and pioneered the study of their properties. Elementary graph theory is a very useful framework for discussing human–computer systems, because it considers systems with discrete states (the nodes or dots in the graph) that are joined together only if certain criteria are met (i.e. if there is an arc joining the points, perhaps with an arrow going in the right direction).

A graph with arrows on it is called a *directed graph* and a graph without loops is called *acyclic*; thus a tree structure (so common in computer science) is an acyclic directed graph (see fig. 6.1).

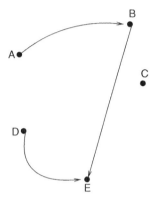

Figure 6.1: A graph is a general assembly of nodes connected by arcs. It is used to describe many situations in science and mathematics.

Definition 24 (Graph) *A graph is a pair* (X, Γ) *that consists of a set of nodes X and a mapping* $\Gamma : X \to X$, *formed by the arcs or lines between the points* $x \in X$.

The degree of any node in a graph is the number of nearest neighbours it has, that is, the number of nodes that can be reached by travelling along those links that are connected to the node.

Definition 25 (Degree of a node) *In a non-directed graph, the number of links connecting node i to all other nodes is called the* degree k_i *of the node. In a directed graph, we distinguish incoming and outgoing degrees.*

Do not be tempted to think of a graph as being necessarily composed of only a sparse set of points with occasional links between them. Even a dense set of points, infinitely close together, such as all the points in a circle form a set and can be mapped onto each other, even if the arcs seem to overlap the points (see fig. 6.2).

A graph may be represented or defined by its *adjacency matrix*[1]. By convention, the adjacency matrix of a network or graph is a symmetric matrix with zero leading diagonal,

---

[1] Working with graphs in adjacency matrix form becomes decreasingly realistic as the size of the graphs grows beyond thousands of nodes; however, this form is very illuminating and is particularly suited to smaller graphs such as those that arise in system administration.

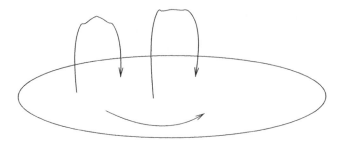

Figure 6.2: A graph mapping points within a disk onto other points within the disk.

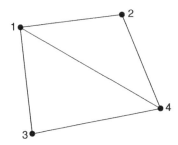

Figure 6.3: A simple network, represented as an undirected graph.

(see, for example, the graph in fig. 6.3)

$$A = \begin{pmatrix} 0 & 1 & 1 & 1 \\ 1 & 0 & 0 & 1 \\ 1 & 0 & 0 & 1 \\ 1 & 1 & 1 & 0 \end{pmatrix} \qquad (6.1)$$

Definition 26 (Adjacency matrix) *An* adjacent matrix *is a square matrix A_{ij} whose rows and columns label the nodes of a graph. If an arc exists from node i to node j, then $A_{ij} = 1$. If the graph is undirected, it must have a symmetric adjacency matrix.*

If we regard the nodes in a graph as possible states in a system, then the adjacency matrix is also a transition matrix between states, where a '1' indicates the possibility of a transition. For non-deterministic or stochastic systems, one could replace '1' by the probability of a transition taking place.

Graphs or networks come in a variety of forms. Figure 6.4 shows the progression from a highly ordered, centralized structure to a de-centralized form, to a generalized mesh. This classification was originally discussed by Paul Baran of RAND corporation in 1964 as part of a project to develop a communications system that would be robust to failure, in the case of a nuclear attack (see Barabási (2002); Buchanan (2002) for a review).

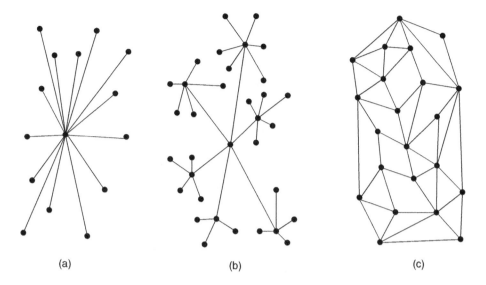

Figure 6.4: Network topologies: (a) centralized, (b) de-centralized or hierarchical and (c) distributed mesh.

6.3 Connectivity

Let $\vec{h}$ be a host or node vector whose components are 1 if a host is available and 0 if unavailable. The level of connectivity within a graph or closed network can be characterized by an invariant scalar value χ:

Definition 27 (Connectivity) *The connectivity, χ, of a network $\mathcal{N}$, is the probability (averaged over all pairs of nodes) that a message can be passed directly between any two nodes. χ may be written as*

$$\chi = \frac{1}{N(N-1)} \vec{h}^{\mathrm{T}} A \vec{h}. \tag{6.2}$$

χ has a maximum value of 1, when every node is connected to every other, and a minimum value of 0 when all nodes are disconnected.

The connectivity of a graph is of great importance to systems. It tells us both how easy it is for information to spread throughout the system and how easy it is for damage to spread. This duality is the essence of the security-convenience conundrum. We explore this issue further in the next section.

6.4 Centrality: maxima and minima in graphs

Where are the best connected nodes in a graph? These are nodes that we would like to identify for a variety of reasons: such nodes have the greatest possible access to the rest of the system. They might be security hazards, bottlenecks for information flows or key components in a system in which we need to invest money and resources to keep the system running smoothly. From the standpoint of security, important nodes in a network (files, users, hosts) are those that are 'well-connected'. We are thus interested in a precise working definition of 'well-connected' (see Burgess et al. (2003a); Canright and Engø (2004)).

A simple starting definition of well-connected could be 'of high degree', that is, count the neighbours. We want, however, to embellish this simple definition in a way that looks beyond just nearest neighbours. To do this, we borrow an old idea from both common folklore and social network theory (see Bonacich (1987)): an important person is not just well endowed with connections, but is well endowed with connections to important persons.

The motivation for this definition is clear from the example in fig. 6.5. It is clear from this figure that a definition of 'well-connected' must look beyond first neighbours. Nodes of equal degree have quite different levels of importance depending on their position within the remainder of the graph.

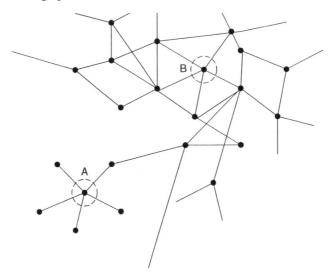

Figure 6.5: Nodes A and B are both connected by five links to the rest of the graph, but node B is clearly more important to security because its neighbours are also well connected.

We can now formulate a precise definition of the importance for non-directed graphs using a concept called *centrality*. We begin by noting that the symmetrical adjacency matrix consists of 1's where a node has neighbours and 0's where it does not. Thus, multiplying a row of this matrix by a column vector of 1's would simply count the number of neighbours for that node. We can use this fact to self-consistently sum the entire graph.

Let v_i denote a vector for the importance ranking, or connectedness, of each node i. Then, the importance of node i is proportional to the sum of the importances of all of i's nearest neighbours $N(i)$:

$$v_i \propto \sum_{j=N(i)} v_j. \tag{6.3}$$

This may be more compactly written as

$$v_i \propto \sum_j A_{ij} v_j, \qquad (6.4)$$

where A is the adjacency matrix. We can rewrite eqn. (6.4) as

$$A\vec{v} = \lambda\vec{v}. \qquad (6.5)$$

Thus, the importance vector is actually an eigenvector of the adjacency matrix A. If A is an $N \times N$ matrix, it has N eigenvectors (one for each node in the network), and correspondingly many eigenvalues. The eigenvalue of interest is the principal eigenvector, that is, that with the highest eigenvalue, since this is the only one that results from summing all of the possible pathways with a positive sign. The components of the principal eigenvector rank how self-consistently 'central' a node is in the graph. Note that only ratios v_i/v_j of the components are meaningfully determined. This is because the lengths $\sum_i v^i v_i$ of the eigenvectors are not determined by the eigenvector equation.

The highest-valued component is the most central, that is, the *eigencentre* of the graph. This form of well-connectedness is termed 'eigenvector centrality' (see Bonacich (1987)) in the field of social network analysis, where several other definitions of centrality exist. We shall use the terms 'centrality' and 'eigenvector centrality' interchangeably.

Example 45 *Consider the graph in fig. 6.6. This has adjacency matrix*

$$A = \begin{pmatrix} 0 & 1 & 0 & 1 & 1 \\ 1 & 0 & 1 & 0 & 1 \\ 0 & 1 & 0 & 1 & 1 \\ 1 & 0 & 1 & 0 & 1 \\ 1 & 1 & 1 & 1 & 0 \end{pmatrix} \qquad (6.6)$$

The matrix has eigenvalues $\lambda = \{-2, -1.2, 0, 0, 3.2\}$. The principal eigenvector is that associated with the last of these, that is, that with the highest value.

$$\vec{P}(A) = (0.43, 0.43, 0.43, 0.43, 0.52). \qquad (6.7)$$

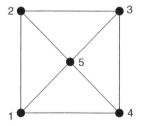

Figure 6.6: An undirected graph with an obvious centre.

This indicates that node 5 is the most central. The remaining symmetrical nodes have symmetrical lower values. Thus, the principal eigenvector maps the topography of the undirected graph.

Figure 6.7 shows a centrality organized graph of the Gnutella peer-to-peer network. Peer-to-peer networks are characterized by having no real centre. The centres here are only marginally higher than their surrounding nodes.

Figure 6.7: A graph of the Gnutella peer-to-peer network using centrality to define local maxima regions. Distinct regions are shown as distinct islands, using the Archipelago tool (Burgess et al. (2003b)).

Definition 28 (Eigenvector Centrality) *The eigenvector centrality of node i is the ith component of the principal eigenvector $\vec{v}$, normalized so that $\max_i v_i = 1$. This is used in importance ranking of nodes.*

We can use this importance ranking to define the *height* of each node in a graph and hence draw the graph as an importance landscape. A local maximum in this landscape defines a 'regional maximum' in the graph or a very important node. Low points indicate nodes of little importance, and, in between tops, we can identify nodes that act as bridges between the local maxima. Eigenvector centrality provides us with a shape for the graph, based on the density of its connections. This serves as a guide for locating hot spots in networks. Note also that, in such diagrams, each dot or node could represent a sub-graph, allowing many levels of detail to be revealed or hidden.

6.5   Ranking in directed graphs

So far, we have considered eigenvector centrality in undirected graphs, where its meaning is unique and unambiguous. The same idea can be applied to directed graphs, but with additional subtleties. The topic of ranking nodes in a directed graph has a history associated with importance ranking by search algorithms for the World Wide Web (see Page et al. (1998) and Kleinberg (1999)).

The arrows on graph edges now identify nodes of three types: sources, sinks and relays.

Definition 29 (Source) *A source is a node from which a directed flow only emerges. In the adjacency matrix, this is represented by a row of 1's with a corresponding column of 0's.*

Definition 30 (Sink) *A sink is a node that is purely absorbing. The adjacency matrix has a column of 1's and a corresponding row of 0's.*

Definition 31 (Relay) *A relay is a node that has both incoming and outgoing flows.*

Principal eigenvector ranking is fraught with subtlety for directed graphs. It does not necessarily have a simple answer for every graph, though it seems to work for some (see section 11.8); nor is there a unique procedure for obtaining an answer for every case. All is not lost, however. One can still use the spirit of the eigenvalue method to learn about graph structure and importance.

There is a fundamental duality about directed graphs that depends on the direction of the arrows. To understand this, look at the two graphs in figs. 6.8 and 6.9.

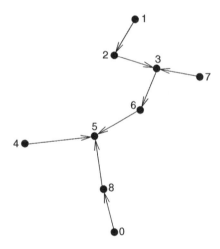

Figure 6.8: A directed graph with a sink, that is, a node (5) that is absorbing of all flows. This is the complement or dual of fig. 6.9.

In the first of these graphs (fig. 6.8), we have a number of nodes with arrows that point mainly towards an obvious centre. This centre happens to be a sink, that is, it absorbs the flows. This diagram has an adjacency matrix A. The dual picture (fig. 6.9) is described by the transpose of the adjacency matrix A^T, since interchanging rows and columns changes the direction of the arrows.

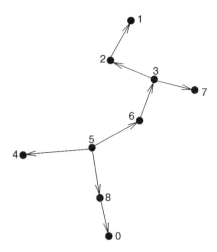

Figure 6.9: A directed graph with a source, that is, a node (5) that is the origin of all flows. This is the complement or dual of fig. 6.8.

Consider now the ranking determined by the principal eigenvector for the numbered nodes. The difficulty one encounters here is that the directness of the arrows prevents any non-trivial solutions because of the absence of paths with loops. All eigenvalues turn out to be zeros; there is no principal eigenvector. However, one can imitate the process of the undirected eigenvalue problem by counting neighbour routes. Consider the table below for the two dual forms of the graph.

Node	$\vec{P}(A)$	Path length from i	Reverse ranking	$\vec{P}(A^{\mathrm{T}})$	Path length from i	Ranking
0	N/A	2	4	N/A	0	6
1	N/A	4	9	N/A	0	9
2	N/A	3	8	N/A	1	7
3	N/A	2	6	N/A	3	3
4	N/A	1	2	N/A	0	4
5	N/A	0	1	N/A	10	1
6	N/A	1	5	N/A	4	2
7	N/A	3	7	N/A	0	8
8	N/A	1	3	N/A	1	3

The iterative equation for the eigenvector sums pathways starting from the node concerned and normalizes the counts according to the self-consistent structure of the graph.

However, this self-consistent sum cannot be made, since the path lengths are finite: routes from each node terminate at different lengths, so total iteration is inconsistent and no solutions exist. The table above sums these values by hand, however, and the ranking is shown in the table. The path length columns show the number of nodes that it could travel starting from the numbered node in the leftmost column, before being absorbed by a sink. If we rank these columns 'backwards' according to 'inverse importance', then the sink node 5 turns out to be the most important (rank 1). Thereafter, the rankings follow the reverse of the path lengths. In this way, we can call this ranking of importance absorbative importance.

Now look at the transposed columns and we find the reverse. The ranking again counts the accessible pathways from each node and ranks them in the opposite order (more or less, up to the self-consistent scalings). There are thus at least two kinds of importance in a directed graph, and these are complementary:

- **Sink importance**: a node is important if it gets a lot of attention from other nodes, that is, if it absorbs a lot of flows. This kind of node is also referred to as an *authority*, since others point to it and hold it in esteem.

- **Source importance**: a node is important if it influences a lot of nodes, that is, if it originates many flows. This kind of node is also called a *hub*, because it shoots out its spokes of influence in all directions.

The problems with this simple use of the principal eigenvector are illustrated by the two diagrams in fig. 6.10. These graphs cannot be distinguished by the basic eigenvector method,

Figure 6.10: These two graphs cannot be distinguished by the principal eigenvalue method.

since there is such great symmetry that all path lengths from the nodes are equal. The eigenvalues are thus all zeros, and there is no way of finding the most important node, despite the central node clearly having a privileged position. An interesting approach to directed graphs that is able to distinguish these graphs has been presented by Kleinberg (1999). The following symmetrized matrices enforce the duality noted above explicitly, by tying together sources and sinks into nearest neighbour 'molecules'. A good source (hub) is one pointing to many good sinks and a good sink is one pointed to by many good sources, at nearest neighbour level. They are guaranteed to have a principal eigenvector.

$$A_A = A^T A \tag{6.8}$$

points to an authority (i.e. its principal eigenvector assigns a high weight to an authority or a sink), while

$$A_H = A A^T \tag{6.9}$$

points to a hub (i.e. its principal eigenvector assigns a high weight to a hub or a source). It is presently unclear what the advantages and limitations of these different approaches are. None of these performs entirely satisfactorily for all graphs (particularly small graphs), and a complete understanding of the identification of roles in directed graphs is still in progress (Burgess et al. (2004)). In cases where both A, A^{T} and the A_{A}, A_{H} yield an answer, they often agree, with approximate correspondence:

$$A \leftrightarrow A_{\mathrm{H}} = A A^{\mathrm{T}}$$
$$A^{\mathrm{T}} \leftrightarrow A_{\mathrm{A}} = A^{\mathrm{T}} A. \tag{6.10}$$

See section 11.8 for an example.

In the hub-authority viewpoint, the matrices attach importance explicitly to local clusters of sources and sinks, but why not longer range dependencies? Another approach is used by the PageRank algorithm (see Page et al. (1998)) in which stochastic noise is added to the actual adjacency matrix in order to generate loops that resolve the eigenvalue problem. The success of the procedure is well known in the form of the search engine Google, but it is also an arbitrary procedure that is now patented and out of public scrutiny. The issues surrounding importance ranking would probably be resolved in future work, allowing methods to be used as organizing principles for systems of all kinds.

6.6   Applied diagrammatical methods

A variety of heuristic diagrammatical forms are in use.

- *Maps*
  Maps (or 'mind maps') are a loose representation of everything in a system. They are used as a basis for identifying relationships and structure, as well as simply documenting all the relevant parts (see fig. 6.11).

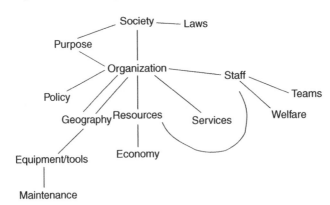

Figure 6.11: A 'mind map' of an enterprise, showing a brainstorming approach to understanding the elements that are important, and their relationship to one another.

Note that this is not a dependency diagram. It is simply a diagram of associations.

- *Flow diagrams*

  These were common in the early days of programming and are still used some-times for representing simple algorithms graphically. Flow diagrams show the causal sequence of actions and the decision branches in simple processes; they are a graphical pseudo-code and thus provide a very low-level picture of a system. For large or complex systems, flow diagrams become unwieldy and more of a hindrance to comprehension than an aid (see fig. 6.12). The Unified Modelling Language (UML) attempts to extend this idea to make diagrams express a strict grammar.

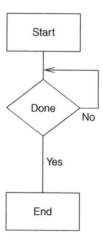

Figure 6.12: Flow diagrams, a graphical pseudo-code that illustrate a sequence of actions and decisions in a simple procedure.

- *Transition diagrams*

  A system of distinct states is called a *Finite State Machine* (see section 5.7.1). It is formally represented as a directed graph. Finite state machines are at the centres of many systems (see fig. 5.5). They represent a coarse type of memory of *context* in a system. Transition diagrams are related to functional structure diagrams. If one labels being in each function as a state of a process, then they represent the same information.

- *Functional structure diagrams*

  A structure diagram is a chart of the independent functions within a process and the flow of information between them. They are sometimes called *Jackson diagrams* (Jackson (1975)).

  A structure diagram describes a chain of command and maps the independent methods in a process, showing how they relate to one another (see fig. 6.13). For instance, the functional declaration

```
begin function A
  do function B
  do function C
end function
```

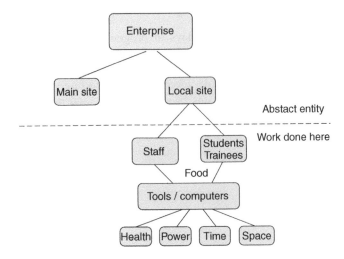

Figure 6.13: An excerpt of a functional diagram showing the economic (resource) organization of an enterprise. In many organizations, this doubles as a chain of command diagram, since control is often chosen to flow with the dissemination of resources, though this need not be the case.

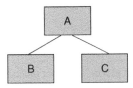

Figure 6.14: A simple Jackson diagram for a function with two dependencies.

would be drawn as in fig. 6.14. By the same token, it charts the flow of resources in a system between the high-level objects within. This diagram limits its view to the top-level structure, so its value is limited; it is most useful for top-down approaches to system design (see also fig. 6.15).

- *Dependency diagrams*

  Dependency diagrams are the basic tool for fault analysis. Each arrow in a dependency diagram should be read 'depends on' (see fig. 6.16 and fig. 6.17). If a component or an object in a system fails, all the components that point to it will also fail. A fault tree is a special case of a dependency diagram.

  In these diagrams, we see the repeated significance of the process of classification and sub-classification of objects. The role of object orientation and type distinction is demonstrated.

- *Entity relation diagrams*

  Entity Relation (ER) diagrams are used in the description of database tables, for example, in the Structured Query Language (SQL). They describe the interrelation-

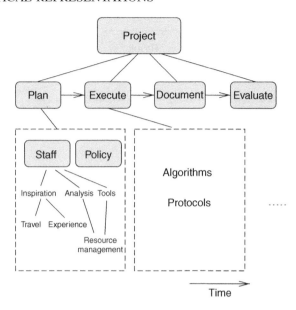

Figure 6.15: An excerpt from a functional diagram showing the resource organization of a project, within a larger enterprise. This diagram takes a two-dimensional view of an organization, by plotting the two degrees of freedom—project development versus dependency structure.

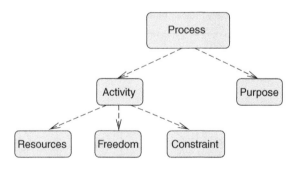

Figure 6.16: A trivial dependency diagram for a process.

ships between objects of different types as well as the cardinality (number of objects that are involved in each type of relationship) that are possible between objects of different types (see fig. 6.18).

Object diagrams that describe object-orientated programming relationships are simplified forms of ER diagrams.

- *Petri nets and stochastic networks*

Petri nets (see David and Alla (1994); Meyer et al. (1985)) are a graphical tool for modelling discrete event systems, for example, network services. They are similar in

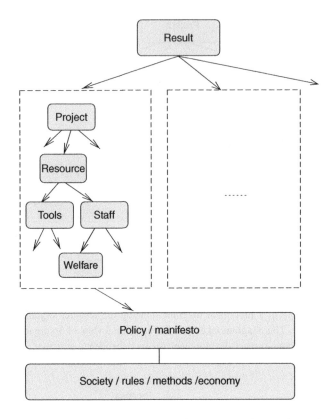

Figure 6.17: An excerpt of a dependency diagram for an enterprise, such as that from figs 6.11 and 6.13.

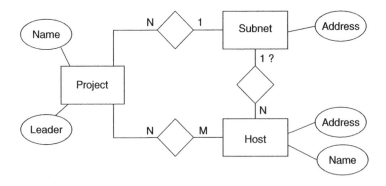

Figure 6.18: Entity relation diagrams describe the structure of tabular data relationships. Each square box is a tabular object with references to fields shown as ellipses. The links show $N{:}M$ relationships

concept to flow diagrams, but are able to describe concurrency, synchronization and resource sharing. Stochastic networks are a related form of network for modelling discrete stochastic events. These topics are beyond the scope of this book.

Diagrammatical methods are only representations of systems, and the systems they represent can often be described more compactly in other forms, such as by algebraic rules. The utility of diagrams lies mainly in human understanding and pedagogy; even graph theory relies more on algebraic construction than on pictures. Some attempts have been made to use diagrams to describe system behaviour in a rigorous way. The Unified Modelling Language (UML) is one such attempt.

Applications and Further Study 6

- *Identifying the main scales and structures in a system.*

- *Classifying system components.*

- *Visualizing relationships and roles between the components.*

- *Evaluating the importance of components within systems.*

- *Finding weak points or inconsistencies in systems.*

7

System variables

To go beyond pictorial representations and be able to make quantitative statements about systems, we must develop a representation of system properties in terms of variables. One of the difficult notions to understand in the science of systems is how to describe their changing properties. In particular, the transition from discrete, digital changes to smooth differentiable functions is unfamiliar in computer science. Both types of variables are required to describe human–computer systems.

7.1 Information systems

Any system can be thought of as an information system (i.e. as the abstract development of a set of variables), simply by describing activity with the help of an abstract representation, that is, by viewing a change performed by the system as being a change in a description of the system's basic resources. This is a one-to-one mapping. The resources themselves describe and are described by variables, which yield information.

Example 46 *A factory that manufactures cars is an information system, because it receives input in the form of steel and changes the pattern of the steel into new shapes, described by a certain amount of information, before delivering cars as output. We can describe this system in terms of the actual physical cars that are produced and the resources that go into them, or we can describe it on paper using numbers to represent the flow of items. Accountants regularly consider businesses as information systems; accountants do not get their hands dirty on the factory floor.*

The converse is also true: in order to represent information, we must encode it in a physical pattern represented in some physical medium (brain cells, paper, computer storage, steel, etc.). To create a pattern, there must be an attribute of some basic resource that can vary, such as colour, height, size, shape and so forth, and these attributes must belong to physical objects. Thus it would be wrong to try to divorce systems from their physical limitations, even though we are interested in abstracting them and speaking only about

Analytical Network and System Administration. Managing Human–Computer Networks Mark Burgess

their information. Many of the limitations in systems arise because of the physical nature of their representation.

7.2 Addresses, labels, keys and other resource locators

In order to speak of change of location, we need to be able to measure and parametrize location. This is not necessarily like measuring things on a ruler. If the locations are named rather than measured, that is, if it is the name rather than the distance that counts, then one does not use a continuously varying parameter like x or t to describe it, but rather a discrete number or label.

Once the medium for representation of information change has been identified, there is the question of where and when the change takes place in that medium. Describing the when and where in a system is called the *parametrization* of the system.

Low-level parameters

Ultimately, any system is bound by physical limitations. In the real world, there are only two variables that can parametrize change in a medium: location (space) and time. Physical objects exist only in space and time. Thus, at the lowest level, a physical dynamical system can only consist of values of the systems' resources at different space–time locations. Mathematically, we denote this as functions that vary in space and time:

$$q(\vec{x}, t). \tag{7.1}$$

Example 47 *Morse code is a pattern of sounds in time. A picture or a shape is a pattern of material in space. A construction site is a pattern in both space and time since it has a definite form in space, but the form changes with time.*

While space and time are *sufficient* to describe any system, they are too low level to be satisfactory in the description of abstract systems. One can create abstract freedoms, such as the *orientation* of a non-trivial shape, or the loudness of a sound, by building on ideas of space and time (this is what physics does), but it is not always necessary to refer to such low-level details.

Derived addresses

In the virtual world of abstract information, and hence of human–computers systems, there are other ways of parametrizing spatial change (location):

- Geographical location of data

- Memory location inside a computer

- Internet address

- Identity of a container

- Ownership.

Each of these examples is an *address*, that is, a label that denotes the location of a resource. We may thus parametrize a change in terms of the value of a resource at an arbitrary address and a particular time. Mathematically, we write this as a function:

$$q(A, t). \tag{7.2}$$

Higher-level patterns: associations

A common generalization of the idea of a resource address is employed in abstract information systems, such as databases, where one does not wish to refer to the physical location of data, but rather to an abstract location. Instead of thinking of a pattern as being a function of a particular location $q(A, t)$ one can use a reference key. The key itself is just a pseudonym label or an alias for the detailed physical address:

$$q(A, t) \rightarrow q(k, t), \tag{7.3}$$

where the key k can be any collection of labels or 'coordinates' in whatever abstract space one cares to imagine, for example,

$$q(k, t) \rightarrow q(\text{building, town, country, floor, project}, t). \tag{7.4}$$

Now, instead of thinking of pattern and structure as varying with address and time, one views it as varying with different values of the abstract key. Thus, we can say that a system is a dynamical function of one or more abstract keys.

The *key* forms an abstract representation of the properties of the data structure within the system; one creates an *association* or an *associative relation* between a *key* and a resource *value* located by that key. Note that, in writing $q(k, t)$, the time at which the change takes place is itself simply a label identifying the time at which the value was true.

Example 48 *The arrival of incoming packets on a network connection is a signal that can be described by a function $q(t)$ that varies according to the chosen representation of the signal, for example, a binary signal with $q \in \{0, 1\}$.*

Example 49 *Consider a pattern of data in the memory of a computer or on some storage medium described by $q(\vec{x}, t)$, where $\vec{x}$ labels the objects of which the pattern is composed. The pattern might be a software package, or an image, and so on. Since the data change over time, at some rate, this function also depends on time.*

Example 50 *Suppose that the software package in the previous example is determined by n criteria or tests that determine its integrity or correctness. The package thus has n state variables, encoding 2^n. Let us give new coordinate labels to all such packages on the system, $\vec{X}$, and consider the variable $Q(X, t)$, where $Q \in \{Q_1, \ldots, Q_n\}$. This variable now describes the state of software packages over time. The value describes the relative integrity of the system.*

Example 51 *A graph is characterized by n nodes linked together. Any property of graph nodes can be represented by vectors of objects $\vec{v} = (v_1, v_2, \ldots, v_n)$. Any property of links in the graph can be represented by a matrix A_{ij} of appropriate objects or values.*

Describing systems formally using variables with particular representations is limited only by imagination. We are free to do whatever is helpful or productive.

7.3   Continuous relationships

An important use for variable quantities is to express relationships between measurables. Comparisons and relationships between quantities are the basic tools one has for expressing policy. To be able to express and enforce what we want, we must be able to compare what is measured with what we specify.

Example 52 *In system performance, configuration and resource usage, we make various comparisons:*

- *Rate of job arrivals* $(<, >, \geq, \leq, =)$ *rate of processing.*

- *State* $q(x, y, z, t) > q(x', y, z, t)$.

Example 53 *Some variables are related to one another in linear combinations:*

- *Total system communications capacity* $C_T = \sum_{i=1}^{N} C_i$.

- *Average traffic at location* x *is the sum of traffics from locations* $x' \in S$: $T(x) = \sum_{x'} T(x')$.

Linear combinations of variables are often useful in parametrizing systems in which hidden relationships occur.

Example 54 *Suppose one finds that the probability of a program crash is a function of the number of users logged on to a computer and the number of processes being executed in separate measurements. One might observe that most processes are started identically for all users (e.g. Window manager processes) and that only negligible differences are measured between users. In this case we notice that* $N_{procs} \propto N_{users}$, *and thus*

$$P_{crash}(N_{procs}, N_{users}) \to P_{crash}(N_{procs}). \tag{7.5}$$

Simplifications arise from a knowledge of relationships.

7.4   Digital comparison

What is the difference between two system configurations C_1 and C_2? This is a question that is frequently asked in connection with device management. Rather than testing whether a machine configuration is consistent with a policy (which normally involves only approximate or *fuzzy* classifiers), it is common to compare a system configuration $C(x, t)$, at location x and time t, to a reference system C_0 and characterize the difference in terms of the number of items that do not agree

$$\Delta C(x, t) = C_0 - C(x, t). \tag{7.6}$$

One can question how useful it is to compare a dynamic system $C(x, t)$ with a static snapshot C_0, but we shall not discuss that here. How shall we make such a comparison? What is the meaning of the difference symbol in eqn. (7.6)? Is it a numerical difference or a difference of discrete sets? The ability to compare configurations depends on the nature of the variables being compared—are they continuous or discrete? Although we shall later argue for the use of continuous variables in chapter 16, most comparisons are made digitally.

Example 55 *Consider two configurations that are coded with symbols $A, B, C, \ldots$:*

$$C_0 = \{A, D, F, G, \ldots\}$$
$$C(x_1, t) = \{B, D, F, A, \ldots\}. \tag{7.7}$$

These configurations can be compared symbol by symbol.

To define a measure of distance between two such configurations, one can take a variety of approaches; there is no unique answer. Instead, we make a definition that suits our purpose. Two strings differ if their symbols do not match. We define a distance function (or metric) to define the distance between differing symbols. The most common distance function is the Hamming distance, which is a count of the number of bits that differ in the binary coding between two strings. A variation on this for higher-level coding is the function for comparing symbol q with reference symbol q':

$$d(q, q') = \begin{cases} a & q = q' \\ 0 & q \neq q' \end{cases} \tag{7.8}$$

This is a linear function; thus the distance grows additively for strings of many symbols. The distance between two configurations is thus

$$D(Q, Q') = \sum_{q \in Q, q' \in Q'} d(q, q'). \tag{7.9}$$

The straightforward comparison of strings is a naive way of comparing configurations that assumes only substitution errors in the symbol string. In general, we can have the following differences:

- Insertions

- Deletions

- Substitutions.

If one relates the differences or 'errors' between configurations to random processes, then one can only speak of the probability of an error. A knowledge of underlying mechanisms can allow the construction of a transition matrix (see section 5.4) that measures the likelihood of a transition from one state to another. This allows us to define a different kind of distance that is symbol-dependent:

$$d_{\text{sub}} = -\log\left(\frac{P(q \to q')}{P(q \to q')}\right)$$
$$d_{\text{del}} = -\log\left(\frac{P(q \to \emptyset)}{P(q \to q)}\right)$$
$$d_{\text{ins}} = -\log\left(\frac{P(\emptyset \to q)}{P(q \to q)}\right). \tag{7.10}$$

This is sometimes called the *Levenshtein distance* (see Bunke and Csirik (1995) and Oommen and Kashyap (1998) for an intelligent discussion of pattern comparison with generalized differences).

Applications and Further Study 7

- *Describing any system or phenomenon quantitatively.*

- *Quantitative analysis allows us to study scalability of systems to changes in parameters.*

- *Comparing systems with different characteristics and determining the 'distance' or metric that measures this distance.*

- *Refining algebraic formulations of a problem in order to better understand its structure.*

- *Manipulating parameter choices in systems and exploring consequences.*

- *Pattern recognition in the layout of and input to systems for identification of system problems (anomaly detection).*

8

Change in systems

Change is probably the most important quality of systems. A system that cannot express change cannot perform a function and is therefore trivial. We need a way to describe changes that occur in human–computer systems; the natural language for this is mathematics, since it is both expressive and precise.

8.1 Renditions of change

Change can be represented in many forms. We choose the mode of description that is most convenient on a case-by-case basis. Some examples include the following:

Time series

The time series is one of the simplest ways of representing change graphically (see Box et al. (1994)). See, for example, figs. 3.11 and 10.5. Time series are most easily plotted for data that are dense or almost continuously varying; however, streams of discrete symbols also form time series. Time series are especially useful in the continuum approximation to discrete systems, since they can be approximated by known functions, for example, in Fourier analysis.

Transitions

Systems that change irregularly between well-defined states are often described in terms of transition tables. Any system can be described in terms of transition tables.[1] Transition tables are essentially like the adjacency matrices of graphs. They tell us what the possible transitions to neighbouring states (nodes) are, given that we are already in a particular state (at a particular node of the graph). Transitions can be deterministic or non-deterministic (see section 8.2).

---

[1] The method of Green functions is the continuous generalization of the transition matrix approach to continuous systems (See Burgess (2002a)).

Analytical Network and System Administration. Managing Human–Computer Networks Mark Burgess
© 2004 John Wiley & Sons, Ltd ISBN 0-470-86100-2

Stability in fixed points

Changes in a system sometimes result in it converging towards a preferred state that is not easily altered again—such a state exhibits some kind of stability and this makes the system easy to predict; other times, it results in chaotic, unpredictable behaviour that has no compact description. If a system wanders into a state from which it does not emerge without outside help, the state is said to be a fixed point of the system. This idea is interesting in management, since it represents a notion of stability. See also section 5.8.

8.2   Determinism and predictability

A *deterministic* process is one that is governed by rules that are always obeyed with complete certainty. For instance, it is generally assumed that the law of gravitation is a deterministic process: if one releases an object in a gravitational field, it falls (every time). In a deterministic system, if a system variable has value $q(t')$ at some earlier time and value $q(t)$ at some later time, then the probability of a transition from the value $q(t')$, given that the earlier value was $q(t')$ is unity:

$$P(t, t') = P(q(t)|q(t')) = 1. \tag{8.1}$$

This certainty about the outcome of the process implies that we can use past knowledge to predict behaviour in the future (see fig. 8.1). The system behaves in the same way under identical conditions, each time it is measured. In practice, this applies only to very simple systems or systems that are isolated from external influences.

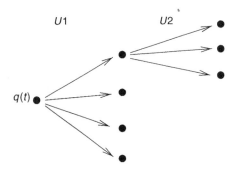

Figure 8.1: The evolution of a function occurs as its value changes at each time step. In a deterministic process, the choice at each time step dt is selected by a predetermined function $U(t, t + dt)$; the result would always be the same, if one rolled back time and repeated the measurements; that is, we can predict the future. In a non-deterministic system, the value of the function is picked at random, so that if one rolled back time and tried again, the outcome could be different.

Non-determinism means that the probability of making a successful prediction about the system $P < 1$. The transition matrix for the process:

$$P(q(t)|q(t')) < 1, \tag{8.2}$$

for any t, t'. At each time step, there is a probability distribution indicating the likelihood of obtaining possible measured values. The distribution $P(q)$ is the probability of value choice q at a given time. Since it is a probability distribution,

$$\int dq\, P(q) = 1. \tag{8.3}$$

Non-determinism implies that we must make an educated guess about what is likely to happen in the future. This means there is uncertainty about what we can expect. The uncertainty is a direct result of what we do not know about the system. Even if there is an underlying deterministic system, it is of such complexity that we cannot realistically predict everything that will happen.

There are many ways in which randomness or unpredictability can enter into systems. One common assumption is that randomness follows the pattern of a Gaussian or 'normal' distribution:

$$P(c) \propto e^{-(c-\mu)^2/\mathcal{C}}. \tag{8.4}$$

This would be equivalent to the assumption that the variable we were measuring had some 'true value' μ that varied at random by about $\pm\sqrt{\mathcal{C}}$ in either direction. This model approximates some phenomena well, but not all.

Example 56 *A transmitted network signal is a variable that has a 'true' or intended value, that is, the signal that we are trying to send. This can pick up random noise along the way. Such noise is often well approximated by a Gaussian random error. Indeed, that is the assumption behind the Shannon formula for the capacity of a communications channel:*

$$C = B \log\left(1 + \frac{S}{N}\right). \tag{8.5}$$

Example 57 *The arrival of packets at a network switch is a random process, but there is no 'correct' or 'true' value to this number. There will be a probability distribution of values from different customers on different arms of the switch, but even this distribution might change slowly over time.*

8.3 Oscillations and fluctuations

Few systems are ever truly constant, as parameters such as time are allowed to vary, but several systems exhibit change that averages out to nothing. Two examples of this that provide a good illustration of the difference between deterministic change and non-deterministic change are oscillations and random fluctuations. An oscillation is a periodic pattern of change that can be expressed as a relatively simple combination of sine and cosine waves:

$$q(t)_{\text{osc}} = \sum_n \sin(\omega_n t + \phi_n), \tag{8.6}$$

for various constant circular frequencies ω_n and phase shifts ϕ_n. The oscillation is deterministic because we have written down its exact functional form, and can therefore predict its behaviour at any time with complete certainty.

Example 58 *Consider a simplistic model of arriving traffic at a Web server over the course of a week. By measuring the actual arrival of requests at the server, we find a complicated pattern that must be described as a random variable, since the requests are sent by a broad number of independent sources at unpredictable times (see fig. 8.2). This behaviour is clearly complicated; but for the purpose of estimating system load, one might try to approximate it by something simpler. We might try a test function of the form*

$$f_1(t) = 2\sin(t) + \cos(2t). \tag{8.7}$$

See fig. 8.3. This function does not really resemble the actual random data, but it has some similar features. We see that there is a general decay in activity towards the end of the week, and we could try to model this by adding an exponential decay term to our approximate model (see fig. 8.4):

$$f_2(t) = (2\sin(t) + \cos(2t))e^{-t/4}. \tag{8.8}$$

The figure starts to take on some of the general features of the actual measurements, but it is still a long way off being a good approximation. On the other hand, the extreme simplicity of the function $f_2(t)$ might outweigh its crude form.

Sometimes, it is useful to model the average properties of a fluctuating function using a deterministic oscillation, as in the example above.

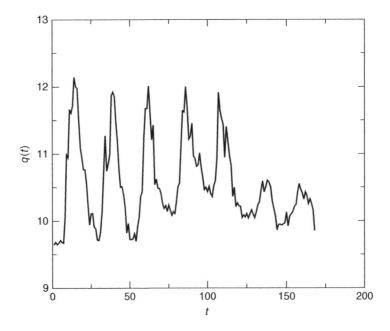

Figure 8.2: A random variable measured from incoming web traffic.

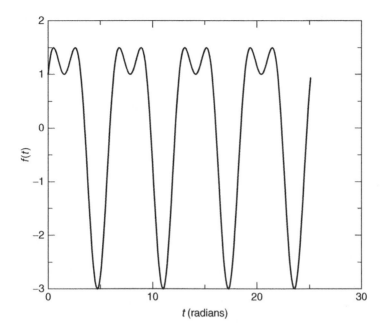

Figure 8.3: An oscillation that is formed by the superposition of two simple waves.

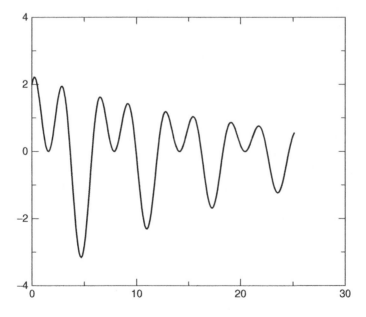

Figure 8.4: An oscillation that is formed by the superposition of two simple waves in which the waves are both slowly dampened out.

Definition 32 (Fluctuation) *A* fluctuation *is a change in a random variable. Fluctuations* $\delta q(t)$ *are sometimes measured relative to the mean value of a variable, that is,*

$$\delta q(t) = q(t) - \langle q(t) \rangle. \tag{8.9}$$

Random fluctuations are characterized by probability distributions $P(q)$, that is, the measured likelihood $P(q)$ that the variable has value q, or the cumulative distribution,

$$F(Q) = P(q \le Q). \tag{8.10}$$

8.4   Rate of change

Rates of change are important for modelling the dynamical interplay between competing processes, so we shall need to be able to describe these changes. For the kinds of random variables that we frequently meet in human–computer systems, there are no smoothly varying quantities in the raw data: the data fluctuate randomly. This makes the description of change more subtle.

In a continuum approximation, the description of rates is an easy matter: we have the derivative or gradient operator, whose effect is given by

$$\frac{\partial q(t, x_i)}{\partial t} \equiv \lim_{\Delta t \to 0} \frac{q(t + \Delta t, x_i) - q(t, x_i)}{\Delta t} \tag{8.11}$$

for the rate of change in time (speed). If there are other approximately continuous dependent labels x_i, or $i = 0, 1, 2, \ldots$, then there will also be partial derivatives for these:

$$\frac{\partial q(t, x_i)}{\partial x_i} \equiv \lim_{\Delta x_i \to 0} \frac{q(t, x_i + \Delta x_i) - q(t, x_i)}{\Delta x_i}. \tag{8.12}$$

For continuous functions, the limit $\lim_{\Delta t \to 0}$ is well-defined, but for stochastic variables it is not. However, one can employ the continuum approximation as described in section 8.5 to approximate the local average behaviour by a smooth function for convenience, or simply use the definition above without the limit for a finite interval Δt.

The error incurred by assuming that these derivatives are actual smooth functions, that is, $\Delta t \to 0$ and $\delta x_i \to 0$ over an interval of time T is of the order $\Delta t / \Delta T$, from the continuum approximation.

Derivatives are used to find the extrema of a function, that is, maxima, minima and inflection points that satisfy

$$\frac{\partial q}{\partial x_i} = 0, \tag{8.13}$$

(see fig. 8.5).

The second derivative or acceleration of the variable

$$\frac{d^2 q(t)}{dt^2}, \frac{\partial^2 q(t, x_i)}{\partial x_i^2}, \tag{8.14}$$

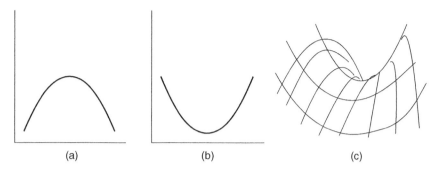

Figure 8.5: Turning points characterize the extrema of functions: (a) is a maximum, (b) is a minimum and (c) is a saddle point or minimax.

describes the curvature of the function, and is commonly used to determine the nature of turning points in a plot.

- At a minimum, the curvature is positive, that is, the function is concave:

$$\frac{dq}{dt} = 0, \frac{d^2q}{dt^2} > 0. \tag{8.15}$$

- At a maximum, the curvature is negative, that is, the function is convex:

$$\frac{dq}{dt} = 0, \frac{d^2q}{dt^2} < 0. \tag{8.16}$$

Generalizations of these for several dimensions can be found in any book on analytical geometry.

We shall have frequent use for the idea of a *saddle point* in describing processes of competition (see fig. 8.5). A saddle point can be thought of as a region of a function in which one parameter is maximized while another is minimized, that is, a saddle is both the top of a hill and the bottom of a valley. This structure occurs in 'tug of war' contests between different processes that share a common resource: one player is trying to maximize gains and the opposing player is trying to minimize the first player's gains. This is a basic scenario in Game Theory (see chapter 19).

8.5   Applications of the continuum approximation

In dealing with probabilities and statistical phenomena, we must distinguish between what is true over short times and what is true over long times. Defining this distinction is central to defining the average properties of systems, as experienced by users. One example is Quality of Service means. If there is a natural separation of timescales, then we can employ an approximation in which we consider the average system behaviour to be smooth and continuous, up to a limited resolution.

- Δt: the interval at which we sample the system. The distribution of possible outcomes is approximately constant over such a small timescale. Even though each measurement contributes to defining the probability distribution of measured values, it would take many more measurements to change the distribution significantly.

- T: the interval over which we can expect the distribution of values to change significantly. This is usually several orders of magnitude greater than the sampling time $T \gg \delta t$.

Do we want quality of service at the level of seconds, minutes or days? This is an important issue. In systems that exhibit approximate stability (i.e. non-chaotic and non-self-similar systems), it is usually possible to separate the deterministic behaviour of the system's average behaviour from the non-deterministic 'fluctuation' behaviour of microscopic details. Suppose that random requests arrive at intervals of the order Δt and that large scale variations in traffic levels occur over times of the order ΔT, where $\Delta T \gg \delta t$ (see fig. 8.6).

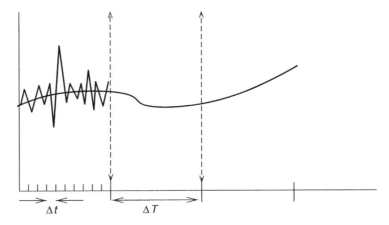

Figure 8.6: Short and long timescales represent actual and averaged variation in service rate. At the microscopic level, there is much noisy variation that averages out over larger periods.

Then, schematically, we may approximate the service rate function $R(t)$, in terms of the average rate R_{av} for all time, as

$$R(t) = R_{av} + R_{av}\, f(t)\, s(t)/2, \tag{8.17}$$

where $s(t)$ is a slowly varying function and $f(t)$ is a rapidly varying function, both of maximum order 1:

$$\max f - \min f \sim 1$$
$$\max s - \min s \sim 1. \tag{8.18}$$

The 'fast' fluctuation function $f(t)$ modulates the average level about the mean value, so its average value over one of the larger time intervals T_n is zero:

$$\int_{T_n}^{T_{n+1}} f(t)\, dt = 0. \tag{8.19}$$

For example, $f(t) = \sin(2\pi N \Delta t / \Delta T)$, for some positive integer N. The slowly varying change in traffic, on the other hand, grows only slightly over the same interval:

$$\int_{T_n}^{T_{n+1}} s(t)\, dt \equiv \Delta s \sim \frac{\Delta t}{\Delta T}. \tag{8.20}$$

Thus, the average service rate over such a long interval is approximately constant (indeed, it tends to a constant as $\Delta T \to \infty$):

$$\langle R(t) \rangle_n = \frac{1}{\Delta T} \int_{T_n}^{T_{n+1}} R(t)\, dt = R_{\mathrm{av}} \left(1 + \mathcal{O}(\Delta t / \Delta T) \right). \tag{8.21}$$

The uncertainty in rate is quantified by the range of values measured over all quality of service time intervals ΔT:

$$\begin{aligned} U &= \sqrt{\sum_n \frac{(\langle R(t)\rangle_n - \langle R(t)\rangle_\infty)^2}{N}} \\ &= \mathcal{O}\left(R_{\mathrm{av}} \frac{\Delta t}{\Delta T} \right). \end{aligned} \tag{8.22}$$

Clearly, this approaches zero as $\Delta T \gg \Delta t$. This simply shows us that we can define any kind of service behaviour as stable and as fulfilling our 'quality' requirements just by choosing a low enough time resolution. Thus, Quality of Service has no meaning unless we specify how large ΔT actually is.

A comment is in order for this reasoning. It has been observed that some service level patterns, such as Ethernet network traffic, follow a self-similar pattern with very large variances over large timescales (see Leland et al. (1994); Willinger and Paxson (1998)). This makes the idea of a quality timescale very hard to implement in terms of averages, because the variances are so large that ΔT needs to be impractically large to get a clean separation of scales. See section 10.9 for further discussion on this.

8.6 Uncertainty in the continuum approximation

If we assume that a system changes smoothly and with infinite resolution, then we must also be realistic about when that assumption will fail to live up to reality. We must estimate the uncertainty in our estimate of the true value.

Example 59 *Many service providers and companies that sell services claim guarantees on the level of service that they can provide, in spite of the inevitable occurrence of random events that limit the predictability of their assumptions. It is important to understand that service is about changes that occur in time, and thus time is an essential element of any service level agreement. If we focus on shorter and shorter intervals of time, it becomes impossible to guarantee what will happen. It is only over longer intervals that we can say, on average, what the level of service has been and what the level of service is likely to be in the future. We must therefore specify the timescale on which we shall measure service levels.*

Example 60 *A Service Level Agreement for UUCP network connectivity could agree to transfer up to 10 MB of data per day. This is an easy goal, by modern standards, and it hardly seems worth including any margin for error in this. On the other hand, a Digital Sub-scription Line (DSL) network provider might offer a guaranteed rate of 350 Mbs (Megabits per second). This is a common level of service at the time of writing. But what are the margins for error now? If each customer has a private network telephone line, we might think that there is no uncertainty here, but this would be wrong. There might be noise on the line, causing a reduction in error-free transmission rate. When the signal reaches the service provider's switching centre, customers are suddenly expected to share common resources, and this sharing must maintain the guarantees. Suddenly, it becomes realistic to assess the margin for error in the figure 350 Mbps.*

Example 61 *A university professor can agree to grade 10 examination papers per day. It is not clear that the level of interruptions and other duties will not make this goal unreasonable. The level of uncertainty is much higher than in a mechanistic network switch. We might estimate it to be 10 ± 3 exam papers per day. In this case, the professor should include this margin for error in the contract of service.*

Uncertainty is an important concern in discussing 'Quality of Service' (QoS); it is calculated using the 'Theory of Errors'. Error theory arises from experimental sciences where one assumes, with some justification, that errors or uncertainties occur at random, with a Gaussian profile, about some true value. The Gaussian property basically ensures that errors are small or do not grow to an arbitrarily large size, compared to the rate of change of the average. However, whether a phenomenon really has a Gaussian profile or not, error-handling techniques can be used to estimate uncertainties, provided there is a suitable separation of timescales. If there is not, the system must be regarded as unstable and therefore no guarantee can be made (see section 10.9). In section 3.9, the method of combining uncertainties is presented.

Example 62 *Consider the rate of arrival of data R, in bytes, from the viewpoint of a network switch or router. The measurables are typically the packet size P and the number of packets per second r. These are independent quantities with independent uncertainties: packet sizes are distributed according to network protocol and traffic types, whereas packet rates are dictated by router/switch performance and queue lengths. The total rate is expressed as*

$$\lambda = rP. \tag{8.23}$$

Using the method of combining independent uncertainties, we write

$$\lambda = \langle \lambda \rangle + \Delta\lambda$$
$$r = \langle r \rangle + \Delta r$$
$$P = \langle P \rangle + \Delta P,$$

and

$$\Delta\lambda = \sqrt{\left(\frac{\partial\lambda}{\partial P}\right)^2 \Delta P^2 + \left(\frac{\partial\lambda}{\partial r}\right)^2 \Delta r^2}. \tag{8.24}$$

Now, Asynchronous Transfer Mode (ATM) packets have a fixed size of 53 bytes, thus $\Delta P_{ATM} = 0$, but Ethernet or Frame Relay packets have varying sizes. An average uncertainty needs to be measured over time. Let us suppose that it might be 1 kB, or something of that order of magnitude.

For a service provider, the uncertainty in r also requires measurement; r represents the aggregated traffic from multiple customers. A service provider could hope that the aggregation of traffic load from several customers would even out, allowing the capacity of a channel to be used evenly at all times. Alas, traffic in the same geographical regions tends to peak at the same times, not different times, so channels must be idle most of the time and inundated for brief periods. To find r and Δr, we aggregate the separate sources into the total packet rate:

$$r(t) = \sum_i r_i(t). \tag{8.25}$$

The aggregated uncertainty in r is the Pythagorean sum:

$$\Delta r = \sqrt{\sum_i \Delta r_i^2} \tag{8.26}$$

The estimated uncertainty is

$$\Delta \lambda = \sqrt{r^2 (\Delta P)^2 + \langle P \rangle^2 (\Delta r)^2} \tag{8.27}$$

Since r and Δr are likely to be of similar orders of magnitude for most customers, whereas $\Delta P < P$, this indicates that uncertainty is dominated by demand uncertainty, that is,

$$\Delta \lambda \simeq \langle P \rangle \Delta r. \tag{8.28}$$

This uncertainty can now be used in queueing estimates.

Applications and Further Study 8

- *Identifying mechanisms and character of changes in systems.*

- *Identifying pathways of causation (cause and effect).*

- *Estimating uncertainties and graininess inherent in changes.*

- *Relate cause and effect by a specific model.*

- *Quantifying the limitations of a description in terms of variables (approximation).*

9

Information

An important concept in describing human–computer systems is the information encoded in changes that take place during their operation. This is one way of measuring the work that is carried out by the system. In the administration of systems, one needs to use the concept of information in several situations to be able to match change with counter-change by sending the opposite information (configuration maintenance), as a principle of maximization or minimization for modelling randomness (disorder or predictability) and as a measure of the wastage in human–computer systems due to the transmission of uncontrolled information (noise).

The formulation of information in symbolic terms is important to system configuration and maintenance because any problem that can be described in this way can be analysed using the tools of stochastic error correction. Information is where the stochastic meets the deterministic.

9.1 What is information?

The idea of information is rather subtle and is used in several different ways. The study of information began in the 1930s and 1940s by mathematicians such as Church, Turing, Von Neumann and Shannon. What one calls information theory today was largely worked out by Claude Shannon (of Bell Labs) and published in 1949 under the title *The Mathematical Theory Of Communication*. He defined the mathematical measure of information, called the *entropy*, and devised many of the core theorems used in information theory today.

Information theory is about the *representation*, *interpretation* and *transmission* of patterns of symbols (data). We attach meaning to patterns of change and call the result information. However, it is vital to distinguish *meaning* from the amount of *information* that represents it.

Example 63 *In Morse code, combinations of the digits '.' and '–' are used to 'mean' letters of the alphabet.*

Analytical Network and System Administration. Managing Human–Computer Networks   Mark Burgess
© 2004 John Wiley & Sons, Ltd   ISBN 0-470-86100-2

Morse code uses strings of up to six dots and dashes to represent every single letter of
the English alphabet. This seems to be rather inefficient, but there are only two symbols that can be communicated in Morse: it is a binary encoding. Clearly, the number
of symbols, of a given alphabet, required to represent the same amount of information is
important.

Example 64 *In the Unix operating system, the symbol* CTRL-C *means 'interrupt program',
while the symbol* # *means 'what follows is a comment and should be ignored'.*

Each of the symbols above conceals a whole series of actions that are carried out, as part
of their interpretation. What is significant is that both meanings could be compressed into a
single symbol, in the appropriate context. Context is very important in coding meaning in
symbols, but symbols can be transmitted by the same rules regardless of their interpretation;
thus they form the basis of information.

9.2   Transmission

Patterns of data are mainly of interest when they are transmitted from a *source* to a *receiver*
over some channel of communication: for example,

- text read from page to brain,

- Morse code sent by telegraph or by lantern,

- communication by memo or letter,

- speech transmitted acoustically or by telephony,

- data copied from hard disk to memory,

- data copied from memory to screen.

In each case, a pattern is transferred from one representation to another and perhaps retrans-
lated at destination. Another way that data are transmitted is to copy data from one place
to another. In computer administration, this is a way of making backups or of installing
systems with software from a source repository. When data are copied, there is a chance
that noise will cause errors to be injected so that the copying is not performed with complete fidelity. The model of transmission again provides a model for discussing this. Data
might be sent

- from one place to another,

- from the past into the future, without moving.

In the first case, data are transmitted by copying, for example, during a system backup.
In the latter case, data do not move, but chance events (cosmic rays, moving magnets,
accidental destruction) can that compromise the integrity of the data, for example, data
stored on a hard disk, or the programs and tasks that work in the system cause the data to
evolve deterministically.

9.3   Information and control

The ability to control a system implies an ability to stabilize it by countering change. In stochastic systems, such change is represented by the information content of the stochastic environment. To control such an environment requires a counter-input of information of comparable complexity. Information thus gauges the likelihood of one's ability to control a system. If the informational entropy of the environment is much greater than the information content of the regulation scheme (e.g. the information content of policy rules), then regulation cannot be guaranteed. We might define the controllability of a system by the ratio of information input from the environment to the information contained in its control policy:

$$C = \frac{I_{\text{environment}}}{I_{\text{policy}}}. \tag{9.1}$$

9.4   Classification and resolution

How shall we define information? We must contend with the following:

- Distinguishing and classifying patterns of symbols.

- Space–time coordinates and units. (Is a long beep the same as a short beep? Quantization/digitization)

- The meaning of redundancy and repetition.

Information arises from the abstract interpretation of changes in a medium.

To build a more precise picture of what information is, we begin with a signal $q(t, x, ..)$, which is a function or a field that changes in time or space. We shall consider only time, as though we are receiving a signal from a stationary antenna, for example, a radio signal. The signal is really a pattern formed by a disturbance in a physical medium. Our everyday experience leads us to believe that there are two types of signal $q(t)$:

- 'Analogue' or continuous functions $q(t)$.

- 'Digital' or discontinuous functions $q(t) = \sum_i \theta(t - t_i) Q_i$.

We shall see that this distinction is artificial but that in this distinction lies the central essence of what information is about. An analogue signal is a limiting case of a digital signal.

In order to say anything about the signal, we have to map it out by placing it on a discrete grid of *coordinates*. At some arbitrary level, one decides not to subdivide space-time any further, and one reaches a limit of resolution. This choice can result in a loss of fidelity in describing the pattern, if the pattern is denser than the coordinate grid (see figures). This coordinatization is called *digitization*, or *coarse-graining* of the signal (see fig. 9.1). Such a process always takes place even if one is not conscious of it. For instance, the eye automatically digitizes images for the brain since there is a finite number of cells on the retina.

Information must be defined relative to this set of coordinates since it is the only means we have of describing change. Let us begin by assuming that the detail in the signal is

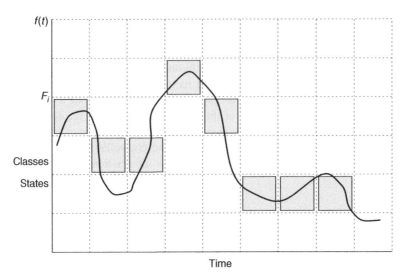

Figure 9.1: Coarse-graining or digitization is a coordinatization of the continuous signal.

greater than the resolution of the grid. We do the following:

- Divide up the time axis into steps of equal width Δt. Here we shall look at an interval of time from $t = 0$ to $t = N\Delta t$, for some N.

- Divide up the q-axis into C *classes* $Q_i = [Q_i^-, Q_i^+]$, which touch such that $Q_i^+ = Q_{i+1}^-$.

Digitization means that whenever the function $q(t)$ is mostly inside a cell Q_i, its value is simply represented by the cell. We have compressed the detail in the square region Q_i into a single *representative* value i over an interval of time.

There are good digitizations and bad digitizations (see figs. 9.2, 9.3 and 9.4). Nyquist's sampling law tells us that the interval widths need to be half the width of the 'finest change' in the signal. In Fourier language, the sampling rate must be twice that of the greatest frequency we wish to resolve.

Example 65 *CD players sample at 44.1 kHz and Digital Audio Tape (DAT) samples at 48 kHz: the limit of human hearing is about 20 kHz when we are young, and falls off to about 12 kHz as we grow old.*

In the physical world, all information is digital if we examine it with sufficient resolution. Electrons are assumed to be indivisible, energy levels are really discrete. Even if we could devise a fully continuous representation, it would not be useful for carrying information because we would have to distinguish between an infinite number of different values, which would be noise.

The digits Q_i are regarded as the basic units of information: they are a strict model for representing change. If $C = 2$, we have binary digits $\{Q_1, Q_2\} = \{0, 1\}$, and so on, or *bits*.

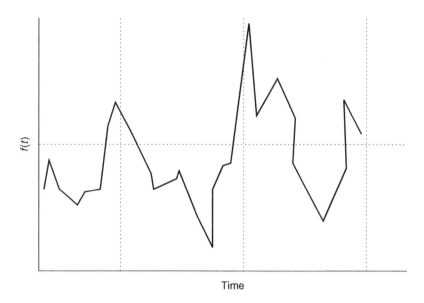

Figure 9.2: A poor digitization cannot sensibly determine the value of the signal within the cells.

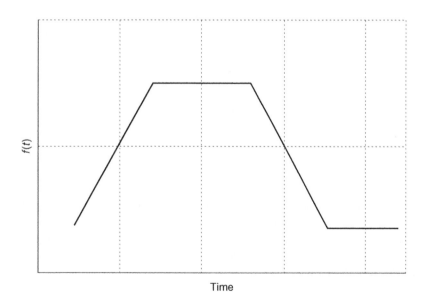

Figure 9.3: A well-suited digitization without loss. This signal can be represented with just two classes, that is, binary digitization.

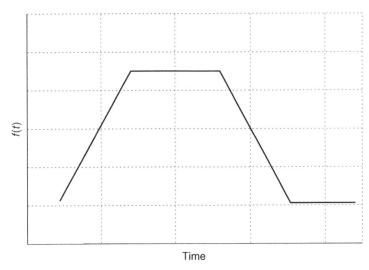

Figure 9.4: The same signal as in fig. 9.3, this time digitized into 6 classes.

9.5   Statistical uncertainty and entropy

In communicating commands and information, we associate meaning primarily to discrete events, or symbols. These might be words, or numbers or even pictograms and glyphs. In western languages, we have grown used to a fixed phonetic alphabet that is used to build up words. In the administration of human–computer systems, communication occurs by several methods:

Example 66 *At the level of the computer, messages are passed to the CPU as instruction opcodes that are read as a stream of words from the memory of the computer. The size of each instruction is known to the processor, so that it always knows where the current instruction starts and ends. The symbol lengths are all the same size, since some instructions contain data and others do not.*

Example 67 *The English language alphabet consists of letters A, B, C, D, E, F, G, H, I, J, K, L, M, N, O, P, Q, R, S, T, U, V, W, X, Y, Z, together with a number of punctuation symbols: , ; . This is a basic set of symbols for forming messages in English. This is the alphabet used to convey information in written communication. In verbal communication, words can be broken down into phonemes or sound digits, though we known from computer speech that this is not always very realistic. Voice is a continuous (non-digital) process.*

Example 68 *In system administration, commands and communication between humans and computers are often conveyed in a specific terminology of code words, almost like a military code book. Special words are used: reboot, format, crash etc. These have special meanings and therefore act as single letters in a command alphabet. If we replaced them by their first* *letters, R, F, C, and so on, the same meaning could be conveyed in the context of system administration[1]. Unix commands have a special alphabetic structure:*

```
command -option1 -option2 ...
```

The set of commands is a finite set that can be labelled by a single digit for each command. Each option (-v, -w, etc.) is also a digital symbol that communicates information to the computer with standard interpretation. Windows commands are often conveyed through menu selections and button selections. In this case, each button or menu item is a digit with an interpreted meaning.

Example 69 *The communication of simple configuration instructions by the Simple Network Management Protocol (SNMP) makes use of an alphabet of configuration codes with standard meanings, defined in Management Information Base (MIB) models for a device. An MIB defines an effective symbolic alphabet for sending streams of instructions with the modifiers: read, write and trap.*

In each of the actual examples above, we see communication—however complex its actual representation—reduced to the transmission of codes or symbols with special meanings. As long as the symbols can be distinguished, each symbol reduces the uncertainty or increases the information of the receiver about the sender. Such a stream of codes or symbols is the basis for nearly all instructions in a human–computer system[2].

We therefore need to recognize each code or symbol in a data stream. As we saw earlier, a signal $q(t)$, is digitized into symbols, or digits, by a digitization procedure (see fig. 9.5). Information is coded into the channel by the variation of the signal. This is digitized into a string of digital symbols (called a message), which occur in a certain order, with a particular frequency. When a digit occurs n_i times out of a total of N digits, where

$$\sum_i n_i = N, \tag{9.2}$$

one may say that the probability of the symbol's occurrence was

$$p_i = \frac{n_i}{N}. \tag{9.3}$$

The probability distribution of symbols in the signal is found by collapsing the image in fig. 3.3 into a histogram, as shown in the figure. Note that it is assumed that only a single symbol can occur at any one time, that is, that the channel is a serial channel.

This probability distribution displays the average amount of *uncertainty* in the symbols being transmitted. If the distribution shows that a message was concentrated entirely around a single symbol, then the uncertainty is very small. If it shows that all symbols were used in equal numbers, then the uncertainty is very large. To convey a lot of information in a

---

[1] Indeed, technical workers have the uncanny habit of replacing all natural language with three-letter abbreviations!

[2] Even where speech is involved, or English text, we tend to reduce the number of possibilities to a few simple symbolic cases by inventing forms to fill out, boxes to tick, or special verbal jargon that reduces the amount of talking required to convey meaning.

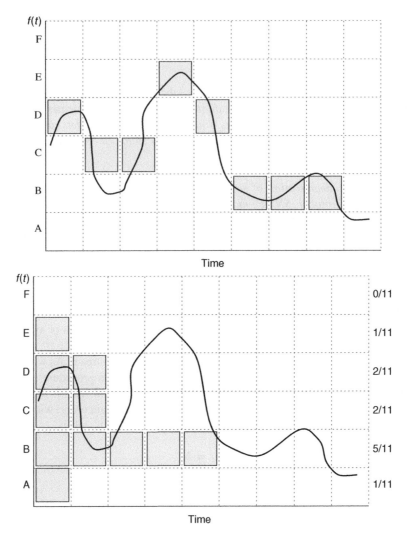

Figure 9.5: Coarse-graining or digitization is a labelling of the signal. First we divide the signal variability into discrete symbols or finite-sized blocks (first picture), and then we approximate the true signal to the coarse representation. Collapsing the digit-blocks into a histogram and counting the number of occurrences (second picture) gives a graphical representation of the average numbers of digits in the signal. By scaling this histogram, one obtains a probability distribution for the signal. This is used to calculate the entropy or average information content.

message, we need to use different symbols; thus uncertainty in the average use of symbols is a measure of information content. This information is measured by the *entropy*:

$$H[p] = -\sum_{i=1}^{A} p_i \log_2 p_i, \qquad (9.4)$$

where A is the number of symbols in the alphabet. The entropy is measured in 'bits', or binary digits, if the logarithm is base 2. Note that the entropy is a scalar functional of the probability distribution.

If we define the expectation value of a function $g = \{g_i\}$, on this discrete space, by

$$\langle g \rangle_p = \sum_i p_i g_i, \tag{9.5}$$

then the entropy may also be written as the expectation value of the logarithm of the number of symbols per message length:

$$H[p] = \langle -\log_2 p \rangle. \tag{9.6}$$

If one further assumes that the transmission of symbols is discrete in time, so that each symbol has the same duration in time, then this has the interpretation of an average rate of bits per second.

Example 70 *If we have a value x, we need $\log_2(x)$ binary digits (bits) to represent and distinguish it from other values, for example, 8 requires at least $\log_2(8) = 3$ bits, 32 requires at least $\log_2(32) = 5$ bits, and so on. If we use base 10 numerals, 8 requires at least $\log_{10}(8) = 0.9$, that is, 1 symbol to represent it, while 16 requires at least $\log_{10}(16) = 1.2$, that is, 2 symbols to represent it. Thus, the amount of information that needs to be distinguished to call a value x includes the values of values less than this that must be coded separately, and the length of the symbol string measures the amount of information required to code the value in some kind of digits.*

The entropy is a single scalar characterization of a probability distribution. Since it collapses all of the changes of the signal into a single number, it cannot distinguish equivalent signals, that is, signals with the same frequency distributions of symbols in them. However, its interpretation as a transmission rate is very useful for considering the efficiency of communication.

Suppose we consider the transmission of information over a communications channel, then there is information both at the input of the channel and at the output. Under ideal conditions, the average information entering the input would be the same as the average information leaving the output; however, this is not necessarily the case if the channel is affected by noise. Noise is simply unpredictable information, or extra uncertainty, which enters the system from the environment and changes some of the symbols.

Figure 9.6 shows how the distribution of symbols in a message can be altered by its transmission over a channel. The entropy may be used to characterize the average information content at the input I and at the output O.

$$H[I] = -\sum_{i=1}^{A} p_i(I) \log_2 p_i(I),$$

$$H[O] = -\sum_{i=1}^{A} p_i(O) \log_2 p_i(O), \tag{9.7}$$

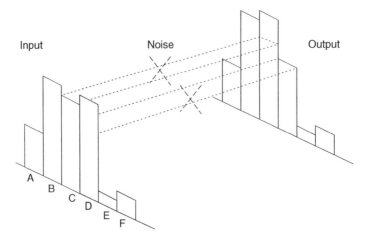

Figure 9.6: Information passing through a noisy channel. The frequency distributions of symbols can be altered by noise introduced by the environment, as the data pass through the channel. Ideally, these two distributions would be equal. In the worst case, they would have nothing in common.

where $p_i(I)$ are the probabilities at the input, and $p_i(O)$ are the probabilities at the output. Note that I and O label *sets* of symbols, whose elements are the alphabet of digits[3]. If transmission of information is perfect, then,

$$H[I] = H[O]. \tag{9.8}$$

This can occur only if the information extracted from the output is predictable using the information at the input, that is, if they are correlated. If $p(I)$ and $p(O)$ are independent events, this cannot be the case since output and input would be unrelated.

9.6   Properties of the entropy

The entropy H has the following properties:

- It is continuous in the p_i.

- When the p_i are equal to $1/C$ it is a monotonically increasing function of C, meaning that information is proportional to resolution.

- If the signal is completely *ordered*, that is, $p_j = 1$, and $p_{i \neq j} = 0$, then the entropy has a minimum value of zero. This tells us that a trivial signal that never changes carries no information.

---

[3] In this picture, the transmission must be a *congruent mapping* of symbols: the same symbol set should be used at the input and output of the channel. Usually, the input and output alphabets are assumed to be the same; however, if the channel is encrypted or otherwise encoded, this need not be the case. For now, we shall assume that the number of symbols is A in both cases, and that the symbols map congruently, in the same sequence.

- If the signal is maximally randomized so that all the p_i are equal to $1/C$, then the entropy takes on its maximum value $\log_m C$. This tells us that a lot of information is needed to characterize a signal with a lot of change.

- It is independent of the route by which the final probabilities were achieved. (Shannon III)

Information is about the length of the shortest message one could send that exactly describes a string of data to someone else, that is, so that they could reproduce it with complete accuracy.

It is important not to confuse information with regularity of a signal, or *orderliness*. We might think that a maximally random, noisy system (like the fuzzy dots on a television screen with no signal) has no information worth speaking of, whereas a system that is very ordered does convey information. This is not true; in fact, the noisy television screen contains so much information that extracting *meaning* from it is difficult. This is a human cognitive problem.

9.7 Uncertainty in communication

To discuss how information is transmitted, we must broaden the discussion to allow for unreliability in copying data. For this, we introduce the joint probability matrix, which describes the probability that sending digit Q_i from A actually results in symbol Q_j being received by B, for any i, j.

Suppose there are two sets denoted, A and B, there is a joint probability matrix

$$p_{ij}(A, B) = p_{ij}(A \cap B), \tag{9.9}$$

which specifies the probability that the ith event will be measured in A in conjunction with the jth event is measured in B. In the case where A and B are mutually independent events, that is, the dual event occurs entirely by coincidence,

$$p_{ij}(A, B) = p_i(A)p_j(B) = p_{ji}(B, A). \tag{9.10}$$

In this case, the probabilities factorize and the combined probability is the overlap of the two sets, that is, the product of the probabilities of the individual events, that is, the two sets have to have some common elements *by coincidence* (see fig. 9.7). Such a factorization is not possible if A and B are more deeply related. In general, one can only say that

$$p_i(A) = \sum_j p_{ij}(A, B) \tag{9.11a}$$

$$p_j(B) = \sum_i p_{ij}(A, B). \tag{9.11b}$$

These are called the *marginal distributions*, and are formed by summing over all other indices than the one of interest. It is natural to apply joint probabilities to the input and output of a communications channel. Since we are interested in communication, the worst-case scenario is when the probability of an event at the output occurs completely independently

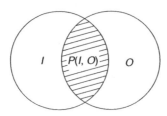

Figure 9.7: The overlap of *independent* sets is the probability of *coincidental* similarity of events measured. Here, we imagine that the two sets are the input and output values. For dependent sets, the joint probability covers the whole of this Venn diagram.

of an event being sent at the input. In that case, the output is purely 'noise', and the only chance of seeing the same data at input and output is when there is a coincidental overlap

$$p_{ij}(I, O)_{\text{worst case}} = p_i(I)p_j(O). \tag{9.12}$$

The ideal case is when a transmission is reproduced with perfect fidelity; that is, $p_{ij}(I, O)$ is a diagonal matrix

$$p_{ii}(I, O)_{\text{best case}} = p_i(I) = p_i(O), \tag{9.13}$$

that is, the probability distributions at input and output are identical.

Using the joint probability of events at the input and output of a communications channel, we can construct a *joint entropy*,

$$H(I, O) = -\sum_{i,j=1}^{A} p_{ij}(I, O) \log_2 p_{ij}(I, O), \tag{9.14}$$

which measures the information in the whole channel, between input and output. As an expectation value, it may be written

$$H(I, O) = \langle -\log_2 p \rangle_{p(I,O)}, \tag{9.15}$$

where the probability set that performs the weighting is specified for clarity. If the input and output are independent (worst case), then one has

$$\begin{aligned} H(I, O) &= -\sum_{i,j=1}^{A} p_i(I)p_j(O) \log_2 p_i(I)p_j(O), \\ &= -\sum_i p_i \log_2 p_i - \sum_j p_j \log_2 p_j \\ &= H(I) + H(O). \end{aligned} \tag{9.16}$$

In general,

$$H(I, O) \le H(I) + H(O), \tag{9.17}$$

that is, the uncertainty (or information) in the joint system is less than that of the two ends combined because some of the information overlaps or is common to both input and output.

Conditional entropy and information flow

The joint entropy measures a kind of correlation of events, but it is not a measure of communication since it does not specify the causal direction of the transmission. We need a way of saying that a certain symbol arrived at the output because it was sent into the input. For this, we define the conditional probability of measuring A, given that we are certain of B:

$$
\begin{aligned}
p_{ij}(I|O) &\equiv \frac{p_{ij}(I, O)}{p_j(O)} \\
&= \frac{p_{ij}(I \cap O)}{p_j(O)} \\
&= \frac{n_{ij}(I \cap O)/N}{n_j(O)/N}. \qquad (n_{ij} \in (I \cap O))
\end{aligned} \qquad (9.18)
$$

This measures the likelihood that the ith symbol was presented to the input, given that the jth symbol was measured at the output. In the case where the input and output are completely independent, the numerator factorizes, and $p(I|O) \rightarrow p(I)$, that is, the knowledge of the output makes no difference.

Another way of interpreting the set $p_{ij}(I|O_j)$, for all i, j, is to ask how many different input distributions give rise to a given output distribution? The sketch in fig. 9.8 shows the regions of the symbol spaces, which give support to this quantity. One may also write,

$$
p_i(A) = \sum_j p_{ij}(A|B) P_j(B) = \frac{n(A)}{(n(A) + n(B))}. \qquad (9.19)
$$

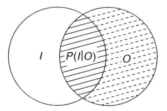

Figure 9.8: The receiver's viewpoint: the conditional probability of input, given that the output is known. This is the space of possible overlaps, given that the space of total possibility is now restricted to those that belong to known outputs. The normalizing factor in the denominator is thus reduced, and the conditional probability is greater for the knowledge of the output.

The conditional probability represents the overlap region, scaled by the space of possible outputs.

The conditional entropy is defined by

$$
\begin{aligned}
H(I|O) &= -\sum_{i,j} p_{ij}(I, O) \log_2 p_{ij}(I|O) \\
&= -\langle \log_2 p(I|O) \rangle_{I \cap O}
\end{aligned} \qquad (9.20)
$$

It is a scalar value, measuring the average variability (information) in I, given that the input overlaps with the output, that is, given that there is a causal connection between input and output. The meaning is slightly non-intuitive. Although the conditional probability is restricted to the output region, on the right-hand side, the averaging is taken over the joint probability space (see fig. 9.9) and thus receives support from the whole region. This

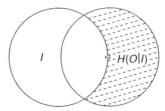

Figure 9.9: The region represented by the conditional entropy is the unrestricted region, plus one point from the conditional region.

measures the average uncertainty that remains about the input, given that the output is known. That uncertainty exists because of the possibility of noise èn route.

We are assuming (hoping) that the probabilities $p(I)$ and $p(O)$ are *not* going to be independent, since that would be an uninteresting system, so the joint probability is not merely the overlap region but the whole space of inputs and outputs. The region that now overlaps with the specific output distribution is only one point, but there is still uncertainty in the input.

The conditional entropy is a measure of information transmitted, because it contains the noise that can be picked up along the channel. To filter out the noise, we consider one final quantity: the relative entropy.

Relative entropy and mutual information

The relative entropy is a measure of the information that is common to two possibility spaces. Consider two spaces P and Q,

$$H(P/Q) \equiv + \sum_P P \log \frac{P}{Q}. \tag{9.21}$$

This is a measure of the distance between P and Q. Although it is not a true metric, it has the property of being zero when P and Q are identical. If we apply this to the input and output of the communications channel, the result is called the *mutual information*, or common entropy.

$$H(I; O) = H(I/O) = + \sum_{ij} p_{ij}(I, O) \log \frac{p_{ij}(I, O)}{p_i(I)p_j(O)}. \tag{9.22}$$

The semicolon is used to indicate that this quantity is a symmetrical function of the input and output. It compares two situations: a source and a receiver that are completely independent (Q), and a general source and receiver that are connected by a partially reliable communications channel. It measures the average reduction in uncertainty about the input that results from learning the value that emerges from the output. In other words, it represents the information in what passes along the channel between input and output.

We can express the mutual information in a number of different ways. Noting that

$$p(I, O) = p(O)p(I|O) = p(I)p(O|I), \tag{9.23}$$

we have

$$H(I; O) = \sum_{ij} p(I, O) \log_2 \frac{p(I|O)}{p(I)}$$
$$= H(I) - H(I|O). \tag{9.24}$$

This has the form of the information (uncertainty plus signal) at the input minus the uncertainty in the input, given a definite output. In other words, it is the part of the input that was transmitted to the output, or the likelihood of transmission (the fidelity). Another representation is

$$H(I; O) = \sum_{ij} p(I, O) \log_2 \frac{p(O|I)}{p(O)}$$
$$= H(O) - H(O|I). \tag{9.25}$$

This is the information arriving at the output (signal plus noise) minus the uncertainty information at the output, given a fixed input (i.e. the noise picked up along the way). Again, this is the likelihood of correct transmission. A third form can be found from

$$H(I, O) = H(I|O) + H(O). \tag{9.26}$$

Using this in eqn. (9.24), we obtain,

$$H(O; I) = H(I) - (H(I, O) - H(O))$$
$$= H(I) + H(O) - H(I, O). \tag{9.27}$$

This is the sum of information entering the input and leaving the output, minus the total amount of independent information in the system. What is left must be the information in the overlap region (see fig. 9.10), that is, the information that is common to both input and output.

9.8 A geometrical interpretation of information

The idea of classification or quantization leads to the idea of *states* Q_i. We say that the signal is in state Q_i at time t. These states are linearly independent: no matter how many times we add together Q_0, we will never get Q_1 or $Q_2 \ldots$ The effect of a message is thus to draw a vector in this space (fig. 9.11):

$$\mathbf{I} = \sum_{i=1}^{C} n_i \hat{\mathbf{e}}_i = n_1 \hat{\mathbf{e}}_1 + n_2 \hat{\mathbf{e}}_2 \ldots n_C \hat{\mathbf{e}}_C. \tag{9.28}$$

This vector summarizes a new kind of state: an average state. It does not preserve the details of the *path* that was taken in order to reach the point. In fact, there is a large number of equivalent paths, of the same length, to the same point.

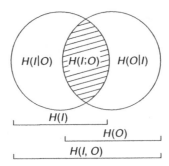

Figure 9.10: The entropies visualized as measures on the space of all possible input and output signals. The Venn diagram shows where the mutual information received support from the input–output probabilities; it does not strictly represent the magnitude of the entropy.

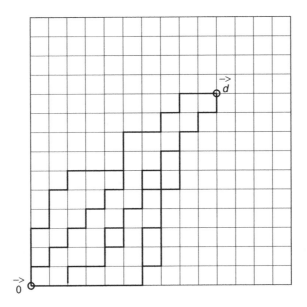

Figure 9.11: A binary message may be drawn as a path in a two-dimensional lattice. A 26-dimensional lattice would be needed to represent a word in English.

The number of equivalent paths is a measure of how much effort one must expend to describe the actual path, that is, it is a measure of the information represented by the path.

I summarizes the number of *units* of information in the signal, since each unit vector is a character or digit in the message. The metric distance of such a vector

$$|\mathbf{I}|^2 = I^i \delta_{ij} I^j = \mathbf{I} \cdot \mathbf{I} \tag{9.29}$$

is a measure of the number of digits, but that is not a good measure of the information. The difference of two message vectors is a measure of how many digits were different (without respect to ordering). This is related to a quantity called the *Hamming distance*, which represents the number of symbols that differ between two messages. We shall return to this when considering how messages are coded.

As digits are received and plotted in the lattice, we build up an average description of the message. If the probability of receiving digit i is p_i, then clearly, after N time units, we have received N digits, and

$$\mathbf{i} = \frac{\mathbf{I}}{N} = \sum_{i=1}^{C} \frac{n_i}{N} \hat{\mathbf{e}}_i = \sum_{i=1}^{C} p_i \hat{\mathbf{e}}_i. \tag{9.30}$$

p_i is the amount of time the signal spends within class Q_i. Suppose we reduced the number of classes of Q_i to one, then there would be only a single path (ordering irrelevant), and only one kind of digit would be possible. Only the number of digits (length) can then convey information. If we had an infinite number of classes or dimensions, then there would be an infinite number of equivalent paths and the amount of detail would be infinite. No two signals would ever be the same in practice, and the signal would be indistinguishable from noise.

Let us call the number of equivalent paths between two points in this lattice h, (the 'hopelessness' or uncertainty of finding the right path, that is, the correct ordering of digits, given only the final state). It is possible to find a formula for this number:

$$h(N) = \frac{\left(\sum_{j=1}^{C} I_j\right)!}{\prod_{k=1}^{C} (I_k!)} = \frac{(I_1 + I_2 \ldots I_C)!}{I_1! I_2! \ldots I_C!}. \tag{9.31}$$

This grows very rapidly with the Euclidean distance $|\mathbf{I}|$ in this message lattice.

$$|\vec{d}| \equiv d = \sqrt{\sum_{i=1}^{C} (I_i)^2}. \tag{9.32}$$

If we consider messages over a fixed number of time, intervals $N\Delta t$, then we can express h as

$$h(N) = \frac{N!}{(Np_1)!(Np_2)! \ldots (Np_C)!}. \tag{9.33}$$

This is a large and awkward number to deal with, so we look at its logarithm H. This represents the length of a string in base m digits that would be able to code the number. This is an important measure because it is the smallest number of digits that one needs to be able to label the exact path and therefore distinguish it from all the other alternative paths:

$$H_N = \log_m h. \tag{9.34}$$

Moreover, by assuming that N is large (high resolution), we can use Stirling's approximation to write this in a simpler form, without awkward factorials. Using

$$\log_m N! \simeq N \log_m N - N$$
$$\sum_i \log_m(Np_i)! \simeq \sum_i [Np_i \log_m p_i] + N \log_m N - N. \tag{9.35}$$

we get

$$H_N = -N \sum_i^C p_i \log_m p_i \equiv N\langle i \rangle. \tag{9.36}$$

To get the fractional uncertainty per digit, we divide by N. Shannon called this quantity the *informational entropy* or average information. He did not derive it in this way, but on general grounds using similar combinatorial properties. It represents the average information needed to distinguish the exact message from a random message with the same probability distribution. It is measured in units of *m-ary digits*:

$$H = H_N/N = -\sum_i^C p_i \log_m p_i. \tag{9.37}$$

This geometrical interpretation is very helpful in interpreting the entropy. It can be extended to the conditional entropies also. Imagine that the picture in fig. 9.11 leads to a slightly different picture at the receiver owing to errors in transmission. Over time, there is an ensemble of similar pictures that might be transmitted with various probabilities. There is thus an additional uncertainty between I and O that is not caused by the information in the message, but by unreliability of the channel. A person receiving a message knows what they have received from the sender, so they are not completely ignorant. That knowledge reduces their uncertainty about the message from a random jumble of digits to one of the possible causes, given our knowledge of the probable distortion by the channel:

$$H(I, O) \to H(I|O) = H(I, O) - H(O). \tag{9.38}$$

The conditional entropy picks out the ensemble of possible causes, that is, the alternative messages that could reasonably have given rise to the message received. From eqn. (9.38), we see that this is the maximum possible uncertainty in the channel ($H(I, O)$) minus the uncertainty that distinguishes a specific message from a random jumble of symbols at the receiver ($H(O)$). The number of possible causes for the message that arrives is therefore of the order

$$E \sim m^{H(I|O)}, \tag{9.39}$$

where m is the base of the logarithm in the entropy ($m = 2$ for binary digits). We should never forget that the entropy makes statements about average properties, not about specific instances.

9.9 Compressibility and size of information

Defining the expectation value of a vector L_i by

$$\langle \mathbf{L} \rangle = \sum_{i=1}^{C} p_i L_i = \mathrm{Tr}(pL), \qquad (9.40)$$

it is possible to view the entropy as the expected uncertainty per digit in the message:

$$H = \langle -\log_m p \rangle = \langle \log_m p^{-1} \rangle = \langle \text{probable digits} \rangle. \qquad (9.41)$$

Example 71 *Consider a message variable that has 16 possible measurable values or states that occur with equal probability. In order to label (and therefore describe) these outcomes, we need 4-bit strings, since $2^4 = 16$. The information/entropy is*

$$H = -\sum_{i=1}^{16} \frac{1}{16} \log_m \frac{1}{16} = \log_m 16. \qquad (9.42)$$

If we take binary digits $m = 2$ (alphabet length 2), we get $H = 4$, showing that the average information per message needed to communicate the behaviour of the random variable over time is 4-bits, as we assumed. This shows that if we choose log-base 2, the answer comes out in bits. This is the uncertainty per digit in the message, since all digits occur with equal probability.

Suppose we encoded the information on a string of DNA (alphabet length 4), then the result would be $m = 4$, $H = 2$, or 2 DNA characters (A, C, T, G). m is the alphabet size of the message.

Example 72 *Consider a human–computer monitoring system in which the probabilities of eight different faults, Δ_i $(i = 1 \ldots 8)$, are found over time to occur with the following probabilities:*

$$p = \left(\frac{1}{2}, \frac{1}{4}, \frac{1}{8}, \frac{1}{16}, \frac{1}{64}, \frac{1}{64}, \frac{1}{64}, \frac{1}{64} \right). \qquad (9.43)$$

Notice the degeneracy of the last four values. The entropy or information of this distribution is

$$H = -\frac{1}{2} \log_2 \frac{1}{2} \cdots - 4\frac{1}{64} \log_m \frac{1}{64} = 2 \text{ bits.} \qquad (9.44)$$

If we want to communicate a message that has this distribution of probabilities on average (i.e. we are looking for long-term average efficiency) to an operator, it looks as though we will need 3-bits in order to label the outcomes of the eight values. However, we do not need to use the same number of bits to code each of the values. It makes sense to code the most probable fault symbol using the smallest number of bits. Suppose we use the bit strings

$$\Delta_1 = 0,$$
$$\Delta_2 = 10,$$

$$\Delta_3 = 110,$$
$$\Delta_4 = 1110,$$
$$\Delta_5 = 111100,$$
$$\Delta_6 = 111101,$$
$$\Delta_7 = 111110,$$
$$\Delta_8 = 111111, \tag{9.45}$$

With this coding, we have distinguished both the distribution of the digits and their positions with a minimum amount of information. The lengths of these strings are $\mathbf{L} = (1, 2, 3, 4, 6, 6, 6, 6)$, *and the expectation value of the length, given the probabilities, is*

$$\langle L \rangle = \sum_{i=1}^{8} p_i L_i = 2 \text{ bits.} \tag{9.46}$$

This illustrates Shannon's theorem that the entropy is lower bound on the average length to which any digit of a message can be compressed. In other words, if we wait until we have received N such fault reports, each coded according to the scheme above, then the total amount of data will be approximately 2N bits long.

9.10   Information and state

Informational entropy tells us about the statistical distribution of digits produced by a system. It measures the average state of the system over an ensemble of measurements, that is, it is a cumulative result. The details of the *current* state are mixed up with all other measurements so that specific information about present state is lost.

One of the surprises of information theory is that one does not usually need to know the precise state of a system in order to be able to describe the system's properties over time[4]. This is an important point, because it means that we do not need to keep infinitely many records or logs of what happened to a system in order to understand its history: it is possible to compress that information into its statistical essence.

If we imagine that a string of symbols, with a certain entropy, is generated by a finite state machine, or other computer program that remembers state, then the entropy also tells us about the average state of the state machine. We cannot tell what state the system is in by looking at the entropy; however, we can tell, at least in principle, that a change of state has occurred within the time resolution of our measurements.

Consider the histogram in fig. 9.12, representing a system with three states: active, waiting and terminated. If a command symbol is transmitted, causing this system to change state to 'active', then the column for the probability for 'active' gets relatively taller and the others get a little shorter. This changes the entropy. If, however, on average we receive equally many commands for each state, then distribution does not change, and we say that the system is in a steady state.

---

[4] This has been one of the most difficult ideas to accept for authors trying to model the system administration process.

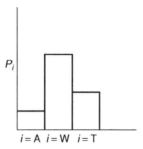

Figure 9.12: The probability distribution of system state, A = active, W = waiting, T = terminated.

Definition 33 (Steady state) *A system is said to be in a steady state if the entropy of the transitions in the system is constant over statistically significant times. Usually, the only stable steady state is one of maximal entropy.*

If we want to change the average state of a system, we need to send it a persistent command signal, consisting of many symbols of the same type (or subset of the whole). This will tend to raise the level of one or more of the histogram columns at the expense of the rest, and 'sharpen' the distribution. In the language of uncertainty, a persistent signal reduces uncertainty about the state of the system, since it forces the signal to be in a known state more often.

This result, although seemingly innocuous, is of the utmost importance to system administration; let us express it in three different ways:

- A stable average state requires no external information to be input to maintain it.

- A random sequence of commands, with maximal entropy, causes no change in the statistical state of a system over long times.

- A sustained system reconfiguration requires the input of statistically significant, low entropy information over long times.

Here, 'long times' means long enough to be statistically significant in relation to any noise or residual randomness. Referring to fig. (8.6), we require a signal of the order of time ΔT to make a significant impact. We shall return to the issue of configuration changes in chapters 15 and 16.

9.11   Maximum entropy principle

The concept of entropy characterizes the uncertainty or 'bluntness' of statistical distributions. If we consider the statistical distributions in fig. 3.4 and in fig. 3.5, then the complete certainty of fig. 3.4 characterizes an absolute minimum of entropy, whereas the completely uncertain fig. 3.5 represents maximal entropy.

If we plot the entropy as a function of the probability of two events, it has the form of fig. 9.13. This shows the shape of the entropy function. More general distributions are

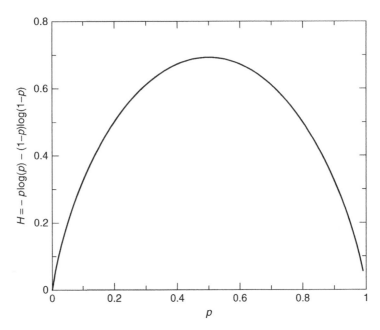

Figure 9.13: The entropy values as a function of p for a two-state system. This is used to illustrate the shape of the entropy function on a flat diagram.

multi-dimensional generalizations of this. The key point is that there is a maximum value for the entropy that is indicated by a turning point in the middle of the graph. There is also a minimum value at the symmetrical points $p = 0$ and $p = 1$, but there is no turning point here. This means that if we try to maximize the entropy by looking for the stationary points of this curve, the result will only find the maximum value, not the minimum (which is always zero, from the form of the function).

Let us work out the maximum entropy distribution for a histogram with C classes. We do this using the method of Lagrange. Let the Lagrangian be

$$L = - \sum_{i=1}^{C} p_i \ln p_i - \alpha \left(\sum_{i=1}^{C} p_i - 1 \right), \tag{9.47}$$

where α is a Lagrange multiplier, or a parameter that enforces the constraint that the sum of probabilities is 1, and we use the natural logarithm for convenience.

$$\sum_{i=1}^{C} p_i = 1. \tag{9.48}$$

Maximizing this with respect to all parameters gives the following:

$$\frac{\partial L}{\partial p_i} = - \ln p_i - 1 - \alpha = 0$$

$$\frac{\partial L}{\partial \alpha} = \sum_{i=1}^{C} p_i - 1 = 0. \tag{9.49}$$

This is solved for p_i by

$$p_i = e^{\alpha - 1}$$
$$\sum_i p_i = Ce^{\alpha - 1} = 1$$
$$\text{i.e. } p_i = \frac{1}{C}. \tag{9.50}$$

In other words, the completely flat distribution is the case of maximum entropy in which all probabilities are the same.

If we maximize the uncertainty about a system, entropy leads us to the distribution of values that contains the least planning (the fewest assumptions) or the most randomness. This is not particularly interesting until we apply additional constraints that tend to make the entropy less than this value. This turns out to be a powerful tool.

Definition 34 (Maximum entropy distribution) *A maximum entropy distribution is that produced by a maximally random variable that is constrained by a function* $\chi(p) = 0$. *It is found by maximizing the Lagrangian function*

$$L = -\sum_{i=1}^{C} p_i \ln p_i - \alpha \left(\sum_{i=1}^{C} p_i - 1 \right) - \beta \chi(p), \tag{9.51}$$

and solving for p_i.

Example 73 *Find the least clustered (most distributed or robust) network, as a function of node-degree, given that we have a fixed number of links L to join the nodes together; that is, where do we place L cables between a number of cities in order to have the best distribution of resources.*

We begin by defining the probability of finding a node of degree k,

$$p_k = \frac{n_k}{N} = \frac{\text{Nodes with degree } k}{\text{Total number of nodes}}. \tag{9.52}$$

To count the number of links in the graph, as a function of k, we note that every node of degree k has k links attached to it, but only half the link is attached since the other half is attached to another node. Thus, the number of links is related to k and n_k *by*

$$L = \sum_k \frac{1}{2} k \times n_k. \tag{9.53}$$

Our constraint is thus

$$\chi(p) = \sum_k \frac{1}{2} k \times n_k - L = \sum_k \frac{1}{2} N p_k - L = 0. \tag{9.54}$$

The Lagrangian is therefore

$$L = -\sum_{i=1}^{C} p_i \ln p_i - \alpha \left(\sum_{i=1}^{C} p_i - 1 \right) - \beta \left(\sum_k \frac{1}{2} N k p_k - L \right), \tag{9.55}$$

Maximizing this function gives

$$\frac{\partial L}{\partial p_k} = -\ln p_k - 1 - \alpha = 0$$

$$\frac{\partial L}{\partial \alpha} = \sum_{k=1} p_k - 1 = 0$$

$$\frac{\partial L}{\partial \beta} = \sum_k \frac{1}{2} N k p_k - L = 0. \tag{9.56}$$

This has the solution from the first two lines in eqn. (9.56)

$$p_k = \frac{e^{-\frac{1}{2} N \beta k}}{\sum_k e^{-\frac{1}{2} N \beta k}}. \tag{9.57}$$

It is an exponential distribution, that is, nodes of large degree are exponentially suppressed ($\beta > 0$ else the distribution is not normalizable for arbitrarily large k) relative to nodes with small k. To express this in terms of the constant L we can perform a final mathematical trick. Suppose we define the generating function

$$Z = \sum_k e^{-\frac{1}{2} N \beta k}, \tag{9.58}$$

then we can express

$$L = -\frac{\partial}{\partial \beta} \ln Z. \tag{9.59}$$

Thus $\ln Z = -L\beta$. The value of $\ln Z$ cannot be evaluated exactly, but it is constant, and we can call it $\ln \equiv -\zeta$ for convenience, so that $\beta = \zeta/L$, thus we have

$$p_k = \frac{e^{-\frac{1}{2} N \zeta k/L}}{\sum_k e^{-\frac{1}{2} N \zeta k/L}}. \tag{9.60}$$

If we express L as a fraction of N, this can be simplified even further. This probability distribution is called the Boltzmann distribution after the physicist L. Boltzmann who discovered its importance in physics. It tells us that most nodes should have small values of $k > 0$, and exponentially fewer large nodes should exist, for maximum entropy or least clustering. Clearly, this makes sense—the fewer large links, the less clustering there will be. The question is why should there be any large degree nodes. The reason is clearly that we have fixed the number of links and the number of nodes to be constants, and there are limits to how many links we can fit into a small number of nodes—some of them will have to be of higher degree if $L > N$. If N/L is small, this exponential falls off only slowly and there will be larger numbers of higher-degree nodes. Thus, this is the least clustered network we can build.

Maximum entropy distributions occur in many situations where one wishes to make the least possible assumption, or invoke the greatest range of possibility in a system.

9.12 Fluctuation spectra

One application of maximal entropy distributions is in the characterization of fluctuating random behaviour in systems.

Definition 35 (Fluctuation spectrum) *A fluctuation spectrum is a probability distribution of values that can be assumed by a random variable, that is, if $P(q)$ is the fluctuation spectrum of the variable, then the average value over all times is*

$$\langle q \rangle = \int dq\, P(q) q. \qquad (9.61)$$

The maximum entropy principle allows us to characterize the likely signal behaviour of random events, such as data requests arriving at a server, given the known constraints under which the system can behave. We do this by assuming maximal entropy for $P(q)$, subject to any boundary conditions.

Example 74 *In (Burgess (2000a)), this method was used to model the observed fluctuation spectra of network services, such as the World Wide Web, using a periodic model for requests based on the working week (see fig. 2.1). The resulting distribution was a Planck exponential distribution (see fig. 9.14). There are many maximum entropy distributions—as many as there are constraints to apply—and they can often be related to simpler forms.*

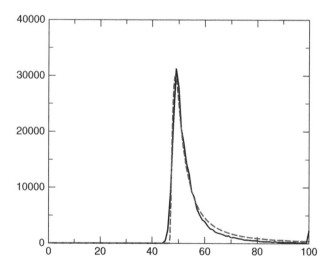

Figure 9.14: The fluctuation spectrum of numbers of World Wide Web requests, corrected for periodic variations has the form of a maximum entropy Planck distribution. The dotted line shows a theoretical curve, calculated using a maximum entropy model, and the solid line is the measured value. The horizontal axis shows the numbers of simultaneous requests, rescaled to a mean value of 50, and the vertical axis is the number of times the system had so many simultaneous requests.

Applications and Further Study 9

- *Quantitative discussion of the flows of information and instruction in a system.*

- *Measurement of workflow in a system.*

- *Gauging the controllability of a system.*

- *Determining how focused, constrained (low entropy) or distributed (high entropy) a process or structure is.*

- *Maximization of entropy for balancing the idea of 'what can happen will happen' with the known constraints.*

10

Stability

One of our fundamental premises about systems is that medium-term stability, allowing for long-term change, is a desirable concept. Systems must be predictable for long enough to perform their intended function. This applies both to the human and machine parts of a system. However, we still need a quantitative description of what such stability means, and on what timescales.

10.1 Basic notions

If we place a ball at the crest of a hill, the slightest movement will cause it to roll down to the bottom. A ball placed at the top of a hill is a mechanical system that is said to be *unstable* to small perturbations, that is, a small push changes the character of the system. By contrast, a ball placed at the bottom of a valley or trough is said to be *stable* to perturbations, because a small push will only take it a short way up the hill before it rolls back down again, preserving the original condition of the system.

Stability is an important idea in systems. If a small change can completely alter a system, then its usefulness is limited. A bomb is a chemical–mechanical system that can be used only once, because it is impractical to reset it to its original condition once it has exploded. Instability affects a great many dynamical systems, from financial systems, to computers, to social systems. The idea of building stability into systems is thus of central importance to human–computer administration.

Modern fighter jets are built with an inherent instability under flight, unlike passenger jets that are inherently more stable. Fighter jets are much more manoeuvrable because of this ability to lapse into instability. The price one pays for this is a much more essential and risky regulation requirement that maintains the system right at the edge of instability, allowing rapid, controlled change, but flirting with rapid, uncontrolled change.

10.2 Types of stability

There are various notions of stability, but they all have in common a search for states that are not significantly altered when we perturb the system by a small amount. We can write

Analytical Network and System Administration. Managing Human–Computer Networks Mark Burgess
© 2004 John Wiley & Sons, Ltd   ISBN 0-470-86100-2

most notions of stability in a generic form:

Definition 36 (Stability) *A stable state is one that is preserved, up to a multiplying factor, when perturbed by some operation. Let Q be a state of a system, and let $\hat{\Delta}_\lambda$ be an unspecified operator that perturbs Q by an amount λ in some parameter. If the generic perturbation leads to the same state Q multiplied by a scale factor $\Omega(\lambda)$, for some function Ω, the state may be described as Ω-stable under this operation; that is, the result of a perturbation*

$$\delta Q \equiv (\hat{\Delta}_\lambda(Q) - \Omega(\lambda)Q) = 0. \tag{10.1}$$

This definition associates stability with stationary variations. This is not a rigorous definition, since the descriptions of $\hat{\Delta}$ and Ω are vague; however, it expresses the general concept behind a broad range of ideas of stability. In some cases, it is desirable to demand restrictions on $\Omega(\lambda)$.

10.3 Constancy

The simplest kind of stability is *constancy* ($\Delta = \Omega = 1$). Mathematically, we express this by saying that a system variable does not change at all with respect to its parameters.

$$q(t, x_1, x_2, \ldots) = \text{const} = q. \tag{10.2}$$

We can change the time and other parameters of the system x_i, but such a change has no effect on $q(t, x_i)$ because the function is a trivial one. Such constancy is a rather simplistic viewpoint that is rarely true of anything but the simplest systems, but it is a convenient approximation in many cases. A slightly more realistic viewpoint is to only expect constancy on average.

Average constancy allows a system to be dynamic and to change over short time intervals in an unspecified way as long as it changes back again so that the average result is zero over longer times:

$$\langle q(t) \rangle_{\Delta t} = \frac{1}{\Delta t} \int_{-\Delta t/t}^{+\Delta t/2} q(t)\, dt = \text{const}. \tag{10.3}$$

An oscillation is a deterministic example of this; random fluctuations are a non-deterministic example. This kind of stability allows us to discuss situations where there are fluctuations in the behaviour of the system that average to zero over time. In this case, we must ask how long do we have to observe the system before the fluctuations will average out?

An enhancement of the previous case is to allow the average of a function to change by a small amount, that is, to exhibit a *trend*, by varying slowly at a rate that is much slower than fluctuations that almost average out. This is called a *separation of scales*, and is discussed in section 8.5.

Finally, a more advanced notion of stability in dynamic systems that exhibit fluctuations is statistical stability. Here we ask the question, in a system that is stochastic, that is, exhibits unpredictable fluctuations, with a particular statistical distribution of values, are there certain statistical distributions that are the natural result of system behaviour? Are some distributions more stable than others if we change the system slightly? Such distributions summarize the stability of the management scale, without disallowing the minutiae of the system.

10.4   Convergence of behaviour

Systems at any scale can exhibit oscillatory or random behaviour. It is sometimes a desirable property and sometimes an undesirable one. An oscillation with fixed frequency (such as a swinging pendulum or a daily task list) is said to be in a steady state, even though it is changing in time in a regular way. Such a steady state is also called a *limit cycle* if it is the result of a process of convergence towards this final steady state from an irregular pattern. Generally, we are interested in systems that enter into a steady state, that is, either a static or dynamical equilibrium.

Another possibility is that oscillations die out and leave a static state. In order to converge to a static state, oscillations must be dissipated by a drain on the system that behaves analogously to the friction in a pendulum.

Example 75 *Circular dependencies often result in oscillations that need to be damped out. Suppose host A requests a result from a database that is located on host B, but the data in the database running on host B is located physically on host A and is shared by a network service. A single request from host A results in traffic*

$$A \rightarrow B \rightarrow A \rightarrow B \rightarrow A. \tag{10.4}$$

Moreover, a timeout or a failure could result in longer oscillations of period A−B. A local cache of data could be used to dampen these oscillatory convulsions and relieve the network of the unnecessary oscillatory behaviour, or the system could be reorganized to alter the flow of communication between the client, the disk and the database.

Example 76 *The build-up of temporary disk files on a computer system can result in an escalating space problem that can be kept under control by introducing a counterforce that makes the behaviour oscillatory. Human work patterns have a daily rhythm, so daily tidying of garbage can keep the build-up within manageable levels. Thus, in this example we are trying to achieve a steady limit cycle from a behaviour that is originally divergent by introducing a countermeasure.*

Example 77 *Tit for tat reprisals between users and/or administrators are ping-pong oscillations that could escalate or, in principle, never terminate. A prolonged bout of tit for tat is warfare.*

We have two possibilities for modelling converging oscillatory behaviour—discrete or continuous models. Discrete changes models use *graphs*, or *chains* of discrete changes can be used to trace the possible changes in the system. Cycles can then be identified by looking at the topology of the graph of the allowed transitions. One way to define convergent behaviour is to consider the transition function T_{ij} of the system that determines what state q_j a system that is in a state q_i will make a transition to. Suppose that, no matter what state we are in, a number n of transitions will bring us into a definite state q_c.

$$(T_{ij})^n \, q = q_c \tag{10.5}$$
$$T_{ij} \, q_c = q_c \tag{10.6}$$

Such a transition function T_{ij} may be called convergent. Such a state q_c is called a *fixed point* of the function (see section 10.13).

If we are willing to ignore the discrete details of the system, over longer times, then another picture of this can be drawn for smooth and differentiable systems (the continuum approximation). Let q be the state of the system and γ is the rate at which it converges to a stable state. In order to make oscillations converge, they are damped by a frictional or counter force γ. The solutions to this kind of motion are *damped oscillations* of the form

$$q(t) \sim e^{-\gamma t} \sin(\omega t), \qquad (10.7)$$

for some frequency ω and damping rate γ. Three cases are distinguished: under-damped motion, damped and over-damped motion. In under-damped motion $\gamma \ll \omega$, there is never a sufficient counterforce to make the oscillations converge to any degree. In damped motion, the oscillations do converge quite quickly $\gamma \sim \omega$. Finally, with over-damped motion $\gamma \gg \omega$, the counter force is so strong as to never allow any change at all.

Under-damped	Inefficient: the system can never quite keep oscillations in check.
Critically damped	System converges in a timescale of the order the rate of fluctuation.
Over-damped	Draconian: changes are stopped before completing a single cycle.

An over-damped solution to system management is rarely acceptable. An under-damped solution will not be able to keep up with the changes made to the system by users or attackers.

10.5 Maxima and minima

The extrema of smooth functions can be used to define stability. The extrema are found at the turning points of functions, where the rate of change of the function is smallest.

$$\frac{dq(\lambda)}{d\lambda} = 0. \qquad (10.8)$$

Thus, both maxima and minima exhibit local sluggishness to change; however, only minima are stable to perturbations.

Maxima and minima can also be defined for non-smooth domains, such as the nodes of a graph, provided there is a function which defines a value ϕ_i on each node i. For instance, if we have a graph formed from a discrete set of nodes, a node is a maximum if it has a greater value ϕ_i than any of its neighbouring (adjacent) nodes.

Definition 37 (Local maximum) *If ϕ is a mapping from a domain i to some range, then a node i in the domain is a local maximum of the mapping if the nearest neighbours j of node i have strictly lower values ϕ_j than node i itself; that is, if A_{ij} is the adjacency matrix of the graph, a local maximum satisfies*

$$\phi_i > \phi_j, \ \forall\{j | A_{ij} \neq 0\}. \tag{10.9}$$

This definition works equally well for sparse or dense sets, that is, for functions of a continuous variable or for discrete graphs.

10.6 Regions of stability in a graph

Another way of talking about the stability of structures is to define regions of persistence in mappings (like the eye of a storm, amongst all the connections). If a mapping leads us into a natural resting place, the graphical equivalent of a basin or minimum, then we can use that to define stability.

There are two distinct notions of stability for graphs that are very useful for us throughout this book.

Definition 38 (Internal stability) *A set $S \in X$ in a graph (X, Γ) is said to be* **internally stable**, *if no two nodes in S are connected, that is, if*

$$\Gamma S \cap S = \emptyset. \tag{10.10}$$

In other words, internal stability says that a region of internal stability consists of points that cannot be exchanged for one another (fig. 10.1). None of them have any arrow of link between them that might identify one as being better than another in the sense of the mapping A. This property makes them 'as good as each other' as end points in a journey through the graph. In the theory of games, that we shall turn to later, this property is used

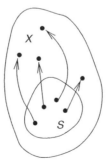

Figure 10.1: An internally stable set is a set of peers within a graph, that are unconnected by any single hop. Once we get into an internally stable state, we cannot get into any other internally stable state without backtracking out of the internally stable region.

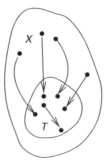

Figure 10.2: An externally stable state is accessible to all points outside the externally stable region, by a single hop.

to mean that none of the points in an internally stable set dominate any of the others, thus they are all equally valid solutions to the game.

Definition 39 (External stability) *A set $T \in X$ in a graph (X, Γ) is said to be* **externally stable** *if every node $x \notin T$, outside of T, satisfies*

$$\Gamma x \cap T \neq \emptyset, \tag{10.11}$$

that is, the image of every node inside of T lies outside of T, or conversely, every point outside of T (in the complement set $X - T$) maps into T by the inverse mapping

$$X - T \in \Gamma^{-1} T. \tag{10.12}$$

External stability tells us that a stable set is a place, like a local minimum, where the graph connections lead us into an end point (see fig. 10.2). In other words, at least one node inside the externally stable set is 'better' than any node outside the set, in the sense of the mapping A.

Definition 40 (Kernel) *A set in a graph that is both internally and externally stable is said to be a* **kernel** *of the graph. The kernel is free of loops and contains all points $x \in X$ for which $\Gamma x = \emptyset$.*

The importance of the kernel for systems is that Γ can be regarded as a mapping of states, that is, of transitions between states of a system—and we are interested in having these transitions converge towards some stable end state. The kernel of a graph contains all the states that can be regarded as being such end states[1]. Internal stability tells us the possible candidates that are at the end of a sequence of arcs, and that once we arrive inside the set, we have chosen one of the sets. External stability tells us that we can always get to one of those points from outside the region, thus the region is accessible to the whole system. If we require both, then the conclusion is that the kernel is the set of states that is accessible to the whole system and is stable under single hop perturbations.

---

[1] Note: the kernel of the mapping is not to be confused with the 'Heat Kernel' generating functional of the graph (see Chung (1997)) that is related to algebraic geometry and field theory.

The kernel is an important concept of equilibrium in games. The notion of a kernel was introduced into game theory by Neumann and Morgenstern (1944) as a proposal for the solution of a game, for finding preferential strategies. Since a point in the kernel is internally stable, no other point is preferable to it; moreover, since it is externally stable, it is preferable to any place outside the kernel. Not all graphs have kernels, but kernels are guaranteed in a number of cases (see for instance Berge (2001)).

Note that although we will always be able to reach a stable node in the kernel, from any point in the system, we will not necessarily be able to find the 'best' node according to some extra criterion, such as centrality or some other 'hidden variable'. The graph mapping Γ itself does not distinguish between the nodes in the internally stable set—they are all valid end states. However, by jumping out of the region and back into it through multiple hops, we can find alternative points in the region that might satisfy additional criteria. This means that other notions of 'preferable' could be allowed to select from the set of states in the kernel of the graph.

10.7   Graph stability under random node removal

The stability of regions and graph structure to node removal has been discussed by various authors. See (Albert and Barabási (2002)) for a review. The stability of local regions in a graph is somewhat unpredictable to the removal of nodes. If we use centrality to define regions (see section 6.4), then it might seem that the removal of the most central nodes in a network would have the most damaging results. However, some graphs with random graph properties are extremely invulnerable to random node removal, including the most central nodes. Peer-to-peer graphs are examples of this because they have no true centre (the maxima are rarely very 'high' above the rest), their structures remain largely unchanged by random node removal. Other structures, such as hub configurations, are extremely susceptible to node removal, since the most central points are of significantly greater importance to the connectivity of the graph. Node removal is clearly more serious to the network as a whole, but perhaps not to individuals (see fig. 10.3).

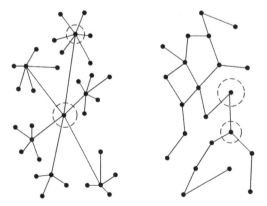

Figure 10.3: Network tolerance to node removal: nodes are more important than connectors.

One might be tempted to use centrality as a measure of stability, however, there is no direct correlation between the centrality values and the connectivities, since the scale-free eigenvector values do not retain information about the total number of nodes in a graph, thus tiny graphs of a few nodes could have similar centrality values to a huge graph. Centrality is only of interest as a relative measure for connected components with a fixed number of nodes.

Susceptibility to node removal can be gauged by examining the *degree distributions* of the nodes

$$P(k) = \frac{n(k)}{\sum_k n(k)},$$ (10.13)

where $n(k)$ is the number of nodes with k connected neighbours. Studies show that random node removal of a fraction f of a graphs nodes has varying effects depending on this degree distribution (see Albert and Barabási (2002) for a review). Large networks undergo phase transitions from states of being connected to being fragmented when critical fractions of nodes are removed.

Scale-free (power law) networks have node degree distributions

$$P(k) \propto \frac{1}{k^\alpha}$$ (10.14)

for some positive constant α. These are especially robust to random node removal, and it is known that peer-to-peer graphs have this structure (see Barabási (2002)). Other reports of self-similar and power law behaviour can be found in connection with the World Wide Web (see Albert et al. (1999); Barabasi and Albert (1999); Barabasi et al. (2000); Glance et al. (1991)).

Attacks by deliberate targeting of the largest nodes are more efficient at breaking up graphs than random failures. Studies typically show numbers at around the order of magnitude of 10% level for fragmenting graphs. A greater fraction of nodes must be destroyed to break up a highly connected graph (see Albert and Barabási (2002)).

10.8   Dynamical equilibria: compromise

A ball sitting at the trough of a valley is said to be in a state of static equilibrium. The forces acting upon it are balanced, and the result is that nothing happens. A 'tug of war', on the other hand, is only in a state of dynamic equilibrium when the two groups pulling on the rope are not moving. Disk storage is a tug of war.

Example 78 *A computer storage system under normal usage is being filled up with data by the actions of users, but there is no counterforce that frees up any of the space again. This is not in equilibrium. A garbage collection service can maintain a dynamic equilibrium if the average amount of storage in use is constant over time.*

Example 79 *In a car park (parking lot), cars come and go. The rates of arrival and departure determine whether the total number of cars is increasing, decreasing or in balance. If the number of cars is in balance, then we say that a dynamic equilibrium has been reached.*

Equilibrium can be mechanical, chemical or statistical. By implication, equilibrium is a balance between opposing forces. In a dynamical equilibrium, there is motion that is continually being reigned in check by other processes.

10.9   Statistical stability

A system that settles into a predictable average behaviour and withstands perturbations to its fluctuation distribution can be said to exhibit statistical stability (see Hughes (1995); Sato (1999)).

Suppose that a system is characterized by a measurable $q(t)$ that varies with time (the same analysis can be applied to any other control parameter). As the state $q(t) \in \{q_1, q_2, \ldots, q_i\}$ varies, it changes value within a set of possible values. In a deterministic system, the pattern of change is predictable at any given moment; in a non-deterministic system it is unpredictable. Either way, if there is sufficient regularity in the behaviour of the system—so that a knowledge of the past can predict an outcome in the future—then we can characterize the change over time by plotting the distribution of values as a histogram (see fig. 10.4) that represents the probability $P(q)$ of measuring a given q_i.

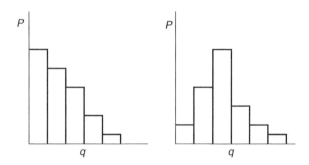

Figure 10.4: The probability distribution of values q_i for a given measurable characterizes the average spread of values that occur over time. If this distribution maintains the same form, as the system develops then it can be called a stable distribution. For systems with random changes, there are only a few possible distributions that are stable over long times: these are the Lévy distributions and the Gaussian 'normal' distribution.

But what happens when we mix different signals together, each with their own distribution of values? The result will surely be a different distribution that is neither the one nor the other, but are there any special distributions that are the stable end result of a lot of mixing of this kind? These would represent the limits of predictability for otherwise unpredictable systems. A. Cauchy and P. Lévy asked this question: what distributions have the property that, when perturbed by mixing, they retain their essential features? Or, is there a $P(x)$ such that the convolution of two signals with the same $p(x)$ results in the same distribution

$$P(q) = \int dq' P_1(q) P_2(q - q')? \tag{10.15}$$

Another way of expressing this is that, if $q_1(t)$ and $q_2(t)$ are random variables with stable distributions, then so is $q_1(t) + q_2(t)$. The Gaussian distribution is one solution to this problem; the other solutions are called the stable Lévy distributions. Some of these have infinite variance and therefore cannot truly represent the behaviour of real systems. However, they do describe systems approximately[2].

The Lévy distributions are denoted by $L_\alpha(q)$ ($0 < \alpha \leq 2$), and most of them cannot be written down as analytical expressions in terms of q. The Fourier transforms (or characteristic functions) of the distributions

$$L_\alpha(k) = \int dq \, e^{iqk} L_\alpha(q) \tag{10.16}$$

(called their characteristic functions) can be written down however. The symmetrical distributions have characteristic functions of the form

$$L_\alpha(k) = \exp(-c_\alpha |k|^\alpha), \tag{10.17}$$

for constants c_α. These allow one to work out the asymptotic behaviour for large q, which turns out to follow a power law behaviour:

$$L_\alpha(q) \sim \frac{\alpha A}{|q|^{1+\alpha}}, \quad q \to \pm\infty. \tag{10.18}$$

for constant A. Two exceptions exist that can be integrated straightforwardly to obtain an analytical form; one is the so-called Cauchy distribution:

$$L_1(q) = \frac{c_1/\pi}{q^2 + c_1^2}; \tag{10.19}$$

the other is the Gaussian

$$L_2(q) = \frac{1}{2\sqrt{\pi c_2}} e^{-q^2/4c_2}. \tag{10.20}$$

For other values of α, the full asymptotic form can be used for large $q \to \infty$,

$$L_\alpha(q) = \sum_{n=1}^{\infty} \frac{(-1)^{n+1}}{\pi} \frac{c_\alpha^n}{q^{1+n\alpha}} \frac{\Gamma(1+n\alpha)}{\Gamma(n+1)} \sin(\pi\alpha n/2). \tag{10.21}$$

and for small $q \to 0$,

$$L_\alpha(q) \sim \frac{1}{\pi} \sum_{m=0}^{\infty} \frac{(-1)^m}{(2m)!} \frac{q^{2m}}{c_\alpha^{\frac{2m+1}{\alpha}}} \Gamma\left(\frac{2m+1}{\alpha}\right). \tag{10.22}$$

The range of values of α that leads to stable behaviour is limited to $0 < \alpha \leq 2$, owing to the scaling behaviour noted in section 10.10; α must exceed 0 in order to have non-negative probabilities, and must be less than or equal to 2 or else there is only short-range dependence of the data, and the distribution has insufficient 'memory' to form a stable distribution. This phenomenon is sometimes called α-stability and the parameter α is related to the Hurst parameter $H = 1/\alpha$ described in section 10.10.

---

[2] Exponentially truncated forms of the Lévy distributions are sometimes used to discuss realistic examples.

Investigating statistical stability allows us to determine whether probabilistic manage-ment policies are sustainable over time. For example, if the fluctuations in a system are not stable, then we can make no guarantees about the system in the future.

10.10   Scaling stability

A related form of stability to the statistical stability is stability under scaling. This is also called scale-invariance, and it asks a more specific version of the same question as in section 10.9: If we have a fluctuating variable $q(t)$, how does the size of the fluctuations depend on how closely we examine the system? If we view the system through a 'magni-fying glass' by focusing on small times, we might see large fluctuations, but if we reduce the resolution by stepping back from the system and examining large time intervals, how does the size of fluctuations change in relation to our resolution. A scale transformation on time is called a *dilatation*.

If the relative size of fluctuations is the same at all scales, the system is said to exhibit *statistical self-similarity*. The scaling hypothesis for a function $q(t)$, under a dilatation by an arbitrary constant s, is expressed by

$$q(st) = \Omega(s)\, q(t). \tag{10.23}$$

In other words, the assumption is that stretching the parameterization of time $t \to st$, leads to a uniform stretching of the function $q(t)$, by a factorizable magnification $\Omega(s)$. The function retains its same 'shape', or functional form; it is just magnified by a constant scale.

This property is clearly not true of an arbitrary function. For example, $q(t) = \sin(\omega t)$ does not satisfy the property. Our interest in such functions is connected with dynamical systems that exist and operate over a wide range of scales. Physical systems are always limited by some constraints, so this kind of scaling law is very unlikely to be true over more than a limited range of s values. Nevertheless, it is possible to discuss functions, which indeed scale in this fashion, for all values of s, as an idealization. Such functions are said to be *scale invariant, dilatation invariant* or self-similar.

Exact self-similarity, for all s, is only a theoretical possibility, and has led to the study of fractals, but a similarity in statistical profiles of functions is a real possibility over finite ranges of s. This is a weaker condition, which means that the behaviour of a complete system S is invariant, but that $q(t)$ itself need not be.

$$S[\Omega^{-1}(\alpha)q(\alpha t)] \to S[q(t)]. \tag{10.24}$$

From eqn. (10.23), the symmetry between $q(t)$ and $\Omega(s)$, tells us that

$$q(x) \sim \Omega(x), \tag{10.25}$$

that is, they must possess similar scaling properties. In fact, $q(t)$ and $\Omega(s)$ must be homo-geneous functions, in order to satisfy this relationship:

$$\begin{aligned} q(t) &= t^H \\ \Omega(s) &= s^H, \end{aligned} \tag{10.26}$$

for some power H. In other words, one has a general scaling law:

$$s^{-H}q(st) = q(t). \tag{10.27}$$

We apply this to locally averaged functions:

$$s^{-H}\langle q(st)\rangle = \langle q(t)\rangle. \tag{10.28}$$

The exponent H is called the Hurst exponent, after the hydrologist H.E. Hurst, who studied such behaviour in the flows of the Nile river. It can be estimated for real data by noting that, over an interval Δt,

$$\langle \max(q(t)) - \min(q(t))\rangle_{s\Delta t} = s^{H}\langle \max(q(t)) - \min(q(t))\rangle_{\Delta t}, \tag{10.29}$$

that is,

$$H = \frac{\log\left(\dfrac{\langle \max - \min\rangle_{s\Delta t}}{\langle \max - \min\rangle_{\Delta t}}\right)}{\log(s)}. \tag{10.30}$$

Note that this estimator will give an incorrect answer for exponential functions $\exp(t^n)$ that increase monotonically; it should be used only on genuine time series. For the Gaussian distribution, we have $H = \frac{1}{2}$, since

$$(\max - \min)\frac{1}{\sqrt{2\pi\sigma^2}}\exp\left(-\frac{q^2}{2\sigma^2}\right) \sim \frac{1}{\sigma}, \tag{10.31}$$

and

$$\sigma(st) = \frac{1}{\sqrt{s}}\sigma(t), \tag{10.32}$$

thus,

$$H_{\text{Gauss}} = \frac{\log(\sqrt{(s)})}{\log(s)} = \frac{1}{2}. \tag{10.33}$$

The parameter H is related to the parameter α from the previous discussion in section 10.9 on α-stability by $H = 1/\alpha$. The Hurst exponent characterizes several properties of the signal $q(t)$'s correlation functions. The auto-correlation is sometimes referred to as the 'memory function' of the system.

For $\frac{1}{2} < H < 1$, processes either have long-range dependence or are correlated significantly with values of the function in the distant past. For $H = \frac{1}{2}$, observations are uncorrelated, and for $0 < H < \frac{1}{2}$ they have short-range dependence and the correlations sum to zero (see Beran (1994) for details).

For $H \geq 1$, the second moment of the probability distribution (the variance σ^2) diverges as a further symptom of long-range dependence. This means that the system can have significant numbers of fluctuations of arbitrarily large size.

Statistical self-similarity has been observed in network traffic (Leland et al. (1994)).

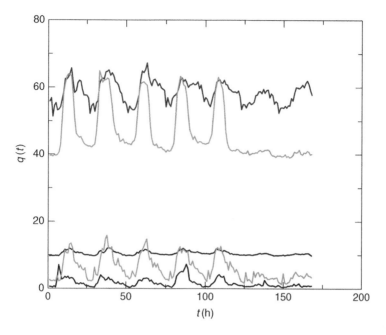

Figure 10.5: Some of the averaged time series resulting in table 10.1

Example 80 *Time series measurements[3] from a local area network measured over an 11-week span at 5 different timescales of up to a week, give the following root mean square values for the Hurst exponent, with 10% tolerance (see fig. 10.5 and table 10.1).*

Here we find no values lower than $H = \frac{1}{2}$, which means that all of the results belong to potentially α-stable distributions. Incoming Web traffic, in particular, has a high value of H, which tends to suggest a long-tailed distribution that signifies very large fluctuations at all the measurable scales. However, this is slightly misleading and is the result of a stable non-equilibrium process that is superimposed on the fluctuating signal as shown in (Burgess et al. (2001)).

The Hurst exponent measures apparent self-similarity. Self-similarity is sometimes linked to long-tailed distributions that are typical of the asymptotic Lévy distributions. However, care is needed to draw any connection between estimates of the Hurst exponent and estimates of the distribution of values measured from limited amounts of data. Heavy-tailed distributions can be caused by monotonically growing fluctuations during the sampling period, for instance, without any scale-invariant behaviour in the fluctuations. It is not just the distribution of values that is important to scale-free behaviour, but the order in which the distributed values occur. Thus, self-similar behaviour *is not the same* as stability of fluctuation distributions. This is the peril of basing the conclusions of average measures.

---

[3] These measurements were made using the cfenvd daemon in cfengine (Burgess (1993)).

Table 10.1: Estimated Hurst exponents for time series measured on a single computer over several weeks. Human activities such as number of users follow a basically Gaussian profile $H = \frac{1}{2}$, while local- and wide-area-network-driven measurements show higher values of the Hurst exponent. Measurements like these are sometimes used to show evidence of self-similar behaviour in computer service patterns, but closer study of the data is required to draw this conclusion.

Variable	$\frac{1}{2} < H < 1$
No. of users	0.5
Root processes	0.6
Other processes	0.7
Diskfree	0.8
Load average	0.6
Incoming netbiosns	1.6
Outgoing netbiosns	1.8
Incoming netbiosdgm	1.4
Outgoing netbiosdgm	1.5
Outgoing netbiosssn	1.6
Incoming nfsd	2.0
Outgoing nfsd	2.3
Incoming smtp	1.6
Outgoing smtp	2.0
Incoming www	2.5
Outgoing www	1.1
Incoming ftp	1.7
Outgoing ftp	2.2
Incoming ssh	1.5
Outgoing ssh	1.4
Incoming telnet	1.2

10.11 Maximum entropy distributions

Maximum entropy distributions are also a kind of stability criterion. By maximizing entropy, we assure a most probable distribution. It is a limiting point. Put in another way, chance will never favour a different configuration once we have arrived at a maximum entropy configuration. See section 9.11 for more about this topic.

10.12 Eigenstates

Eigenvalues are especially stable solutions of simultaneous linear equations. For an $N \times N$ matrix M, eigenvectors $\vec{v}_\lambda$ and their associated eigenvalues λ satisfy the matrix equation

$$M\vec{v}_\lambda = \lambda \vec{v}_\lambda, \tag{10.34}$$

or in component form

$$M_{ij}\vec{v}_\lambda^j = \lambda \vec{v}_\lambda^i. \tag{10.35}$$

Put in another way, when the matrix M acts on certain vectors, the vectors can become longer or shorter (by a factor λ) but they still point in the same direction. If this were to be true for any vector $\vec{v}$, it would only be true for diagonal matrices; however, every non-singular $N \times N$ matrix has this property for a special set of N linearly independent vectors, and for special values λ. These vectors are the *eigenvectors* and *eigenvalues* of the matrix.

Eqn. (10.34) has the form

$$\delta \vec{v}_\lambda = (M\vec{v}_\lambda - \lambda \vec{v}_\lambda) = 0, \tag{10.36}$$

and thus defines a set of stable vectors. For $\lambda = 1$, the vectors are invariant, or constant under the perturbation by M.

Example 81 *Consider the adjacency matrix of a simple graph (fig. 10.6). The adjacency*

Figure 10.6: A three-node graph with symmetry about node 2.

matrix plays a dual role for a graph both as a representation of the connectivity between adjacent nodes and as a recipe for summing over nearest neighbours (see section 6.4). Consider the simple three-node graph, where nodes (1,2) and (2,3) are joined by two links. The adjacency matrix is given by

$$A = \begin{pmatrix} 0 & 1 & 0 \\ 1 & 0 & 1 \\ 0 & 1 & 0 \end{pmatrix} \tag{10.37}$$

This has eigenvalues $\lambda = \{0, \pm\sqrt{2}\}$ and corresponding eigenvectors

$$\begin{pmatrix} v_1 \\ v_2 \\ v_3 \end{pmatrix} = \left\{ \frac{1}{\sqrt{2}} \begin{pmatrix} 1 \\ 0 \\ -1 \end{pmatrix}, \begin{pmatrix} -\frac{1}{2} \\ \frac{1}{\sqrt{2}} \\ -\frac{1}{2} \end{pmatrix}, \begin{pmatrix} \frac{1}{2} \\ \frac{1}{\sqrt{2}} \\ \frac{1}{2} \end{pmatrix} \right\}. \tag{10.38}$$

The structure of the eigenvalues and eigenvectors reflects the symmetry of the graph, and the principal eigenvector (that belonging to the highest eigenvalue) has its highest value for node number 2, indicating that it is the most 'central' node in the graph.

Sometimes, eigenvectors and eigenvalues have more subtle meanings that arise from the mathematical structure that underlies a problem. One must always be careful in interpreting mathematical results, in terms of the assumptions that are entered at the start.

Example 82 *Consider two service departments or servers d_1 and d_2 that work together and share the load of work between them, with manager m (fig. 10.7).*

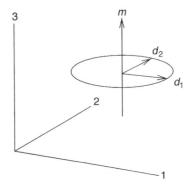

Figure 10.7: Two redundant servers and a manager, represented in three-dimensional space. Redundancy is represented as a circular (rotational symmetry) between the departments d_1 and d_2.

Both departments have a limited capacity C and can only just cope with the total workload alone. If both worked flat out, they could muster a total of

$$C_{\max} = \sqrt{C_1^2 + C_2^2}, \tag{10.39}$$

as independent units. However, in order to provide for full redundancy, we must not use up this total capacity, but instead require $C_1 = C_2 = C_{\text{tot}}$. Now, to balance their load and provide redundancy, they split the workload between themselves in such a way that either one could take over at a moment's notice, by changing a management parameter θ.

$$d_1 \rightarrow d_1 \cos\theta + d_2 \sin\theta \tag{10.40}$$
$$d_2 \rightarrow -d_1 \sin\theta + d_2 \cos\theta. \tag{10.41}$$

Are there any stable or preferred solutions to this problem? It is helpful to draw this situation geometrically by defining a vector:

$$\begin{pmatrix} d_1 \\ d_2 \\ m \end{pmatrix} \rightarrow \begin{pmatrix} v_1 \\ v_2 \\ v_3 \end{pmatrix} \tag{10.42}$$

in three-dimensional space.

Consider a matrix M representing a two-dimensional rotation about the m axis. The manager can choose to rotate the load from one department to another by changing the angle θ. *What are the privileged vectors for this perturbation matrix?*

$$M = \begin{pmatrix} \cos\theta & \sin\theta & 0 \\ -\sin\theta & \cos\theta & 0 \\ 0 & 0 & 1 \end{pmatrix} \tag{10.43}$$

The characteristic equation for the eigenvalues is

$$\det|M - \lambda I| = 0, \tag{10.44}$$

giving

$$\lambda = 1, \exp(\pm i\theta), \tag{10.45}$$

and eigenvectors

$$\begin{pmatrix} v_1 \\ v_2 \\ v_3 \end{pmatrix} = \left\{ \frac{1}{\sqrt{2}}\begin{pmatrix} 1 \\ i \\ 0 \end{pmatrix}, \frac{1}{\sqrt{2}}\begin{pmatrix} 1 \\ -i \\ 0 \end{pmatrix}, \begin{pmatrix} 0 \\ 0 \\ 1 \end{pmatrix} \right\}. \tag{10.46}$$

In this example, two of the eigenvalues are complex numbers, but this should not distract from the interpretation of the result. For $\lambda = 1$, we find an eigenvector along the management rotation axis. This tells us that the management axis has a privileged status in this system; it is independent of any change by varying θ. The remaining two eigenvalues and eigenvectors appear to be lines pointing at $45°$ into the complex plane. If we substitute these values into the eigenvalue equation $M\vec{v} = \lambda\vec{v}$, we find that it is identically satisfied, that is, no conditions are placed on θ or a specific configuration. This makes sense from our original system design; the rotational symmetry was introduced to provide redundancy, or independence of configuration, so any balance is as good as any other[4].

This last example illustrates an important point: eigenvectors and eigenvalue solutions fall into two categories: solution by linear *identity* and solution by invariance (symmetry), so eigenvalue solutions will always tell us the intrinsic, stable or invariant properties of the perturbation matrix. Eigenvectors are important in dynamics because systems have often preferred configurations that are described by their eigenvectors.

10.13   Fixed points of maps

In the theory of dynamical systems, stationary points where derivatives vanish are sometimes called fixed points; however, there is a more fundamental and interesting definition of fixed points that is of great importance to average stability of human–computer systems.

---

[4] The presence of $i = \sqrt{-1} = \exp(i\pi/2)$ in $(1, \pm i)$ signals a phase shift between the preferred solution for d_1 and d_2 of $90°$ ($\pi/2$ radians). This is exactly reflected in the placement of cos and sin in eqn. (10.41), since sine and cosine are phase shifted by $90°$ with respect to one another.

Any function $x' = f(x)$ defines a mapping from some domain of values to a range of values. In a continuum description, the domain and range are normally subsets of the Euclidean space of real numbers R^n. A fixed point x^* of a mapping $f(x)$ is any point that maps onto itself.

$$f(x^*) = x^*. \tag{10.47}$$

In other words, if a system finds itself in x^* by iteration of this mapping, it will remain there. This is clearly a definition of stability, since

$$\delta Q^* = (f(Q^*) - Q^*) = 0. \tag{10.48}$$

We are particularly interested in functions that determine the time development or running of a system, since Q^* is a natural choice for a configuration of a system that is required to be stable.

In this section, we describe, without proof or detailed explanation, two notions of fixed point that relate to continuous functions: the Brouwer fixed point theorem and the Kakutani fixed point theorem[5].

Theorem 10.13.1 (Intermediate values) *Let $f : [a, b] \to R^1$ be a continuous function, where $[a, b]$ is non-empty, compact, convex subset of R^1 and $f(a)f(b) < 0$, then there exists an $x^* \in [a, b]$ such that $f(x^*) = 0$.*

This theorem makes the simple point that if two points lie in the range of a continuous function, and one point is positive and the other is negative, then the function has to cross the $f = 0$ axis by virtue of its continuity (see fig. 10.8). The only way that a function would not satisfy this property is if it were broken into disjointed pieces. Although the theorem uses the value $f = 0$ in this construction, the result remains true if we re-label the axes to any values. The important point about this theorem is a corollary. If we restrict the domain and range to the unit interval (for convenience):

Corollary 10.13.2 *Let $f : [0, 1] \to [0, 1]$ be a continuous function, then there exists a fixed point $x^* \in [0, 1]$ such that $x^* = f(x^*)$.*

This tells us that for a function that is convex over the unit interval, and therefore covers the range $[0, 1]$, the function must cross the $45°$ line $x = f(x)$ at at least one point x^*. This leads us to Brouwer's theorem, which is a generalization of this idea for more general sets.

To describe the two fixed-point theorems, we need to use mathematical terms describing sets: *convex* and upper semi-continuous[6].

Theorem 10.13.3 (Brouwer) *Let $f : S \to S$ be a continuous mapping from a non-empty convex, compact set $S \subset R^n$ into itself (see fig. 10.9). Then there exists an $x^* \in S$ such that $x^* = f(x^*)$.*

---

[5] Several excellent explanations of these results exist, often in books on the Theory of Games. The World Wide Web also has numerous helpful pages on these theorems.

[6] Also called upper hemi-continuous in some texts.

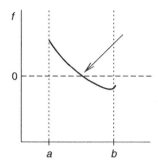

Figure 10.8: The Intermediate Value Theorem says that a continuous function must cross a line that passes between two points that lie in the function's range. The figure makes it obvious why this must be true.

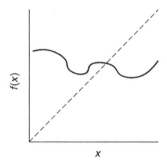

Figure 10.9: Brouwer's Fixed-Point Theorem. At some point, a function that satisfies the convexity conditions must cross the diagonal line. Thus, a mapping from one axis to the other must have a fixed point.

The convex property is important here; if the upper and lower portions of the curve F were not connected and convex, then the mapping would not necessarily intersect the 45° line and thus there would be no fixed point. (The set of rotations of a circle by irrational angles has no fixed point, since the set has gaps.)

In system administration, we are often interested in procedures that map some input state to some output state. Arbitrary discrete sets of operators need not map convex sets of states. This places the limits on what kinds of operator can have a stable outcome. One of the challenges of system administration is determining ways of *ensuring* that a system has a fixed point.

Example 83 *Discrete operations do not necessarily have fixed points, but a continuum approximation created by a local averaging procedure can interpolate a non-convex set of operations into a convex one, allowing a virtual fixed point to be defined, on average. This is one of the main themes of this book. It is the essence of the maintenance theorem (see section 16.8) and it is a major reason for abandoning a strictly discrete formulation of systems.*

Kakutani's generalization of Brouwer's theorem allows us to map not just points to points, but sets to sets or points to sets. A *correspondence* is a mapping from a point to a set. This allows us to have not just stable points for simple mappings from point to point, but also in mappings that are formed by sewing together disjointed pieces into a consistent union.

Theorem 10.13.4 (Kakutani) *Let $F : S \rightarrow S$ be an upper semi-continuous correspondence, or mapping from a non-empty convex, compact set $S \subset R^n$ into itself (see fig. 10.10), such that, for all $x \in S$, the set $F(x)$ is convex and non-empty, then $F(\cdot)$ (which can now have a multitude of values as it is a set) has a fixed point $x^* \in S$ such that $x^* \in F(x^*)$.*

The importance of this theorem is in being able to identify fixed points even in functions that map sets that are more complicated than one-to-one mappings; we can talk of identifying the regions in parameter spaces.

The theorem tells us that there is at least one point $x = P$ within the values laid out along the axes that map, which meets itself within the regions that are defined by the correspondence. In fig. 10.10, there is a range of values where the $45°$ line intersects the region that are fixed points. Thus, we have a *fixed subset*.

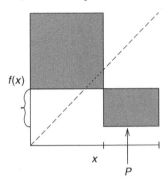

Figure 10.10: Kakutani's Fixed Point Theorem. The Brouwer theorem can be generalized to regions, provided the regions touch and have the same convexity properties. A point on the horizontal axis now maps to a range of values within itself indicated by the brace. Had the two regions not touched and been intersected, the system would have been doomed to oscillate between the two regions and never stabilize.

The importance of this final theorem is that it applies to objects (sets) that are general enough to be able to describe policy in a continuum approximation. Although the figure illustrates this for a subset of R^1, that is, a single parameter, the regions can be multi-dimensional, so we can identify stable sub-spaces. It is seldom that we can identify a single 'correct point', but a correct region is a more likely prospect in a complex system of many variables. Note, however, that these theorems apply to continuous functions, and thus apply to the continuum approximation of human–computer systems, which in turn applies on average. That is another reason why regions are important—we must allow for uncertainty. Discrete systems do not necessarily have stable fixed points.

Example 84 *An important equilibrium for defining system policy is based on the notion of a fixed point of the graph of all rational strategy preferences (see below). This is not* *stability of the system's evolution, but stability of the choice of end point in the kernel under different policy decision criteria. This equilibrium is best known as the Nash equilibrium or Kakutani fixed point (see Myerson (1991) for an excellent introduction) of the preference graph. It looks for a subset of states that can be regarded as a limit point of competing decision criteria. In a two-person zero sum game, this corresponds to the minimax solution that is used below. The idea of an ideal configuration Q^* for a system can be defined as a fixed point of the 'response matrix' for mapping non-ideal states onto ideal ones. This matrix is readily defined in terms of convergent mappings (Burgess (1995); Burgess and Ralston (1997)),*

$$Q' = R(Q), \tag{10.49}$$

However, the convergence property is not enough to select a stable base state for a convergent process, because convergence can be applied to any state. In order to prevent configuration loops and find the set of self-consistent fixed points that can be identified with policies, we must solve

$$Q^* = R(Q^*). \tag{10.50}$$

This condition is the essence of the Nash equilibrium in Game Theory (see chapter 19).

10.14   Metastable alternatives and adaptability

If a system has a single global minimum, or stable end point, it will seek out this point and never emerge from it. Few systems benefit from being this stable, since changes in their environments usually demand a greater flexibility.

A system that can adapt to new conditions needs to be able to get out of its state of stability, by perturbing it with a large enough perturbation of an appropriate type. We would then like it to settle into a new and 'better' stable state that is more appropriate to the new conditions.

Systems that have some kind of symmetry (a choice that does not matter), will naturally have several equivalent stable states, each of which is equally good. The archetypal model for such a system is a system that has several internally stable regions (fig. 10.11), with barriers in between. If we provide a sufficient perturbation, we can end up in a different minimum that is as good. This is called *tunnelling*.

This kind of multi-stability will not help the system to adapt, unless the criterion separating them suddenly becomes important in making one state better than the other. In this case, we say that the symmetry has been broken.

Example 85 *Consider the set of states for access permissions to a computer password database. In fig. 10.11, the first minimum is the end state in which read-only permissions are granted to all; in the second minimum, read-write permissions are granted to all; in the third minimum, privileged access is given to the administrative user. Now suppose that, initially, only a single administrative user has physical access to the database so that the permissions on the database are irrelevant. If conditions change, so that several users are introduced to the system, then the symmetry is now broken under the criterion of 'security', where security means access to restricted data.*

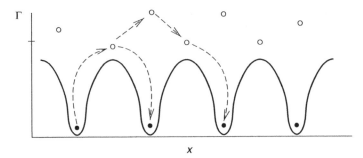

Figure 10.11: Multi-stability means that there are several alternative stable states that are equally good. These are said to be symmetrical. As environmental conditions change, one of these states might become preferable to all of the others. Then the symmetry is said to be broken. To tunnel from one state to the next, we must jump out of the region of internal stability, and back into it in a different location.

The idea of metastability or multi-stability is connected to the notion of the kernel of a graph. The kernel consists of stable end points of a system's policies, or preference criteria. If conditions change, then we want to be able to break down the internal stability or allow transitions out of it, backtracking out and then transferring into a new state. If we are interested in adaptability, then it is wise to arrange for several of these alternative states to exist, so that a rapid change of policy can select the new 'best choice' without a major architectural change.

10.15 Final remarks

Stability is a powerful concept, with many interpretations. We have considered a few possible interpretations in this chapter, but this by no means covers the full repertoire. The stability of mappings, for instance, are quite general—they are generic behaviours that one would expect to find in any directed mapping, and thus we can expect these concepts to emerge in a variety of situations.

Equilibria of games, discussed in chapter 19, are a natural extension of the idea of stability to decision making. In collaborative networks, such as peer-to-peer networks, cooperation or conflict between rational members of the collective can be modelled as persistent coalitions between individuals pursuing selfish interests. The concept of *imputations*, or coalition that strictly increase a player's benefit arises in this case (see Rapoport (1970) for a lucid introduction); ψ-stability can then be used to describe the likelihood for lasting cooperation.

Stability provides a set of concepts that we can draw on to analyse the behaviour of human–computer systems—both stochastic and some to rule-based systems. Concepts of stability prove to be particularly important when defining the meaning of a sustainable policy.

Applications and Further Study 10

- *Characterizing and quantifying stability.*

- *Relating stability to predictability.*

- *Using stability as a basis for choosing arbitrary policy in a system.*

- *Looking for controllable or maintainable pathways of change.*

11

Resource networks

This chapter considers the relationship between a system and its critical dependencies: the description of basic resources that permit the system to function. It was commented earlier that any system can be thought of as an information system, because anything that happens to its resources must be described with the help information; thus, one is never more than one step of abstraction away from talking about the change of information about objects, rather than talking about the objects themselves.

11.1   What is a system resource?

The word resource is often used to describe the *assets* of a system or an enterprise that is often used by businesses to describe what things are valuable to the business. Resources or assets are the objects that describe the make-up of a system; they are also the freedom to change in the system. The word resources is preferable to assets, since it does not imply something that is automatically good.

Processes manipulate a system's resources, leading to a change in their organization or perhaps to an exchange of one resource for another. It is the bookkeeping or accounting of resources that is described by the dynamics of a system. In short, all systems follow a kind of 'economy' where the accounting parameters have names and properties that go beyond money and goods.

A whole plethora of words has been introduced to describe assets and resources. Some of them refer to tangible, physical items and others are more abstract qualities. Here are some examples.

- Raw materials (stock)

- Processing power (CPU or human)

- Storage space (memory or archive)

- Movement in space

Analytical Network and System Administration. Managing Human–Computer Networks   Mark Burgess
© 2004 John Wiley & Sons, Ltd   ISBN 0-470-86100-2

- Time

- 'Potential' for reward

- Personnel

- Property (real estate)

- Intellectual capital

- Respect, status

- Privilege.

Resource availability is something that results in both freedoms and constraints. The availability of space allows a system's expansion; conversely, the limited size of a space is a constraint on what can develop.

There are two issues with resources: how resources change and how they are organized, for example, do we have access to them? Another way of saying this is that there are both local and global issues. Resources can be modelled as a number of nodes within a network topology, where the network indicates the pathway for interaction between the resources.

11.2   Representation of resources

We want to be able to talk about resources, their distribution and their usage formally. In mathematical or formal terms, a resource is a quantity that can be measured about the system. A *measurable* must be represented as a *variable* that depends on *parameters* that describe its distribution and change with respect to time.

Let us separate space and time as resources, and think of these rather as parameters that are perhaps restricted by boundary conditions.

Definition 41 (Resource variables) *The resources of a system are functions of time, space and other parameters that describe their distribution and patterns of change. They are written as variables of state*

$$q_i(t, x, \ldots) \qquad \text{for } i = 0, 1, 2, \ldots. \tag{11.1}$$

Definition 42 (Degree of freedom II) *The tuple of parameters $(x, t, \ldots) \in X$, represents* degrees of freedom; *they are labels that describe the different alternatives or possible arrangements of resources: for instance, a particular time, a location in space, a given person, a colour etc.*

The parameter space X is sometimes called the configuration space of the system, because it describes how resources are arranged or configured.

Resources take values that are measured in some form of currency (see section 4.9), and they represent the valuables of the system, both actual, potential, material and social. A set of resource variables describes a partial state of the system. The amount or availability of a resource is described by a value, measured in its own form of currency, which represents the balance of payments that have led to the current state of resources.

A description of resources involves a set of values at various locations $q(t, x)$. The set of all such values, $\{q(t_0, x)\}$, at a given moment t_0 is called a configuration of the system. Managing system resources includes managing their configurations over time.

11.3   Resource currency relationships

Resources come in many flavours, and depend on various location parameters. A resource can also have a value measured on any number of different currency scales. If one wishes to express a statement saying that a resource has a value with respect to more than one system of values, then one must specify a relationship between those values, that is, the value systems are not independent. This applies to any relationship between values in the system.

Example 86 *'Time is money' is a functional relationship between a parameter t and a value M belonging to a state space consisting of all the possible values that money can take. It is written M(t) saying that time and money are related by a formula, for example, in the simplest case,*

$$M(t) \propto t = k\,t. \tag{11.2}$$

This example is often quoted frivolously, but it expresses an economic truth about systems, in which the constant of proportionality k is the average sum of money that can be earned by the system per unit time. Since human–computer systems are frequently driven by economic interests, this kind of relationship will be used frequently, and will feed into models of the system.

Example 87 *Consider social scales that drive the human components in a system. Intellectual capital is a potential for innovation within a human system. If intellect leads to respect and status, it could also lead to privilege or even money. Each of the emphasized words is a social value system. By identifying a relationship between those above, we are saying that they are all related. Thus, we might choose to measure intellect by IQ, but if intellect leads to respect, then respect must be measured on a scale that is a function of intellect (and other inputs), calibrated to the respect scale. Similarly, if this leads to status, then status must be a function of respect (and other inputs) calibrated to the status scale. This dependency chain continues.*

In practice, particularly in the West, human social systems are organized around money, and all other scales can be measures in terms of money, according to some elaborate formula. While this is certainly a cynical view of things, and somewhat of an oversimplification, it is a pragmatic reality precisely because it leads to a concrete, measurable value for what is happening in society. A similar situation exists in computer systems, where most measurements can be associated with a certain amount of processing time, or memory consumed.

In making informed decisions about human–computer systems, we are often forced to make value judgements of the kind described above. The key to making rational, rather than ad hoc, decisions lies in quantifying that value system in relation to the other valuables in the equation.

Example 88 *Suppose we are interested in implementing a security system for online banking (see section 19.9). A security consultant has determined the relative security of different available technologies and how much they will cost to be implemented, but how do we decide whether it is worth paying more, for a better solution or not? We can do this by defining the 'payoff' or utility of the solution, and define it by*

$$Payoff\,(\Pi) = Security\ evaluation\ (S) - cost\ of\ implementation\ (C).$$

At this stage, however, security and cost of implementation are measured using quite different scales, so we need to relate them to one another. Let us measure payoff on a scale from 0 to 100, and the security evaluation on the same scale. The cost of implementation can be measured in Euros, so we must relate Euros to security level. Let us suppose that the cost/security ratio is constant ('you get what you pay for'); then:

$$\Pi = S - f(C) = S - kC, \tag{11.3}$$

for some constant k. There is no rational way of determining k. It is a human value judgement: determining it must be a part of policy. *Policy necessarily intrudes on the rational process through the need for certain ad hoc judgements.*

11.4   Resource allocation, consumption and conservation

Resources are both the machinery and the fuel in a system; thus, they fall into two categories.

Definition 43 (Reusable resources) *These are resources that are allocated temporarily to a particular task. Once they are no longer needed, they can be passed on to another task. Reusable resources include people, computers, storage and communications lines.*

Example 89 *A car park (parking lot) consists of a number of parking spaces that are occupied for a certain time and then are freed for reuse. A computer disk has a number of sectors that are used to store data for a certain time and are later freed for reuse.*

Definition 44 (Consumable resources) *These are resources that can be used only once. Once they are used, they disappear from the resource pool. The resource pool can be refilled if a new quota of resources becomes available.*

Example 90 *Time is a resource that cannot be reused. Money, electricity, oil, write-once memory and damaged equipment are other examples.*

In the physical world, resources do not simply disappear or get used up; rather, they are converted from one form into another, or bound in some role, where they cannot be freed for reuse. In practice, however, it is convenient to think of resources as disappearing from the pool of resources. One says that the system is an open system. The mathematical expression of conservation, as expressed in the continuum approximation is given by the formula:

$$\frac{\partial}{\partial x^i}\,J^i = -\frac{\partial \rho}{\partial t}, \tag{11.4}$$

for some vector $\vec{q}(t, x^i) = (\rho, J^i)$. It expresses the idea that if there is regional conservation in a system, the density of a system property $\rho(t, \vec{x})$ cannot change except by spreading out into a current or flow $J^i(t, \vec{x})$ from that location (see for instance Burgess (2002a)).

Another consequence of open systems is that the distribution of resources is not always under the control of the system itself. Some resources are distributed at random or by external agents.

Example 91 *The number of passengers arriving at a bus stop is not determined by the bus transport system, but the arrival of buses is. The number of e-mail messages arriving at a computer system is not determined by that system, but the availability of CPU and memory to handle the requests is.*

The problem of utilizing available resources that are under local control involves two issues: allocation and reclamation.

Definition 45 Resource allocation *is the association or assignment of a resource from a shared pool to a process within the system. The concept of ownership of a resource is achieved by attaching an extra parameter to the tuple X.*

Example 92 *When a process is defined, it is built up of an assembly of resources.*

Definition 46 Resource reclamation *is the freeing of a reusable resource back into a shared pool.*

Example 93 *When a process is terminated, the resources that were attached to it are returned to a state of disuse. When cars leave a car park (parking lot), the spaces become available for others to use, that is, the spaces are not reserved for private use. Disk space that is allocated by* quota *to a given computer user does not become free for other users when it is no longer used by the private user.*

11.5   Where to attach resources?

A system does not function without resources. Resources come in many forms, both literal (e.g. processor time, memory, floor space, raw materials) and abstract (e.g. goodwill, permission, credit). Materials, workforce, creativity, tools and equipment are all resources. We describe these resource by variables $q(t, A)$ that vary in time and address.

For a system architecture to work optimally, resources must be made available at the right places and times. Once we know where the resources are needed, this becomes a scheduling problem. Methods for approaching these matters have been considered in the previous chapters.

- Have the resource requirements been targeted correctly?

- Will resources reach their target or disperse before arriving?

In order for resources to be available to other parts of the system, there has to be a route connecting them to their point of consumption. Resources may be provided by direct injection, or by controlled delivery through a mediator (a metaphorical or literal valve or regulator). There are often administrative overheads involved with regulation.

Direct access to resources allows maximum efficiency, but sometimes resources have to be transported by an unreliable channel that is shared by several tasks (see fig. 11.1).

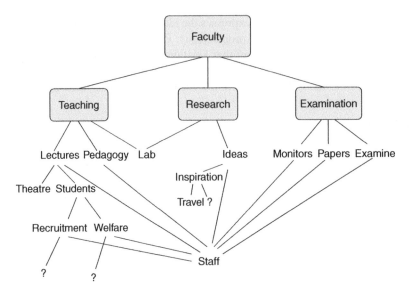

Figure 11.1: A simplified functional dependency structure diagram for a university faculty department. The level of dependency increases with distance down the page. Note that while the goal of the department is to furnish society with knowledge, the dependencies point to the staff as being the central element on which almost everything above depends. This indicates that investment in this resource is important to the system. The dependency chart continues downwards, of course, with food, power and so on. A formal method for this analysis is given in section 6.4.

Example 94 *The monitoring of resource usage (e.g. system accounting, or bureaucracy) can result in inefficiency, because it relies on sharing the same resources as the system itself. Thus, it becomes overhead that takes resources away from the main purpose of the system. We must decide whether this overhead is worth the loss of resources.*

The distance from resources supply to the point of usage should be minimized in order to minimize delay and other losses. Food, power and basic freedoms are at the bottom of any system involving humans or machines. If access to these is restricted or constricted, the system performance will suffer. There are two extremes for injecting resources:

- Bottom up: Resources are inserted directly at the location within the functional tree, closest to the action (direct access, but no control or accountability);

- Top down: Resources are inserted at the top of the functional tree: disseminated by percolating down from above (has to pass through many junctures—this is inefficient, but accountable).

If two processes share a common dependency, resources can be inserted at the dependency for maximum efficiency, that is, we can make the dependency a formal service that supplies resources to the processes. This is a form of *system normalization* (see chapter 14).

11.6   Access to resources

The ability for information and resources to spread through a system depends on the availability of free channels of communication within the system. Percolation is the phenomenon that occurs when a fluid manages to penetrate a porous medium. For a network, the term is used in the following way.

Definition 47 (Percolation transition) *A network is said to* percolate *if it is possible to reach any node from any other by at least one route.*

This means that information or resources can be communicated to the parts of the system where they are needed; conversely, it might mean that an attacker can reach any part of the system from any other. Either way, the percolation transition in any network is important to understand. It tells us both about security and availability.

Since connectivity has such basic ramifications for a system, it is of considerable interest to be able to measure it. This is a particularly difficult task in large organizations where we might have only partial information about. In a dynamic system, pathways of communication open and close at different times, perhaps for security reasons, and perhaps for efficiency reasons.

Since it is not always possible to measure a system completely or directly, we are interested in gauging the *probability* for percolation in two distinct cases:

- *Perfect information*: If we know precise details about the graph of a system, it is possible to work out whether there is a route connecting any two nodes. The All Pairs Shortest Distance Matrix, as defined by standard algorithms by Floyd or Dijkstra (Baase and Gelder (1999); Cormen et al. (2001)), for instance, is an example algorithm for evaluating the possibility of communication between nodes in a graph. Another measure is based on the connectivity that comes from asking how many pairs of nodes, out of all possible pairs, can reach one another in a finite number of hops. We thus define the ratio R_C of connected pairs of nodes to the total number of pairs that could be connected (Burgess et al. (2003a)):

$$R_C = \sum_{i=\text{clusters}} \frac{\frac{1}{2}n_i(n_i - 1)}{\frac{1}{2}N(N - 1)} = 1. \qquad (11.5)$$

  This is simply the criterion that the graph be connected. Normally, it is only possible to evaluate this quantity for theoretical models, or for very small organizations under tight control, with perfect information, so we need other ways of estimating percolation with only partial information.

- *Partial information*: A system administrator is not always aware of every detail of the system or its users. Real systems are inherently probabilistic. In human–computer systems especially, there is the possibility of covert channels of communication that link together parts of the system in a non-obvious fashion. For instance, a married couple working in different fragments of a system could easily leak information to one another. Conversely, an artificial barrier between nodes might be introduced by sickness or accidental disconnection of a node. In such cases, we must admit to possessing only incomplete or probabilistic information about a system and make do with an estimate of the likelihood for percolation.

In both cases, there are methods for determining the likelihood of complete connectivity within a part of the system.

In Appendix C, the results for percolation thresholds of approximately random graphs are derived. These are based on the work of Newman et al. (2001) for huge random graphs, and were adapted for small graphs in (Burgess et al. (2003a)). Random graphs are a useful measuring stick for actual graphs because they tend to percolate very easily, with only a small number of connections. They are very efficient at covering the nodes with available routes.

A connected part of a graph is called a *cluster*. We would like to find the sizes of clusters in a graph and see when they become large, that is, of the same order of magnitude as the size of the graph. The giant component or cluster is thus defined to be a cluster that is of order N nodes. If such a cluster exists, then other smaller clusters of order $\log N$ might also exist (see Molloy and Reed (1998)). The large-graph condition for the existence of a giant cluster (of infinite size) is simply

$$\sum_k k(k - 2) \, p_k \geq 0. \tag{11.6}$$

Here, the sum is over k, the degrees of nodes in the graph, and p_k is the probability of finding a node of degree k in the graph, that is, it is the number of nodes of degree k divided by the total number of nodes n_k/N (see section 6.2).

This provides a simple test that can be applied to a human–computer system, in order to estimate the possibility of complete penetrability. If we determine only the p_k, then we have an immediate machine-testable criterion for the possibility of a systemwide security breach, or efficient transmission.

The problem with the above expression is that it is derived under the assumption of there being a smooth differentiable structure to the average properties of the graphs. This is really only a good approximation in the infinite graph limit, so it is mainly of interest to huge systems like the entire Internet. For a small graph with N nodes, the above criterion for a giant cluster is inaccurate. Clusters do not grow to infinity, they can only grow to size N at the most, hence we must be more precise and use a dimensionful scale rather than infinity as a reference point. For a small graph, the size of a giant cluster is N and the size of below-threshold clusters is $\log(N)$ (see Molloy and Reed (1998)). An improved criterion was found in (Burgess et al. (2003a)), and is given by

$$\langle k \rangle^2 + \sum_k k(k - 2) \, p_k > \log(N). \tag{11.7}$$

This can be understood as follows. If a graph contains a giant component, it is of order N and the size of the next largest component is typically $O(\log N)$ (see Molloy and Reed (1998)); thus, according to the theory of random graphs the margin for error in estimating a giant component is of the order $\pm \log N$. We thus add the magnitude of the uncertainty in order to reduce the likelihood of a false positive conclusion.

Ad hoc networks are dynamically *random graphs* in which connections are initiated and broken in a non-deterministic way. The above criteria are useful for estimating the penetrability of mobile and other *ad hoc* networks on the fly.

11.7   Methods of resource allocation

Since resources are crucial to the ability for a system to function, resource allocation
is a critical dependency. There are two sides to the problem: one is the organization of
resources given to a single process, the other is how several processes can coexist within
shared resources. There has to be a pool of resources that acts as a constraint on the
operation. Methods of assignment often need to make certain compromise.

Example 95 *Suppose you need to draw a complicated map on a small piece of paper. How
will you allocate the space on the page?*

Since a dynamical system requires the flow of work or information, restrictions on the
flow can limit the performance of the system. Time resources must be assigned to process
incoming information so as to maintain the flow. Mismatches of flow rates between parts
of the system can be a problem. If one part of a system has more resistance to the flow of
work than another, it becomes a bottleneck that delays the entire collaborative network.

Example 96 *Many systems employ resources to monitor the usage of other resources, for
example, bureaucratic controls or SNMP monitors. Too much monitoring or verification
communication ties up resources that could be used to do useful work. In such a case, a
resource becomes an overhead.*

11.7.1   Logical regions of systems

Systems are well described by graphs, and graphs have properties that allow them to be
classified into regions. A fitness landscape view of a graph can be created using eigenvalue
centrality to define the 'height' of any node above an imaginary sea level or null point of
zero centrality. Local maxima, the tops of the mountains, in this landscape provide a notion
of maximum local importance. The nodes that are connected to these local maxima then
define a *region* connected to that local centre. If we trace regions by travelling outwards
from a local maximum and find nodes and meet a node in between two maxima, then
it lies in a valley between the tops, and we can call it a *bridge* between regions (see
fig. 11.2).Thus, we classify regions into maxima, bridges and regions. This will prove to
be a widely used tool for identifying distinct regions in systems.

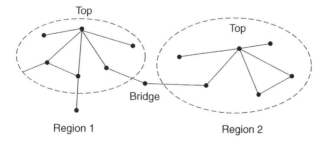

Figure 11.2: The classification of a graph into regions and bridges, using eigenvalue centrality as a
sort criterion.

11.7.2   Using centrality to identify resource bottlenecks

The graph theoretical technique above allows us to identify the places in a network where workflow resources are most crucial to the continued functioning of the system. Eigenvector centrality (see section 6.4) is a relatively easy and straightforward way of identifying the confluences of information flows in a network. It can be applied with various levels of approximation to take account of different aspects of the system. Let us consider how to identify the 'hot spots' in a network, that is, the places where resources will need to be invested in order to provide optimum fuel for the system.

Let the graph of the systems (X, Γ) with configuration space X and arcs Γ be a representation of any aspect of a system involving association.

1. Draw a graph of the human–computer system, with all information flows or associations represented as arcs between the nodes.

2. For each connected fragment of the graph, construct the symmetrical adjacency matrix for the non-directed graph, setting a constant value, for example, 1 for a connection and 0 for no connection.

3. Calculate the principal eigenvector of the adjacency matrix, that is, the eigenvector belonging to the highest eigenvalue.

4. Normalize the elements so that the maximum value is $+1$.

5. We now find the regions and bridges as described in section 11.7.1.

Each of the elements in the principal eigenvector now rank the importance of the nodes in the graph to the workflow.

Example 97 *Consider the human–computer system for Internet commerce depicted in fig. 11.3. This graph is a mixture of human and computer elements: departments and servers. We represent the outside world by a single outgoing or incoming link (node 5). The*

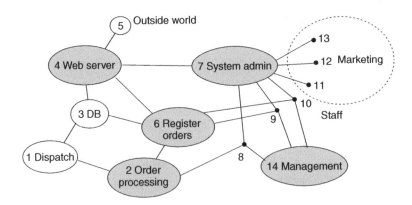

Figure 11.3: Unstructured graph of a human–computer system—an organization that deals with Internet orders and dispatches goods by post.

organization consists of a web server connected to a sales database, that collects orders that are then passed on to the order registration department. These collect money and pass on the orders to the order processing who collect the orders and send them to dispatch for postal delivery to the customers. A marketing department is linked to the web server through the system administrator, and the management sits on the edge of the company, liaising with various staff members who run the departments.

Let us find the central resource sinks in this organization, first assuming that all of the arcs are equally weighted, that is, contribute about the same amount to the average flow through the organization. We construct the adjacency matrix, shown in eqn. (11.8):

$$
A_{ij} =
\begin{pmatrix}
0 & 1 & 1 & 0 & 0 & 0 & 0 & 0 & 0 & 0 & 0 & 0 & 0 & 0 \\
1 & 0 & 0 & 0 & 0 & 1 & 0 & 1 & 0 & 0 & 0 & 0 & 0 & 0 \\
1 & 0 & 0 & 1 & 0 & 1 & 0 & 0 & 0 & 0 & 0 & 0 & 0 & 0 \\
0 & 0 & 1 & 0 & 1 & 1 & 1 & 0 & 0 & 0 & 0 & 0 & 0 & 0 \\
0 & 0 & 0 & 1 & 0 & 0 & 0 & 0 & 0 & 0 & 0 & 0 & 0 & 0 \\
0 & 1 & 1 & 1 & 0 & 0 & 0 & 0 & 1 & 1 & 0 & 0 & 0 & 0 \\
0 & 0 & 0 & 1 & 0 & 0 & 0 & 1 & 1 & 1 & 1 & 1 & 1 & 0 \\
0 & 1 & 0 & 0 & 0 & 0 & 1 & 0 & 0 & 0 & 0 & 0 & 0 & 1 \\
0 & 0 & 0 & 0 & 0 & 1 & 1 & 0 & 0 & 0 & 0 & 0 & 0 & 1 \\
0 & 0 & 0 & 0 & 0 & 1 & 1 & 0 & 0 & 0 & 0 & 0 & 0 & 1 \\
0 & 0 & 0 & 0 & 0 & 0 & 1 & 0 & 0 & 0 & 0 & 0 & 0 & 0 \\
0 & 0 & 0 & 0 & 0 & 0 & 1 & 0 & 0 & 0 & 0 & 0 & 0 & 0 \\
0 & 0 & 0 & 0 & 0 & 0 & 1 & 0 & 0 & 0 & 0 & 0 & 0 & 0 \\
0 & 0 & 0 & 0 & 0 & 0 & 0 & 1 & 1 & 1 & 0 & 0 & 0 & 0 \\
\end{pmatrix}. \tag{11.8}
$$

Some elements are marked bold here, for later removal. The eigenvalues of this matrix are now computed and the transposed principal eigenvector is

$$
\vec{v}^{\mathrm{T}} = (0.29, 0.49, 0.54, 0.75, 0.21, 0.89, 1.00, 0.58, 0.69, 0.69, 0.28, 0.28, 0.28, 0.55) \tag{11.9}
$$

Node 7 is clearly the most central. This is the system administrator. This is perhaps a surprising result for an organization, but it is a common situation in which many parts of an organization rely on the basic support services to function, but at an unconscious level. This immediately suggests that system administration services are important to the organization and that resources should be given to this basic service. Node 6 is the next highest ranking node; this is the order registration department. Again, this is not particularly obvious from the diagram alone: it does not seem to be any more important than order processing. However, with hindsight, we can see that its importance arises because it has to liase closely with all other departments.

Using the definitions of regions and bridges from section 11.7.1, we can redraw the graph using centrality to organize it. The result is shown in fig. 11.4. The structure revealed by graph centrality accurately reflects the structure of the organization: it is composed largely of two separate enterprises: marketing and order processing. These departments are bound together by certain bridges that include management and staff that liase with the departments. Surprisingly, system administration services fall at the centre of the staff/marketing part of the organization. Again, this occurs because it is a critical dependency of this region of the system. Finally, the web server is a bridge that connects both departments to the outside world—the outside hanging on at the periphery of the systems.

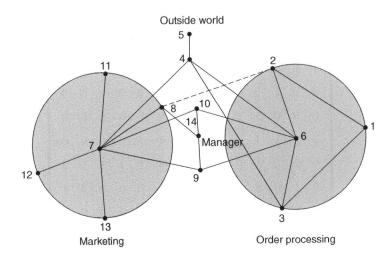

Figure 11.4: A centrality self-organized graph showing the structure of the graph centred around two local maxima or 'most important' nodes, that are the order registration department and the system administrator. There are also four bridge nodes and a bridging link between the regions.

The results of the example above could be enhanced by weighting the different connections in the adjacency matrix to reflect the workload. In a more realistic graph, one would measure the volumes of communication between the different nodes and create a more detailed model. The inclusion of every person and computer as a separate node would then automatically provide the appropriate weighting.

Resource centrality is a powerful method for identifying the stable regions of resource networks, as a guide to the appropriate investment of resources. Taking account of the directed nature of a graph can also affect the identification of regions (see section 6.5).

11.8   Directed resources: flow asymmetries

As noted in section 6.5, the centrality of directed graphs is much harder to interpret; indeed, the concept falls apart for graphs that do not have enough return paths to make a self-consistent picture. As long as there is a well-defined principal eigenvector, we can use directedness in the centrality method to find the important hot spots in systems.

Consider the direction of the flows in fig. 11.3, and let us suppose that some of the flows are directional, as in fig. 11.5. The adjacency matrix is now made asymmetrical, by setting to zero the bold connections in eqn. (11.8).

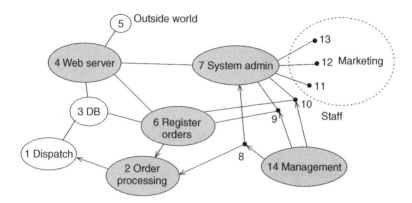

Figure 11.5: The organization in fig. 11.3 with some directed flows. Although management becomes a 'non-listening' source, the stable structure does not break down and importance ranking proceeds as for the non-directed case.

Both A and A^{T} have a principal eigenvector, with the following ranking:

i	$\vec{P}(A)$	$\vec{P}(A^{\mathrm{T}})$	$\vec{P}(AA^{\mathrm{T}})$	$\vec{P}(A^{\mathrm{T}}A)$
1	0.1	0.2	0.1	0.1
2	0.03	0.14	0.03	0.2
3	0.3	0.3	0.3	0.3
4	0.4	0.4	0.3	**0.4**
5	0.1	0.1	0.13	0.1
6	0.4	0.4	0.5	0.3
7	**0.5**	**0.5**	**0.5**	0.4
8	0.2	0.0	0.2	0.1
9	0.3	0.3	0.2	**0.4**
10	0.3	0.3	0.2	**0.4**
11	0.2	0.2	0.1	0.2
12	0.2	0.2	0.1	0.2
13	0.2	0.2	0.1	0.2
14	0.2	0.0	0.3	0.0

The highest values in each eigenvector are labelled in boldface. As expected, A and AA^T tend to identify the most sink-like centres, or 'authorities', while the two transposed quantities pick out source-like centres or 'hubs'. In this graph, there is no clear distinction, however; node 7 is both a source and a sink as far as its role in the system is concerned. In directed graphs, understanding the structure requires more consideration than in the undirected case. In particular, nodes that have both incoming and outgoing, unidirectional and bidirectional connections to their neighbours can play multiple roles simultaneously.

Applications and Further Study 11

- *Organizing and planning the deployment of system resources.*

- *Formalizing the economic aspects of a system, both for humans and machines.*

12

Task management and services

Never fear the event

—Horatio Nelson (1801)

In the previous chapters, we considered how resources fit into the scheme of a system. Now we consider how to optimize their allocation according to quotas or the allocation criteria. The resource we focus mostly on is time, since this is the fundamental currency underlying all dynamical systems; however, this chapter is really an extension of the previous chapter on resource usage. Time allocation applies both to humans and computers and effects a strong dependency on the ability of the system to process information.

12.1 Task list scheduling

Scheduling is a way of parsing a tree of tasks as efficiently as possible. The techniques for scheduling are well known from parallel processing and operating system design (Ahmad and Dhodhi (1996); Kasahara and Narita (1984)). An important aspect of configuration management is how management operations are scheduled within a system, both in response to specific events and as a general matter of the maintenance of the system.

Example 98 *A service help desk receives calls at a peak rate during the middle of the day, in its time zone, and at a slower rate at other times. The number of jobs, with a certain level of difficulty, follows a rough pattern. The number of available case handlers is also maximal during the day, but the numbers are not quite certain because of possible sickness. What is the optimum approach to completing all the incoming tasks?*

Example 99 *A list of routine maintenance operations, including system backup, security checks and software updates, must be completed each day to satisfy policy. Which ordering of tasks causes the least disruption to normal activities and leads to the most up-to-date, best-maintained system.*

Analytical Network and System Administration. Managing Human–Computer Networks   Mark Burgess
© 2004 John Wiley & Sons, Ltd   ISBN 0-470-86100-2

Scheduling clearly encompasses precise comparisons and heuristic value judgements. Specifying a schedule naturally becomes a part of any policy.

Scheduling takes many forms, such as job-shop scheduling, production scheduling, multiprocessor scheduling, human time management and so on. It can take place within any extent of time, space or other suitable system parameter. The two main classes of scheduling are *dynamic* and *static* scheduling. Dynamical schedules can change their own execution pattern, while static ones are fully predetermined. In general, solving static scheduling problems belongs to the class of NP problems (it is presumed to be computationally intensive).

If we represent scheduling as a graph theoretical problem, it involves assigning the vertices (task nodes) of an acyclic, directed graph onto a set of resources, such that the total time to process all the tasks is minimized. The total time to process all the tasks is usually referred to as the *makespan*. An additional objective is often to achieve a short makespan while minimizing the use of resources. Such multi-objective optimization problems involve complex trade-offs and compromises, and good scheduling strategies are almost based on a detailed and deep understanding of the specific problem domain.

Most approaches belong to the family of priority-list scheduling algorithms, differentiated by the way in which task priorities are assigned to the set of resources. Traditionally, heuristic methods have been employed in the search for high-quality solutions (Kasahara and Narita (1984)). Over the last decade, heuristics have been combined with modern search techniques such as simulated annealing and genetic algorithms (see Ahmad and Dhodhi (1996)).

12.2   Deterministic and non-deterministic schedules

There is a distinction between the arrival of a random event (a stochastic task such as a telephone call or a server request) and a planned or structured task with long-term predictability, (such as periodically executed tasks like daily cleaning, regular software updates, system backups). This is reflected in two possible ways of initiating a system task:

- A random starting point in the system graph,

- A random arrival time at some point in the graph.

Whichever method is used to parse the graph of tasks, the result of the process must end up with n ordered lists, to be carried out by n 'servers'. A server might be a person, a computer or an enterprise.

Event handling

System administration, employing agents or software robots, is an administrative method that potentially scales to large numbers of hosts in distributed systems, provided each host is responsible for its own state of configuration (see Burgess and Canright (2003)). However, the interdependencies between networked computers makes cooperation essential, and the distributed nature of the system makes the timing of events, for all intents and purposes, random.

Example 100 *Policy-based configuration languages associate the occurrence of specified events or conditions, with responses to be carried out by an agent. Cfengine accomplishes this, for instance, by classifying the state of a host, at the time of invocation, into a number of classifiers. Some of these represent the time of invocation, others the nature of the environment, and so on. For example,*

```
files:

 (linux|solaris).Hr12::

  /etc/passwd mode=0644 action=fixall inform=true
```

The class membership is described in the second line. In this instance, it specifies the class of all hosts that are of type Linux or Solaris, during the time interval from 12:00 hours to 12:59 (Hr12). Tasks to be scheduled are placed in classes that determine the host(s) on which they should be executed, or the time at which they should be executed. Actions are placed in such classes and are only performed if the agent executes the code in an environment that belongs to one of the relevant classes. Thus, by placing actions in judiciously chosen classes, one specifies actions to be carried out on either individual machines or on arbitrary groups of machines that have a common feature relating them. This is a scheduling procedure. We thus have

- *Scheduling in time,*

- *Scheduling by host attribute (location, type, etc.).*

Task scheduling

A set of ordered tasks or precedences can be represented by a directed graph, (V, E), with vertices, V, and directed edges, E. The collection of vertices, $V = \{v_1, v_2, \ldots, v_n\}$, represents the set of n tasks to be scheduled and the directed edges, $E = \{e_{ij}\}$, define the precedence relations that exist between these tasks (e_{ij} denotes a directed edge from task v_i to v_j). The graph might contain loops or cycles. These can cause unwanted repetition of tasks by mistake. Edges can be removed from the graph to convert it into an acyclic graph, or *spanning tree* that avoids this problem.

The task management process can be understood as scheduling in several ways. A graphical representation allows modelling of task management (see fig. 12.1).

Within a single set of policy rules there is a set of schedulable tasks that is ordered by precedence relations (arrows in the graph). These relations constrain the order in which policies can be applied, and thus how the graph has to be parsed.

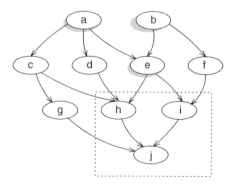

Figure 12.1: Scheduling of constrained policies.

A second way by which scheduling enters is through the response of the system to the arriving tasks. Should the server agents activate once every day, hour or minute, in order to check for scheduled tasks, or immediately? Should they start at random times or at predictable times? Should the policies scheduled for specific times of day, occur always at the same times of day, or at variable times, perhaps random?

Finally, although scheduling is normally regarded as referring to extent over time, a distributed system also has two other degrees of 'spatial' extent: h and c. Scheduling tasks over different hosts or changing the details of software components is also a possibility. It is possible to confound the predictability of software component configuration to present a 'moving target' to would-be attackers (see Burgess and Sandnes (2001); Sandnes (2001)).

12.3   Human–computer scheduling

Configuration management is a mixture of a *dynamic* and *static* scheduling. It is dynamic in the sense that it is an ongoing real-time process in which policies are triggered as a result of the environment. It is static in the sense that all policies are known *a priori* during any work interval. Policies can be added, changed and removed arbitrarily in a dynamical fashion. However, this does not really interfere with a static model because such changes would typically be made during a time interval in which the configuration tool was idle or off-line (in a quiescent state).

The separation of timescales is an issue that we harp on continually in this book. It is important here too. The management of time can only be successfully prosecuted by separating independent timescales from one another. Using the method of local averaging (see appendix B), it can be used to identify timescales over which average behaviour is only slowly varying. These are the independent levels of scheduling.

Example 101 *A simple analogy is helpful for understanding this. Suppose we are interested in planning the time of an enterprise that deals with selling furniture on the Internet (see fig. 11.3). The timescale at which products are redesigned is on the order of a year, so we expect to have to update web pages on this timescale. We must therefore allocate a number of man-hours per year to this task. (It is not necessary to check this every few minutes.) However, the timescale at which the sales requests arrive is about one per hour, so we need to allocate sufficient order processing power on the scale of an hour to check and process these orders. The number of web page accesses is on the order of several hits per second, so a web server must schedule resources to cope with this demand.*

12.4   Service provision and policy

In networks and systems alike, we are interested in maintaining predictable levels of service. Service providers would, after all, like to sell these services to customers and thus need to be able to offer guarantees about what will be delivered and what service will cost. This brings up two issues:

- *Quality of Service (QoS)*: Quality of Service is a goal of all service providers, whether the service is a network transmission rate or a system up-time level. Although one often has the impression of service quality as being a solved problem, this is far from the case. Guaranteeing service levels requires one to address the uncertainties in service provision at all levels, and these are often complex.

- *Service Level Agreements (SLA)*: A Service Level Agreement is a contract of service levels that is offered to a customer by a provider. The provision of service cannot be guaranteed with complete certainty, as we shall see below, so Service Level Agreements are above determining acceptable margins of behaviour in the system, and what recompense will be offered to a customer if the levels are not met.

A strategy for meeting service level guarantees is *over-provision* to provide a margin that can absorb sudden demand. Clearly, no system can absorb any sudden change in demand, thus maximum limits are placed on expectancy.

12.5   Queue processing

Whenever there is a confluence of information at some point in a system, there is a serialization of processing that bottlenecks information into a queue. In stochastic systems and queueing theory, these processes are called *birth–death processes*. In physics, they are referred to as *creation–annihilation processes* or *prey–predator models*.

Queueing networks are representations of resource-consuming systems on graphs, such that information or work flows from node to node, often with cyclic repetition. A queue is a generic processing model: jobs arrive as events, or they are pre-allocated and some form of human–computer processing must be applied to eliminate them from the 'to-do' list.

Queueing models have been devised with all levels of complexity (see Jain (1991) for an introduction); they are statistical models that describe the steady-state properties of stochastic task systems. At the coarsest level of approximation, the mean value theory can be used to obtain order of magnitude estimates of processing efficiencies. Queueing models consist of a number of choices or parameters:

- *Arrival time distribution process*

  If service requests occur at times $t_1, t_2, \ldots$, the values of the random variable $\tilde{t}_i = t_i - t_{i-1}$ are called the *inter-arrival times* of the process. It is commonly assumed that these form a Poisson (exponential) distribution; however, many network processes have long-tailed distributions (see section 10.10).

- *Process time distribution*

  The time that each client is engaged in requesting a service.

- *Number of servers*

  The number of humans, computers or other entities responsible for processing requests. The work rate is a function of this number.

- *System processing limit*

  If the system has a maximum throughput, or processing rate, this limits the behaviour of the system as a whole.

- *Maximum population size*

  The maximum number of clients that can ask for services.

- *Scheduling policy*:

  Various queueing policies are used to try to empty the queue as quickly as possible. The most common is First Come First Served (FCFS), which is a first-in-first-out (FIFO) structure. Round robin scheduling is a way of sharing time between multiple jobs. Shortest Job First picks the task whose estimated completion time is least. This potentially suffers from *starvation* of some tasks, that is, long jobs never get executed because resources are saturated with an influx of new small jobs. Shortest Remaining Time First is a variation on the previous policy. Clearly, there are many possible strategies for processing requests. In a human context, a policy that is often used is Loudest Voice, First Served, that is, those clients who make the biggest nuisance of themselves are disposed of quickly. The efficiency of these different policies is often difficult to evaluate, and depends on the nature of the system. In general, experimental analysis is required to determine an appropriate choice, or combination of choices.

12.6   Models

Queues are denoted in Kendall notation in the form $A/S/c(/B/K/P)$, where A is the inter-arrival distribution that usually takes one of the following values:

M Memoryless (exponential/Poisson)

E_k Erlang with parameter k

H_k Hyper-exponential with parameter k

D Deterministic

G General.

A deterministic distribution has constant inter-arrival times, with no variance, and 'general' means that the model's results apply for any distribution. The term *memoryless* implies that the arrival process is a steady-state process, in which the current state of the distribution does not depend on what happened in the past; that is, if the arrival time has a given form now, then it will have the same form for all subsequent arrivals, that is, it is statistically stable.

S is the processing time distribution, with the same values as for A. c is the number of servers or service entities. B is the number of buffers (the system processing limit), K is the maximum population size and P is the policy, if these are specified.

Example 102 *The basic queue in section 12.7 is referred to as an M/M/1 model.*

The Machine Repair-man Model (Scherr (1967)) is a simple queueing model that considers the problem of assigning the repair of machines to a repair queue. When a machine breaks, it is put into a queue until a repair-man can service it.

The Central Server Model (Buzen (1973, 1976)) is an event-driven model, with polling. A central server schedules a visit to a device (i.e. it polls a number of devices). If a request is pending, it is serviced; after polling is finished, the device returns to processing other tasks until the next event.

Example 103 *The Simple Network Management Protocol (SNMP) uses essentially the Central Server Model. When an event or a 'trap' occurs from a monitored device, a manager can poll the devices in the network and attend to any configuration changes in turn.*

12.7   The prototype queue M/M/1

At a service centre (server), the incoming traffic has to balance with the outgoing traffic, or a queue will grow. In the worst case, incoming jobs will have to be dropped.

Suppose we have an average of $n - 1$ tasks already in a queue and tasks are arriving at a rate of λ per second, then as soon as there are n tasks in the queue, some tasks must be forwarded at the rate μ, otherwise the average number of tasks in the queue will not stay constant. We can write this

$$\lambda \, p_{n-1} = \mu \, p_n,$$
$$p_n = \rho p_{n-1},$$

for any n, where $\rho = \lambda/\mu$. ρ is called the traffic intensity. If $\rho > 1$ then the incoming rate is higher than the outgoing rate. Now since the recurrence relation above holds for all n, clearly

$$p_1 = \rho p_0$$
$$p_2 = \rho^2 p_0$$
$$p_n = \rho^n p_0.$$

The sum of probabilities is always 1, so

$$\sum_{n=0}^{\infty} p_n = 1.$$

This is a geometric series, so we can find p_0:

$$\sum_{n=0}^{\infty} p_n = \frac{p_0}{1 - \rho} = 1.$$

Thus we have, for any n,

$$p_n = (1 - \rho)\rho^n.$$

Although this is clearly an idealization (there is never an infinite number of tasks, even on the whole Internet), this gives us a distribution that we can use to estimate the average number of tasks in the queue. The expectation value of the number of tasks is

$$E(n) = \langle n \rangle = \sum_{n=0}^{\infty} n p_n.$$

Substituting for p_n,

$$\langle n \rangle = \sum_{n=0}^{\infty} n(1 - \rho)\rho^n$$

$$= \sum_{n=0}^{\infty} n\rho^n - \sum_{n=0}^{\infty} n\rho^{n+1}$$

Relabelling $n \to n + 1$ in the second term gives us another geometric series:

$$\langle n \rangle = \sum_{n=0}^{\infty} n\rho^n - \sum_{n=1}^{\infty} (n - 1)\rho^n$$

$$= \sum_{n=1}^{\infty} \rho^n$$

$$= \frac{\rho}{1 - \rho}$$

Thus the mean number of tasks is

$$\langle n \rangle = \frac{\rho}{1 - \rho}$$

$$\to \infty \qquad (\rho \to 1)$$

The variance can also be worked out

$$\langle (n - \langle n \rangle)^2 \rangle = \frac{\rho}{(1 - \rho)^2}$$

This gives us an estimate of the size of queue that we need in order to cope with a normal traffic rate.

The system is busy whenever there is at least one job in the system. We can use this to characterize how busy a server is.

Definition 48 (Load average or Utilization) *The probability of finding at least one job in the system is called the load average. It is 'NOT' the probability of finding no jobs in the system, that is, $1 - P(n = 0)$.*

For the M/M/1 queue, we thus write the load average, also called the Utilization, as

$$U = 1 - p_0 = 1 - (1 - \rho) = \rho = \frac{\lambda}{\mu}. \tag{12.1}$$

The above assumptions that a fixed steady-state (memoryless) rate equation holds imply that the arrival times are Poisson distributed. However, research over the last ten years has revealed that Internet traffic does not satisfy this pattern. Voice traffic on the telephone system has always been well modelled in this way, but packet switched traffic is 'bursty'—it has no scale that can be averaged out.

Example 104 (Sharing and contention) *Some resources can be used by only one client at a time. If several clients try to use these simultaneously, there is contention. Is the solution to put these requests in a queue, with some scheduling policy, or to make them try again later? A scheduling of requests requires a protocol or even a service to exist, to manage the queue. Without this, clients will compete and contend for the resource. To fully appreciate the nature of contention, we need to combine the queueing theory with the contention theory. See section 19.8 for a more complete method of analysis.*

System loads that do not even out over reasonable timescales is said to be *long-tailed* and sometimes *statistically self-similar* (see section 10.10). This is often associated with a 'burstiness' or a power-law clustering of events. In self-similar traffic, there is a long-range dependence, and theoretically infinite (at least unpredictably large) fluctuations. This means that the queue size estimates need to be re-evaluated to avoid the queue length from growing uncontrollably.

Garbage collection and overflow

Not all systems have procedures in place to reclaim resources that are no longer in use. The reclamation of reusable resources and the discarding of consumable by-products is called *garbage collection*. In a queue, there is automatic reclamation as items are removed from the queue, but in memory or space allocation that is not necessarily true. Resource reclamation is crucial for the survival of systems in the long term[1].

12.8   Queue relationships or basic 'laws'

Basic dimensional analysis, or linear rate equations, provide a number of basic 'laws' about queues. These are rather simple relationships that barely deserve to be called something so elevated as laws, but they express basic truths about rate-flow systems. They are nicely summarized in (Jain (1991)). We mention a few examples here by way of illustration. The laws express basic linear relationships about service flow rate. These form the basis of subsequent approximations in section 18.3, for instance. We use the following definitions for the ith server:

$$\text{Arrival rate } \lambda_i = \frac{\text{No. of arrivals}}{\text{Time}} = \frac{A_i}{T} \tag{12.2}$$

$$\text{Throughput } \mu_i = \frac{\text{No. of completions}}{\text{Time}} = \frac{C_i}{T} \tag{12.3}$$

$$\text{Utilization } U_i = \frac{\text{Busy time}}{\text{Total time}} = \frac{B_i}{T} \tag{12.4}$$

$$\text{Mean service time } S_i = \frac{\text{Busy Time}}{\text{No. of completions}} = \frac{A_i}{T}. \tag{12.5}$$

---

[1] Humans eventually die because DNA does not perform (inverse) garbage collection of telomeres that are involved in DNA replication.

The Utilization law

The utilization law tells us the mean levels at which resources are being scheduled in the system. The law notes simply that

$$U_i = \frac{B_i}{T} = \frac{C_i}{T} \times \frac{B_i}{C_i} \qquad (12.6)$$

or

$$U_i = \mu_i S_i. \qquad (12.7)$$

So, utilization is proportional to the rate at which jobs are completed and the mean time to complete a job. It can be interpreted as the probability that there is at least one job in the system (see eqn. 12.1).

Example 105 *Suppose a web server receives hits at a mean rate of 1.25 hits per second, and the server takes an average of 2 milliseconds to reply. The law tells us that the utilization of the server is*

$$U = 1.25 \times 0.002 = 0.0025 = 0.25\%. \qquad (12.8)$$

This indicates to us that the system could probably work at 400 times this rate before saturation occurs, since $400 \times 0.25 = 100\%$.

Although the conclusion in this example is quite straightforward and only approximate, it is perhaps not immediately obvious from the initial numbers. Its value therefore lies in making a probable conclusion more obvious.

Example 106 *A university teacher complains that most of the terminals or workstations in the terminal room of the University are idle when he looks in the afternoon, so there must be too many terminals and money can be saved by at least halving the number. The students, on the other hand, complain that they can never find a free terminal when they need one. Who is right? The system administrator decides to look through the logs and apply the Utilization law. She finds that the time an average student spends at a terminal is $S = 1$ hour, and the average number of users using a given terminal per working day (of 8 hours) is $\mu = 4/8 = 0.5$ per hour. The utilization is thus $U = 0.5 \times 1 = 0.5$. In other words, the terminals are in use about half of the time.*

    The system administrator realizes that the peak load on the system is time-dependent. Around midday, many students arrive looking for a terminal and cannot find one. In the early mornings and the late afternoons, the students are all sleeping (or in lectures, or both), so there are many spare terminals. By reorganizing their time, the system administrator concludes that the students could make use of the machines that are there but that an overcapacity of double is acceptable for covering the peak load.

Series and parallel utilization

When dealing with arrangements of queues, working in parallel (independently) or in series (waiting for each other), it is sometimes helpful to replace the array of queues with a single effective queue. We can rewrite the formulae to obtain effective formulae for these cases by assuming conservation of jobs: what goes in must come out. This leads us basically to the analogy with Ohm's law in electrical circuits, or Kirchoff's laws.

If a number of queue servers is arranged in series (see fig. 12.2), then the utilization law applies to each component. System components in series have the same average flow rate μ_i through each component, assuming that the system is running in a steady state—otherwise the pressure in the system would build up somewhere in one component, which would eventually cause a failure (like a burst pipe). Thus, the utilization of each component is

$$U_i = \mu S_i. \tag{12.9}$$

The utilization for the whole series of queues is thus

$$U_{\text{tot}} = \sum_i U_i = \mu \sum_i S_i$$
$$= \mu S_{\text{serial}}. \tag{12.10}$$

Thus, we see that the average service time for the whole series is

$$S_{\text{serial}} = \sum_i S_i. \tag{12.11}$$

Figure 12.2: System components in series have the same average flow rate through each component, assuming that the system is running in a steady state—otherwise the pressure in the system would build up somewhere in one component, which would eventually cause a failure (like a burst pipe).

It is simply the sum of the service times of each sub-queue. If, on the other hand, we couple the queues in parallel (see fig. 12.3), then it is now the total utilization that is shared equally between the queues, that is, U is common to each queue.

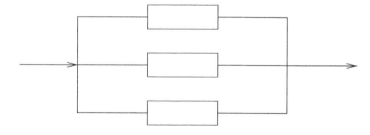

Figure 12.3: System components in parallel share the flow between the different server queues. Here it is the average utilization that is common to each server, since each experiences the same average incoming load.

Thus,

$$\mu = \sum_{i=1}^{N} \mu_i = \sum_{i=1}^{N} \frac{U}{S_i} = \frac{U}{S_{\text{parallel}}}, \tag{12.12}$$

and we can write

$$\frac{1}{S_{\text{parallel}}} = \sum_{i=1}^{N} \frac{1}{S_i}. \tag{12.13}$$

If we have N identical servers $S_{\text{par}} \to \langle S \rangle_{\text{par}}/N$, that is, the service time is reduced by a factor of N on average. Alternatively, throughput can be increased by a factor of N. Note, however, that this does not tell us how long a client task will have to wait for completion. To determine that, we need a more advanced analysis (see section 12.9).

Example 107 *Suppose a student terminal room has 100 computers to share between 500 students. If the number of students increases now by 60, how many extra computers do we require to maintain the same level of utilization? The simplest answer to this is a naive ratio estimation, using the ratio of students to machines.*

$$\frac{\text{Students}}{\text{Terminals}} = \frac{500}{100} = \frac{560}{x} \tag{12.14}$$

Thus $x = 112$ terminals. However, this result assumes that students actually arrive as a predictable flow process. In fact, they arrive as a random process, and things are not as bad as this. Utilization of overlapping random processes can be more efficient than for predictable processes; the events of one process can slot into the gaps in another. See section 12.9.

Little's law

Another self-evident consequence of rates is Little's law of queue size. It says that the mean number of jobs in a queue Q_i^n, in queue i, is equal to the product of the mean arrival rate λ_i (jobs per second) and the mean delay (seconds) R_i incurred by the queue:

$$Q_i^n = \lambda_i R_i. \tag{12.15}$$

If one assumes that the queue is balanced, as in eqn. (12.1), then $\lambda_i \propto \mu_i$, and we may write

$$Q_i^n \propto \mu_i R_i \tag{12.16}$$

or

$$Q_i^n = c\mu_i R_i. \tag{12.17}$$

Another way of explaining this equation is to say that the amount of information in the queue is proportional to both the throughput rate and the amount of time each job takes to execute.

Before leaving this section, note that this equation has the generic form $V = IR$, similar to Ohm's law in electricity. This analogy is a direct consequence of a simple balanced flow model; it is not so much an analogy as an isomorphism. We shall make use of this result again in chapter 18.

Example 108 *In the M/M/1 queue of section 12.7, it is useful to characterize the expected response time of the service centre. In other words, what is the likely time a client will have to wait in order to have a task completed? From Little's law, we know that the average number of tasks in a queue is the product of the average response time and the average arrival rate, so*

$$R = \frac{Q_n}{\lambda} = \frac{\langle n \rangle}{\lambda}$$
$$= \frac{1}{\mu(1 - \rho)}$$
$$= \frac{1}{(\mu - \lambda)}. \tag{12.18}$$

Notice that this is finite as long as $\lambda \ll \mu$, but as $\lambda \to \mu$, the response time becomes unbounded.

Response time law

A situation that parallels the coupling of Ohmic resistances in a series is to consider the coupling of a number of queues in a series. This is common in systems of all kinds; once one part of a system is finished with a task, it is passed on to another part for further processing. What then is the response time of the whole system?

Let Q_{tot} be the total number of jobs in the system. Then clearly,

$$Q_{\text{tot}} = Q_1 + Q_2 + \cdots + Q_n. \tag{12.19}$$

Also, the total time can be written by Little's law as a product of the average throughput and the total response time as

$$Q_{\text{tot}} = \langle \mu \rangle R_{\text{tot}}. \tag{12.20}$$

The law applies to each of the component queues also, so eqn. 12.19 becomes

$$\langle \mu \rangle R_{\text{tot}} = \sum_{i=1}^{n} \mu_i R_i, \tag{12.21}$$

where $\langle \mu \rangle$ is the average throughput of the system. Thus, we have the response time

$$R_{\text{tot}} = \sum_{i=1}^{n} \frac{\mu_i}{\langle \mu \rangle} R_i = \sum_{i=1}^{n} x_i R_i, \tag{12.22}$$

where x_i is the capacity of the component, or the fraction of jobs that are held in flow on average in component i:

$$x_i = \frac{\mu_i}{\langle \mu \rangle} = \frac{\text{Jobs/s in component}}{\text{Average total jobs/s}} = \frac{C_i}{\langle C \rangle}. \tag{12.23}$$

This tells us that the total response time is simply a weighted sum of the individual response times at any moment. It allows us to see where bottlenecks are likely to occur, so that processing resources can be reallocated.

12.9   Expediting tasks with multiple servers M/M/k

How does the ability to process input change if we add more servers to accept jobs from the input queue? Intuition tells us that this must be more efficient, but, as always, the answer should be qualified and quantified to be certain about what we mean.

Suppose we have k servers in parallel, removing jobs from a single input queue. We assume that the servers are identical components (computers, humans etc.), so that the input rate is still λ and each server has a processing rate of μ. Clearly, the maximum processing rate is now $k\mu$ when all of the servers are busy. If we assume that there is no overhead incurred in allocating tasks to servers, then this is also the rate at which work is expedited. If there are fewer than $n < k$ jobs waiting, then the average service rate will be $n\mu$. Thus, our balance equations are now

$$
\lambda p_{n-1} =
\begin{cases}
n\mu\, p_n & (0 < n \le k) \\[2mm]
k\mu\, p_n & (n > k)
\end{cases}
\tag{12.24}
$$

These can be solved to give

$$
\lambda p_n =
\begin{cases}
p_0 \left(\dfrac{\lambda}{\mu}\right)^n \dfrac{1}{n!} & (0 < n \le k) \\[3mm]
p_k \left(\dfrac{\lambda}{\mu}\right)^{n-k} \dfrac{1}{k^{n-k}} = p_0 \left(\dfrac{\lambda}{\mu}\right)^n \dfrac{1}{k!k^{n-k}} & (n > k)
\end{cases}
\tag{12.25}
$$

and p_0 is found by normalizing

$$
\sum_{n=0}^{\infty} p_n = p_0 \left(\sum_{n=0}^{k-1} \left(\frac{\lambda}{\mu}\right)^n \frac{1}{n!} + \sum_{n=k}^{\infty} \left(\frac{\lambda}{\mu}\right)^n \frac{1}{k!k^{n-k}} \right) = 1,
\tag{12.26}
$$

that is, if we now let the traffic intensity per server be $\rho = \lambda/\mu k$,

$$
p_0 = \left(1 + \sum_{n=1}^{k-1} (k\rho)^n \frac{1}{n!} + \frac{(k\rho)^k}{k!(1-\rho)} \right)^{-1}.
\tag{12.27}
$$

The probability that a task will have to wait to be performed κ is the probability that there are k or more tasks already in the system,

$$
\kappa \equiv P(n \ge k) = \sum_{n=k}^{\infty} p_n = \frac{(k\rho)^k}{k!(1-\rho)} p_0.
\tag{12.28}
$$

Similarly, the average number of jobs in the system is the expectation value of n:

$$
\langle n \rangle = \sum_{n=0}^{\infty} n p_n = k\rho + \frac{\kappa \rho}{1-\rho}.
\tag{12.29}
$$

Little's law again gives the average response time for the system in responding to a task as

$$
R = \frac{\langle n \rangle}{\lambda} = \frac{1}{\mu}\left(1 + \frac{\kappa}{k(1-\rho)} \right).
\tag{12.30}
$$

Readers can verify that $\kappa(k = 1) = \rho$ and that the expressions above agree with the single server queue for $k = 1$.

The k dependence of the above expression is rather complicated; in terms of power-counting, κ is approximately neutral to changes in k, but decreases slightly for increasing k. Expression (12.30) therefore tells us that as the number of servers k increases, the response time for incoming jobs falls off slightly faster than $1/k$. Thus, the response time of a single queue with k servers is slightly better than k separate queues, each with a $1/k$th of the tasks to complete. This assumes that all of the jobs and servers are identical, of course. Why should this be? The reason is that single queue servers make all jobs wait even when the load is low, whereas a parallel server strategy can keep the level of incoming jobs below the queueing threshold a greater percentage of the time. The difference becomes most noticeable as the traffic intensity increases.

Example 109 *A web hotel company has five customers who need their sites hosted. The company must decide whether to use a separate computer for each web server, or whether they should host all sites as virtual domains on a single site with a load balancer or multiple CPUs. Considering the performance aspect, they note that traffic arrives most of the time at a rate of 10 hits per second ($\lambda = 0.01$ per millisecond) and that the web server has a service rate of 20 ms ($\mu = 0.05$ per millisecond). Let us model this as an M/M/5 queue.*

The traffic intensity is $\rho = \lambda/k\mu = 0.04$, and thus the probability that the server is idle is

$$p_0 = \left[1 + \frac{(5 \times 0.04)^5}{5!(1 - 0.04)} + 5 \times 0.04 + \cdots \right]^{-1}$$
$$= 0.82. \tag{12.31}$$

Clearly, normal traffic levels are low, and this is a high probability. The probability that all the terminals are busy is

$$\kappa = \frac{(k\rho)^k}{k!(1 - \rho)} p_0 = 1.1 \times 10^{-5}. \tag{12.32}$$

Hence the average response time is

$$R = \frac{1}{\mu} \left(1 + \frac{\kappa}{k(1 - \rho)} \right) = 20 \text{ ms}. \tag{12.33}$$

Now, suppose we have used five separate machines each with an M/M/1 queue. The effective arrival rate λ can be divided evenly between them, so that $\lambda' = 0.01/5$. The probability of a given queue being idle is

$$p_0' = (1 - \rho') = 0.96, \tag{12.34}$$

and the average response time is

$$R' = \frac{1}{\mu - \lambda} = 20.8 \text{ ms}. \tag{12.35}$$

This shows that the multiple queues lead to a slightly larger result, even at this low level of utilization.

Suppose we now compare the behaviours of these two alternatives at peak times, where the number of arrivals is ten times as much. Substituting $\lambda = 0.1$ and $\lambda' = 0.1/5$, we find *that the probable response time for the M/M/5 queue is largely unchanged*

$$R = 20 \text{ ms}. \tag{12.36}$$

However, the response times of the multiple M/M/1 queues are

$$R' = 33.3 \text{ ms}. \tag{12.37}$$

We verify that a multiple server handling of a single queue is at least as good as multiple queues, because the probability of a multi-server handler being idle is much lesser. Multiple queues with separate servers force even the shortest jobs to wait unnecessarily, whereas a multiple server scheduling would have cleared these jobs quickly, leaving a greater chance of being able to handle incoming tasks immediately. This phenomenon is related to the 'folk theorems' in section 18.3, eqn. 18.24.

12.10   Maximum entropy input events in periodic systems

One of the themes that we return to repeatedly in this book is the idea of systems that respond to random events. When these events change measurables in an unpredictable way, we call them fluctuations. Characterizing fluctuations is important because it is a way of representing the spectrum of input.

The problem with random events is that they come from an environment that is complex and that we know little about. By definition, the environment is that which is outside our control, so how should we describe the unknown? One way of approaching this problem is to assume the greatest level of randomness, given a constraint that says something about the interaction of the system with the environment.

If we characterize input as an alphabet of symbols labelled by i and with probability distribution p_i, then the assumption of maximum randomness can be accomplished by using a maximum entropy principle (see section 9.11). This is a form of 'constrained fair weighting' that is sometimes used as a way of scheduling events between different service centres (servers).

Example 110 *Consider the problem of modelling the input event stream of a server, with expected input current I symbols per second, formed from the weighted sum of individual symbols rates I_i:*

$$\langle I \rangle = \sum_{i=1}^{N} p_i I_i. \tag{12.38}$$

Thus, we maximize entropy given this 'constraint' on p_i. The result is the well-known Boltzmann distribution $p_i \propto \exp(-\beta I_i)$.

The most common solutions of the maximum entropy hypothesis are the completely flat distribution (all input events equally likely) and the exponential Boltzmann distribution that follows from a constant sum constraint.

How good is this assumption of maximal entropy fluctuation? This was investigated partially in (Burgess et al. (2001)), where it was found that the assumption is a good representation, but only if a modification is made to the maximum entropy signal. The modification is a periodic scale transformation that follows the daily and weekly rhythm of system behaviour.

If we fit a maximum entropy distribution to actual data, it will have constant variance, or the moments of the distribution will remain fixed. However, this would imply that the environment was a steady-state process, which we know to be false. Observations show that there is a distinct periodicity, not only in the expectation values of environmental signals but also in their variance over a periodic representation of several weeks (see fig. 2.1).

Some authors maintain that this is evidence for self-similar behaviour, and the modified distribution is in fact a power-law spectrum, rather than a maximum entropy distribution. One possible explanation for this is the similarity in form between the asymptotic form of the stable Lévy distributions and the form of the periodically corrected maximum entropy model (Sato (1999)).

12.11   Miscellaneous issues in scheduling

Numerous additional factors about human–computer systems can be incorporated into time management: humans get bored, and they have different specializations and preferences, different skill levels, and so on. Finding the optimal use of time, given these additional constraints requires a more sophisticated model, with heuristic currencies.

One model that describes human scheduling of interruptions or intervals of concentration is the game theoretical payoff model described in section 19.8. In this model, which concerns cooperation versus competition, we can think of time cooperation as giving humans space to work in fixed schedules and competition for time as being continual interruptions at random. In other words, tasks cooperate in orderly scheduled quotas, or they demand resources at random. This model addresses many situations, and we can use it for human scheduling too. The basic result from game theory, although simplistic, tells us that random event competition works well as long as the number of jobs is small, that is, as long as there is plenty of time and no one is taxed to the limit. However, as someone becomes very busy, it is better to schedule fixed-quota time slices for jobs, otherwise all jobs tend to suffer and nothing gets done. The overhead of swapping tasks at random can lead to great inefficiency if time is short.

The potential for harvesting the vast computing resources of the Internet has led to the notion of Grid computing, or massive distributed parallelism. To accomplish worldwide cooperation of computing resources, models extend their timescales to include batch-style processing as part of a larger task. Examples of this include the screen-saver processing methods used by the Search For Extraterrestrial Intelligence (SETI) project. Each of these cases can be studied with the help of parallel and distributed scheduling models.

Applications and Further Study 12

- *Designing workflow patterns and schedules.*

- *Resource deployment.*

- *Measuring workflow and efficiency.*

- *Modelling of generalized queueing processes.*

13

System architectures

In the foregoing chapters, we mainly considered systems that could be described by simple scalar variables, with no larger structure. That was a useful simplification for considering the basic effects of change and dynamics, but few human–computer systems, worthy of investigation, are this simple. In technology, we normally want to design a system with a specific function in mind. System design requires a number of elements as follows:

- Strategy

- Policy

- Procedure

- Activity.

These fit together into flows of information that govern change within the system.

13.1 Policy for organization

No system has to have a purpose in order to function. Biological organisms do not have a purpose *a priori*, they exist only to reproduce and start over; if they are lucky they find an ecological niche. It has been said that bureaucracies are self-sustaining, in the same way as biological life is, and that they will happily consume resources without ever outwardly progressing beyond an inward communication. Normally, when we speak of systems, however, we are interested in what effect the systems have on their resources and environment, since systems consume resources and are thus expensive.

The term *policy* is often used in the management of human–computer systems to mean a specification of the goal of the system, together with decisions limiting the behaviour of processes within the system. Policy has potentially several goals as follows:

- To maximize production of something

- To minimize the extent of a problem

- To identify and distinguish input (accountability).

Analytical Network and System Administration. Managing Human–Computer Networks   Mark Burgess
© 2004 John Wiley & Sons, Ltd   ISBN 0-470-86100-2

These are ad hoc decisions about a system; they cannot be derived. Policy exists at the high level and the low level. High-level policy might include ethical considerations and other human issues. Low-level policy constrains the details of work processes, for example, in the placement of regulating valves (administrative overseer) for control, convenience measures, security measures, consistency and the overall flow of information.

13.2   Informative and procedural flows

Dynamical systems generate and manipulate information. They require input and they generate output. The information flows around the system as the system carries out its function, and the final result is usually a synthesis of products that have been accumulated and assembled from different parts of the cooperating system.

It is important to distinguish two types of flow, or development within systems:

> **Definition 49 (Algorithmic flow)** *This is a map of the way the actions taken by the system interact and how control flows from instruction to instruction within the processes of the system. The instructions within a system are often laid out from start to finish in a control list, or a control loop. It is the flow of activity and authority in the system.*

An example of this is a computer program.

Example 111 *A computer program has a functional structure, with each function composed of a linear sequence of instructions. Some of these instructions require the evaluation of subroutines or subordinate functions. A university is a system that accepts students and churns them through a learning process, emitting them at the other end hopefully invigorated with new ideas and abilities. There is a definite flow within this system, from lecture to lecture. Event-driven systems, on the other hand, respond to specific happenings. A fire service is an example of an event-driven system.*

Another kind of flow is that taken by the resources that are manipulated by the processes.

> **Definition 50 (Resource flow)** *The produce of a system, or the work it actually carries out, also has a rate of change. This is the rate of resource and information flow.*

The flow of information does not necessarily mirror the flow of algorithms and instructions that produces it.

Example 112 *The command structure of an army might involve a flow of decision-making by radio to and from a command centre, during a military operation. The movement of the troops (i.e. the resources) is not related to this information flow.*

Example 113 *In a user-support desk or help desk, a flow of control information passes between the user and those helping the user. The actors, who help the user, manipulate resources in order to solve the user's problem. These two information flows are different.*

The distinction between algorithmic and resource flow is important to the design of a system, because one should not be tempted to organize these two flows by the same standards. Both flows are important to the overall function.

Table 13.1:   Object/entity names in human–computer systems.

Passive	Active	Human
Data structure	Program	Site/enterprise
Entity	Process	Department
Database	Sub-process	Work group
Record	Thread	Project

13.3   Structured systems and *ad hoc* systems

In most cases, data and processes are arranged in components, objects or *entities* that allow specialization and collaboration. These cooperate in the execution of the system, by communicating via channels of communications. These might be 'word of mouth', written on paper or electronic; the means of communication is unimportant.

Consider the names in table 13.1. These names represent logically distinct elements within different kinds of system. The units are often arranged in a hierarchical fashion, that is, a site contains several departments, which in turn contains several groups composed of individuals; a database contains many records, which in turn consist of many sub-records.

The structure of systems that perform a stable function is usually fixed for the duration of that function. However, this limits their ability to adapt to slight changes in their environments. Change can therefore be allowed to take place in a deterministic way by having a continual re-evaluation of the system at regular intervals. An alternative procedure is to allow systems to perform randomly.

An *ad hoc* system has no predictable structure over long intervals of time. We can define it as follows:

Definition 51 (*Ad hoc* system) *A system is said to be* ad hoc *if its structure is periodically re-determined by random variables.*

Example 114 *Mobile ad hoc networks are networks for radio communication that are formed by randomly moving mobile devices with transceivers. These devices link up in an opportunistic manner to form a relaying network. Messages can be routed and relayed through such networks if a sufficient number of active devices are in the range of one another.*

It is tempting to associate ad hoc systems with the concept of a random graph; however, an ad hoc system need not be a random graph—it could be based on a predetermined structural plan, but with only a finite probability of being connected at any given moment.

13.4   Dependence policy

All systems have components that depend on other components; that is implicit in the definition of a system. This dependence might be a strong dependence, with dramatic consequences if a part fails, or because of a merely weak influence. Some factors in a system enable the system to function, while others merely enhance or amplify its abilities. In the next chapter, we shall consider how to rationalize and analyse the effects of dependencies; for now, we introduce only the concepts.

Definition 52 (Strong dependence) *This is a dependence on another part of the system, in which the removal of the dependency (the tie between the dependent and the dependee) leaves the dependent (that which depends) unable to function.*

Definition 53 (Weak dependence) *This is a dependence in which the removal of a dependency does not prevent the functioning of the dependent, but merely alters its possible behaviour.*

Example 115 *An aircraft depends strongly on its fuel; without fuel, it cannot function. The same aircraft depends weakly on the weather; this can affect its performance, but does not prevent it from functioning.*

A computer system depends strongly on its hardware and operating system; without this, it cannot function. The same computer depends only weakly on its third-party software.

A system must be analysed in terms of its *dependencies*. Dependence on other parts of a system has many implications for a system, including its efficiency and its ultimate limitations. As implied above, a *critical dependence* on some component can affect the ability of the system to perform its function. If such a crucial part becomes damaged or unavailable, then the whole system can fail to function; the system is then said to have a *single point of failure*. Note that this phrase does not necessarily mean that there is only one of them, but that one is enough to cause complete failure. In other words, the failure of the system at a single point is enough to halt it completely. This is clearly a precarious situation to be in, and is usually a sign of poor design. A strategy for avoiding such problems in system design is *redundancy*, or the use of back-up systems.

Definition 54 (Redundancy) *The duplication of resources in such a way that there are no common dependencies between duplicates, that is, so that the failure of one does not lead to the failure of the others.*

Here is an example of a lack of redundancy.

Example 116 *A clustered structure of computers centred on a single server forms a hub-centric structure. If the computer at the centre of the hub should fail, all of the dependent computers will cease to function properly. There is no back-up.*

Example 117 *Developed societies are almost completely dependent on electricity for their functioning. If a bad storm, or flood were to take out power lines to a major community, it would cease to function. This is the reason why military operations usually target an enemy's critical infrastructure first.*

Redundancy is an important strategy, but too much redundancy can lead to inefficiency.

Example 118 *Systems depend on the flow of information through channels of communication. Unstructured communication often results in too much communication (repetition and little progress). Regulation of dependencies is a strategy for minimizing unnecessary uncertainty.*

13.5   System design strategy

There are many ways to attempt the design of a system. One looks for a strategy for breaking a complex task down into manageable pieces in order to execute it. Experienced practitioners rely on their experience to suggest solutions, but sometimes intuition can fail and a more rational enquiry is needed to make a decision. We shall deal with the requirements for rational decision-making in the remainder of the book. Some common alternatives are described here.

Note that the graphical view of systems expounded in this book implies no particular need for hierarchy. There is a long-standing tendency of imposing hierarchies on systems, even where they might be detrimental to functioning, because hierarchies are ingrained into most societal structures. One of the hardest lessons to learn in system analysis is that decision-making does not necessarily imply a directed branching tree-like structure; it can also arise by competition, by voting, by ad hoc timing or by random events. Our main concern is whether systems are appropriately stable in their decision-making.

Modular design

Different authors use this expression in different ways. Modular design is, first and foremost, a strategy of breaking up a system into smaller parts, in order to promote comprehensibility, adaptability and extensibility. Is a single, closed function a module? The answer is clearly yes, but one normally reserves this expression for larger entities than this.

Top-down versus bottom-up

Two design strategies for building systems have emerged, and have developed with increasing refinements and compromises between the two viewpoints. Although it is too much to ask for a consensus of naming, broadly speaking, these are represented by the following:

- *Top-down*: The name 'top-down' is motivated by the traditional way of drawing hierarchical structure, with high-level (low detail) at the top, and increasing low-level detail at the bottom. A top-down analysis is *goal driven*: one proceeds by describing the goals of the system, and by systematically breaking these up into components, by a process of *functional decomposition* or *normalization*.

- *Bottom-up*: In a 'bottom-up' design, one begins by building a 'library' of the components that are probably required in order to build the system. This approach is thus *driven by specialization*. One then tries to assemble the components into larger structures, like building blocks, and fit them to the solution of the problem. This approach is useful for building a solution from 'off-the-shelf', existing tools.

The difference between these two strategies is often one of pragmatism. A top-down design is usually only possible if one is starting with a blank slate. A 'bottom-up' design is a design based on existing constraints, namely, the components or resources that are to hand. An advantage of building from the bottom-up is that one solves each problem only once. In a top-down strategy, one could conceivably encounter the same problem in different branches of the structure, and attempt to solve these instances independently. This could lead to inconsistent behaviour. The process of 'normalization' (see chapter 14) of a system is about eliminating such inconsistencies.

Example 119 *(Computer system - 'top-down') In the design of a new computer system, one examines the problem to be solved for its users (banking system, accounts), then one finds software packages that solve these problems, then one chooses a platform on which to run the software (Windows, Macintosh, Unix), and finally one buys and deploys the hardware that will run those systems.*

(Computer system - 'bottom-up') First one buys reliable hardware, often with the operating system already installed, and installs it for all the users; then one looks for a software package that will run on that system and installs that. Finally, the users are taught to use the system.

Example 120 *(Enterprise - 'top-down') In the organization of a maintenance crew, one looks at the problems that exist and breaks these down into a number of independent tasks (plumbing, electrical, ventilation). Each of these tasks is broken down into independent tasks (diagnosis, repair) and finally individuals are assigned to these tasks.*

(Enterprise - 'bottom-up') The enterprise bosses look at everyone they have working for them and catalogue their skills. These are the basic components of the organization. They are then grouped into teams that can cooperate to solve problems like diagnosis and repair. Finally, these teams are assigned tasks from the list of plumbing, electrical work and ventilation.

In system administration, especially configuration management, these two strategies are both widely used in different ways. One may either implement primitive tools (bottom-up) that are designed to automatically satisfy the constraints of a system, or one can use trial and error to find a top-down approach that satisfies the constraints.

Functional design and object design

Similar to the ideas of top-down and bottom-up design are the design strategies widely used for building computer software, based on functional and object decomposition. Computer software is simply a system of algorithms and data structures that operate on the resources of a computer. The principles of software design are the same as those of any system design.

A functional design is an algorithmic (top-down) design, in which the goals of the system motivate the architecture of the components within the system. The system is geared around the evaluation of its function, that is, performing its task. This approach is illustrated in figs. 13.1, 13.2 and 11.1. The system state is usually centralized in a monolithic data structure, whereas operators are distributed throughout the different functional entities within the system.

In an object-oriented design (bottom-up), one begins by isolating logically independent objects or tasks within the whole system, that is, tasks that are able to carry out a specific function without outside help (see fig. 13.3). One then constructs the system around these objects by sending messages between them. Any details that only apply to the particular object are contained with in it, that is, hidden from public view. Whereas a functional model reflects a flow of activity and information between the levels of the boxes, an object model requires a separate plan for action, at each level of its operation. Private algorithms are usually referred to as the *methods* of the object. An object model requires at least one public 'method', or control function that binds the objects together. The state of an object is private to that object, and all of the operators that change that state are private

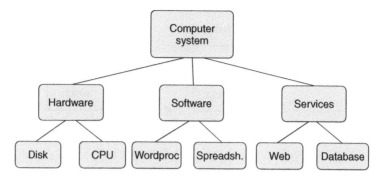

Figure 13.1: A functional structure diagram for a computer system, somewhat simplified, shows how each level of the computer system depends on other parts.

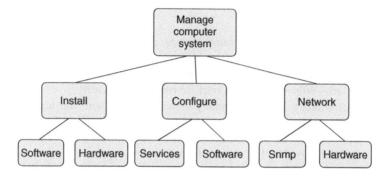

Figure 13.2: A functional structure diagram for the meta system that manages (administers) a computer system. The diagram, somewhat simplified, shows how each task level of the system depends on sub-tasks. This diagram is easily drawn and easily used to over-simplify the issues in system management. For instance, it gives the impression that management is a one-off process and moves from left to right in the diagram whereas, in fact, it is a dynamical process in which all of the parts are called upon repeatedly.

to the object. In order to function in concert, one often requires the guidance of a simple functional system. The total state is decentralized, and is the sum of the private states of each of the objects.

It is sometimes said that 'objects' communicate by sending messages (such as a peer-to-peer network). While this is the idealized viewpoint taken in pure object-oriented design, it is often an exaggeration of the truth. In practice, there is some guiding functional super-structure that manages the objects; seldom are they able to work independently. We can imagine this as a principle.

Principle 7 (Separation of management and work structures) *A higher-level management framework is required to bind low-level operations together, and guide them towards the larger goal of policy, both in human and machine parts of any system. The management framework is formally separable from the low-level operations.*

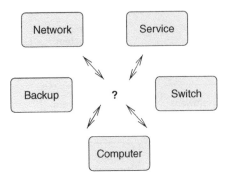

Figure 13.3: In an object design, one bases the structure around the objects that need to be visible at the functional level of the system. Any dependencies that belong only to the object are concealed within the object concerned.

Each object has a private notion of state, and its own internal methods for changing that state. The object can communicate with the outside world, through authorized channels, but is otherwise closed. From a security or privacy perspective, the object model is based on the Clark–Wilson security model, or role-based security.

Object strategy is conventionally mixed together with the idea of classification also. Classification has to do with object *types* or *name conventions*. For example, if one defines an object that is a bank, then there might be several instances of a bank system in use, at different locations. If all of those instances share the same design, then they are then all said to be members of a *class* of banking systems.

In a rough sense, a pure, functional decomposition is rather like the dreams of the communist states in which every part of the system is steered from above by the orders propagating down the hierarchy, and in which every part contributes in its own unique way to the larger goal of executing the function of the system. A pure object model, on the other hand, is somewhat like a system of decentralized control, in which different objects work independently, but cooperate in order to achieve larger goals. These are two different strategies for building systems. The relative efficiencies of the two strategies are hotly disputed, but there are no unilateral conclusions. At best, we can say that the ability to control requires the system to be well behaved, or sufficiently stable. Then, 'control' implies change by external influence.

Principle 8 (Control) *Objects must either compete freely with one another, or be guided by a superstructure in order to be controlled. This does not require hierarchy, only persuasive interaction.*

The advantage of functional decomposition is that it has a structure that directly reflects the task it is designed to perform. It is therefore easy to understand in a causal sense. The advantage of the object model is a more rigorous separation of logically independent tasks and resources, which makes the reuse of those separate items in other contexts easy. The lack of an obvious structure of algorithmic flow makes understanding the causality of an object system harder, however. Current trends tend to favour the object model because, if nothing else, it provides a tidy way of housekeeping the parts of the system and their dependencies.

In practice, it is rare to find either of these strategies in a pure form. Both strategies need each other in order to succeed: an object model has no implied algorithmic flow, so it needs a functional model alongside in order to guide the flow of production. Similarly, any function within a functional design is a rudimentary object, and benefits from a disciplined logical decomposition. Thus, one must view these two philosophies as being complementary.

Client model and peer model

Another way of looking at the centralized versus decentralized debate is through the paradigm of *services*. In a functional decomposition, one could say that each function requests services of its subordinate functions. Similarly, in an object decomposition, each object contains internal services (methods), but can also opt to send messages to other objects in order to request information, by delegation: this request is a service performed by one object for the other.

Example 121 *A building contractor might hire a subcontractor to perform part of a job, in which he lacks skills. The subcontractor does not work 'under' the main contractor in actuality; he works alongside it. Thus, while the conceptual flow was initiated by one contractor, there is nothing intrinsically subordinate about the subcontractor.*

Rather than focusing on issues that are subordinate to others, in the manner of a hierarchy, one can also view a system from the perspective of its actors. This is particularly appropriate to systems that are already distributed in space, either by geographical considerations, or by network.

- *Client-service model*: In this viewpoint, a system is broken up in an orderly fashion into a service-providing part and a service-requiring part. This is the traditional model of shop and customer.

- *Peer model*: This viewpoint is a more anarchistic viewpoint of the actors, as skilled individuals. A peer model has no 'political' centre. Everyone can do something for everyone else. No one is intrinsically a client or a service provider; anyone can play any role. In this way, one hopes to maximize access to individual resources, without shielding those resources from view, by layers of organization. This model lacks a clear organization and is vulnerable to loss of focus. On the other hand, it is highly robust to total failure.

These models tend to apply in different contexts.

Example 122 *Computers that are 'politically independent', that is, that have independent policies can interact but often wish to protect themselves. There is no automatic trust. A client-server type of service request opens up one party (the server) to a situation of vulnerability in which clients might actually attack. In (Burgess et al. (2003c)), the idea of voluntary cooperation is used, in which each individual device in a network collaborates by prior agreement, and then only from behind a 'hands-off' firewall system, like the one-way drawers that are used to pass money over the counter in banks and post offices.*

Open source software development is an example of a peer model, for instance. A peer model also applies in the larger landscape of commerce, for instance, where many companies and individuals require each other's services, and they are all equally important to the running of the 'total system'. This is the way that the society works, as a whole. The peer model is also very much an object viewpoint: one defines objects on the basis of their particular attributes without attempting to evaluate which objects are subordinate to others (at least to a first approximation). All objects are equally important, but there is cooperation between them. The peer model can be very efficient, in local groups, but it presents new problems in scaling to large groups.

The client-service model is a more focused strategy for a single part of a larger process. It is more restricted, but more orderly. It is more susceptible to bottlenecks and efficiency problems, because one focuses the execution of all work at the service provider. Many clients to a single service provider can easily overwhelm a poorly designed service provider.

13.6   Event-driven systems and functional systems

Systems can be characterized by whether they are designed only to respond to the requests of clients (i.e. they are services), or whether they have a pre-programmed procedure that evaluates some function (they are stand-alone programs).

- A system that exists only to service others, needs to be able to schedule work flow to cope with demands at the behest of others.

- A system that produces, manufactures or evaluates something is freer to organize its time and resources according to its own scheduling plan.

In large organizations and computer systems, there is often a mixture of these two types of system and the resource requirements for the two processes can conflict.

Example 123 *A help desk that is constantly being interrupted by telephone calls cannot produce information, courses or solve real problems that require more functional work.*

Example 124 *A computer that is acting as a database server and as a numerical analysis engine must share resources between these tasks. The server requires intensive access to disk and interrupt driven resources, whereas the numerical calculation requires constant periods of CPU concentration to advance. These two requirements are not strongly compatible.*

There are the following two ways to cope with this kind of conflict:

- Separate the two parts of the system into independent units;

- Interleave the scheduling of resources in a single unit.

The latter strategy is more difficult for humans to accept than for computers, as humans are slower to switch context than machines. The advantage of the second strategy for human organization, if it is performed as a long-term job rotation, is a broader experience. Humans are, after all, learning systems and can benefit from a broader insight into a system's operation. However, the timescale of the interleaving must be appropriate for humans.

13.7   The organization of human resources

Organization includes many issues as follows:

- Geographical organization

- Psychological organization

- Process decomposition: How the organization is split up into different activities (whether these overlap, leading to consistency and integrity issues).

The geographical organization of human–computer systems has become less important with the arrival of information networks and telephone systems, but still there are issues where geography can play a role, such as the need for face-to-face communication. To analyse resource organization, we need a value system or currency of organizational structures.

1. Efficiency (cost/speed)

2. Convenience

3. Comprehensibility.

This is a task for observational verification over time.

Humans are known to be good at decision-making and creative thought, but poor at discipline and consistency. It seems sensible to assign humans to creative work and machines to repetitive, precision work[1]. The role of human beings in systems, be they human–computer systems or other man-machine liaisons, has been studied in a variety of frameworks, using fuzzy logic, symbolic interpretation and other cognitive hooks. The extensive literature can be navigated starting from a few bases (see Endsley (1995); McRuer (1980); Rasmussen (1983); Rouse (1989); Sheridan (1996), for starters).

Socio-anthropological research suggests that these two human faculties require different organizations. Decisions are made quickly by individuals, but individuals do not always have impartial interests or complete information, thus one generally involves several individuals in decision-making. Peer review of decisions is performed by committees. For effective committee work, where decisions need to be made quickly, a group size of no more than six is found to be a limit, beyond which ordered decision becomes chaos. Conversely, for brainstorming and creative thought, a larger group size is an advantage (Dunbar (1996)).

Research into humans' abilities to collaborate is based on the brain size to group-size hypothesis. Human social group sizes are observed to be limited to 150 people in an organized society; this is about the number of friends and acquaintances that we can relate to. For animals with smaller neo-cortices, the number is smaller. This suggests to anthropologists the hypothesis that organizations that grow beyond about a hundred individuals are likely to become unmanageable.

---

[1] It has been suggested to the author, informally, that skill-based management is a waste of time for companies. In terms of economics, one has more to lose today from the consequences of expensive licence agreements and contract clauses than from learning particular technical skills. The losses due to incompetence are negligible compared to the cost of expensive licence agreements for a company's software base. Readers may calculate for themselves whether this might be true in their own enterprise.

Another datum is of interest here. Hunter-gathering tribes of humans have evolved to work in groups of more than about 30 before they become unwieldy and break apart, even though they might regroup in a social context to numbers of up to a hundred and fifty, or so it is thought that busy groups of workers are limited by this order of magnitude (Dunbar (1996)). In other words, when we are preoccupied with work, we have less aptitude for dealing with other persons. Thus, if we take our evolutionary heritage seriously, it makes sense to pay attention to any hints provided by our genetic heritage, and perhaps limit the sizes of working groups and enterprises with these numbers in mind.

13.8   Principle of minimal dependency

The minimization of dependencies is a principle that attempts to reduce the possibility for failure and inefficiency. In any system, there is a conflict of interest between tidiness of structure and efficiency of operation.

Example 125 *A traditional hierarchical structure, stemming from a military past leads to great orderliness, but many resources are lost in passing down the many layers.*

Dependency carries with it a *functional inertia* or *resistance* that makes systems inefficient. It also introduces points of failure, since it increases the number of parts of a system that need to be connected.

Definition 55 (Point of failure) *In a graph (X, Γ), any node or edge of the graph is a potential point of failure.*

The principle of minimal dependency sometimes conflicts with the need for system normalization (see chapter 14). The balance between order and efficiency is a *game*, in the Game Theoretical sense (see chapter 19).

13.9   Decision-making within a system

Systems have to make decisions in many contexts. Decisions affect changes of process, strategy or even policy, based on information that is available to a part of the system. The question of where decisions should be made is a controversial one, both in human-centric and computer-centric systems. The decision requires both access to the information and a knowledge of the goals, that is, the policy of the system.

13.9.1   Layered systems: Managers and workers

Traditionally, system decision-making has been a task assigned to 'managers' or 'commanding officers'. This stems from two reasons: first, autonomous machine decision has not always been technologically possible, so it has fallen to humans to guide every detail as a matter of history; second, is the prejudice that managers or commanders have superior intelligence compared to lower-level workers or components[2]. Consider these strategies:

---

[2] This latter assumption has its roots in history, where the upper classes placed themselves at the helm, and were usually better educated than the working classes who carried out their orders. In our present technological

- Move decision-making as close as possible to the part of the system that is affected by the decision. Input from higher levels is applied by globally available policy constraints, to avoid the need for directed message passing, since message passing would consume resources.

- High-level decisions act as constraints on low-level decision-making. Authority to act is delegated as far as possible to the part of the system that is affected.

- By definition, high-level processes depend on low-level ones. Thus, high-level processes need to search for strategies that are implementable using the low-level components.

Too much policy constraint and monitoring from higher levels stifles the freedom of the lower levels to perform their function. Too much communication back and forth leads to inefficiency.

Example 126 *A symphony orchestra is a system executed by humans, following mechanical instructions, with a human interpretation. Here, the players are technically specialized in their individual instruments, whereas the conductor's job is to look at the broader picture. The conductor does not have detailed skills in the individual instruments, nor does he have the resources to follow every detail of each instrument's part, but he has a unique perspective on the broader picture, which is not available to the individual players, because they are shielded from the full sound by their neighbours. The conductor's role is therefore to support the individual players, and orchestrate their collaboration. Decisions about how to play the instruments and interpret the music are made by the players, with high-level hints from the leadership.*

The next example takes a more questionable approach to its task:

Example 127 *A security monitoring system is usually a collection of cameras and alarms, linked to a control centre, which is monitored by a human. The human can respond to alarms and data from the camera, by locking doors, and by going out to investigate. This reliance on a human manager introduces a communication channel and a dependency for each alarm. If one of these fails, the system could be delayed or fail to function.*

13.9.2   Efficiency

Decision-making is an example of dependency management. Systems are dependent on decisions being made, and thus the decision-making process is an area that needs to be crafted carefully.

From the viewpoint of efficiency, a brief inspection of the figs. 6.15 or 6.17 is clear that autonomous decision-making, in the individual parts, is optimal because it avoids communication and thus introduces no further dependency or delay. If information has to be passed up or down through the system in order to be analysed, then significant overhead can

---

society, the association between education and position has been dismembered and replaced by an association of personalities and interests with position. This tends to invalidate the assumption that 'high level' means more qualified to decide on every issue. What emerges, in a skill-based system, is a layered approach to decision-making.

be incurred. In human-centric systems, however, it is a matter of policy that decisions are made by 'management'. Management can have a bird's-eye view that lower-level agents cannot.

Many companies in the technology sector have reorganized themselves from a hierarchical model to an object model because object models are built around specialization, rather than control (see Tapscott and McQueen (1995)).

Specialized knowledge about the system is found within specialized components at a low level, but the command decisions are usually made from the top. How does the information get from top to bottom to top again? That is, how does decision-making circulate around the system?

A global decision is the most resource-consuming kind of decision, usually only required in cases of global policy review.

1. Data collection from all levels

2. Data interpretation by all levels

3. Re-examination of goals

4. Strategy options

5. Policy adjustment, system changes.

13.10   Prediction, verification and their limitations

Maximizing predictability is a key aim of system design. Similarly, one would like to verify that a system design is in accordance with expectation and policy. The idea that systems have well-defined states with deterministic behaviour is common in many branches of computer science and has a limited validity. Only simple systems are deterministic. In complex systems, particularly those immersed in an environment, information is being injected into the system from an unpredictable source all the time. This means that absolute determinism is unrealistic. Verifiability is thus not only useful when designing a system but also during its operation, in order to evaluate its behaviour.

There are several levels of verification. To verify correctness in a system that generates non-ordered behaviour, one can use a checksum or hash function to map results to a single scalar value that can easily be compared.

Example 128 *MD5 and SHA checksums are used to verify the correctness data transmitted through routed network systems, compared to original control-values, after being reformatted and encapsulated many times to ensure correct transmission.*

To compare more structured operations, in which dependent ordering is involved, we use the notion of language and grammar.

Example 129 *OSI model network packets that have multi-layer encapsulation have a simple grammatical structure. Each layer of the TCP/IP system has a header, followed by a payload that includes the previous layer. The IEEE 802 data-link layer protocols, such as Ethernet, have both a header and a trailer. These encapsulations have a simple grammatical structure that is verified by the unpacking process.*

Example 130 *The configuration rules of a routing policy protocol, such as BGP, form a structured language of relationships with dependencies. This forms a recursively enumerable graph (for instance, see Griffin and Wilfong (2002); Qie and Narain (2003)).*

In chapter 5, we reviewed grammars as a way of describing structural complexity in strings of operations. If the syntax of a system, that is, the list of all legal operation strings, is described by a known grammar, then its correctness can easily be verified by an automated procedure or automaton that attempts to parse it. Grammar works as a sophisticated operational checklist for system correctness. Errors in system functioning can easily be identified by comparing the actual behaviour of the system to the legal strings of the language. Correctness can then be evaluated as true or false.

This is the idea behind software engineering tools such as the Unified Modelling Language (UML), which attempts to apply the methods of algorithmic rigour to the complexities of human–computer system interaction[3]. The problem with grammatical structure is that it is only a guide to structure in many systems. The rule-based part of a system is often at a low level. The part that resists such formalization is the high-level behaviour. An obvious example of this is in biology; at a low level, simple rules of chemistry tell us how molecules fit together, but as we put molecules together to form cells, and cells together to form tissue, and tissue to form organisms, the idea of simple structural rules becomes absurd.

13.11 Graphical methods

Graph theory is a mode of structural expression that is closely related to grammatical methods. It can be used to describe and discuss various aspects of architectural and structural efficiency. For example:

- Connectivity and robustness

- Optimal constructions

    - Minimum weight spanning trees
    - Maximum weight branchings

- Shortest path problems.

These topics find a natural language of expression in graph theory, and there are plenty of theorems and results that can be used to assist in the design and understanding of systems. For an excellent introduction to the graph theory of these issues see (Balakrishnan (1997)).

Applications and Further Study 13

- *Combining the stuff of the previous chapters into a plan for connecting components.*

- *Collaboration and partitioning.*

- *Understanding hierarchy versus flat weblike structures.*

---

[3] In the opinion of the author, UML has been rather unsuccessful in this task. Although it offers a formal framework, it provides no way of dealing with the scales of complexity that are characteristic of actual systems of substance

14

System normalization

Normalization of a system is a process by which one verifies the implementation of its goals and possibly corrects its organization with respect to guiding principles (see Date (1999)). Normalization principles are meant to avoid certain problems of design and operation in systems and to simplify the semantics of system operation as far as possible. This has consequences for maintenance and long-term consistency. For instance, systems should not contain elements that actively oppose one another in the performance of a task; such a system would be wasteful of resources, and possibly even harmful.

Data structures form the foundations of any system. The purpose of a data structure is to catalogue and organize variables, that is, the changing configurations and records that form the substance of a system. These are then arranged in a pattern that can be navigated in a manner most conducive to the efficient functioning of the system.

In any system, the law of causality applies. For every effect, there must be one or more causes. Determining the causes of an effect becomes increasingly difficult with the increasing complexity of the system. One says that the effect is *dependent* on the cause.

14.1 Dependency

In a dynamical process, it is not only the information that has relational patterns but also the functional components, or 'subroutines' of the total process. In either case, guidelines exist for breaking up larger data-structures and processes into smaller, optimal types of parts, the purpose of which is to isolate 'repeatedly used information' for optimal reuse and maintenance. In this way, one avoids unnecessary repetition that can lead to inconsistency and dependency conflicts. Seeking the optimal decomposition into layers is the process referred to as *normalization*, and is commonly discussed in relational databases (see for instance Date (1999)).

This book is an example of a passive data structure that is not fully normalized. Complete normalization would lead to an obnoxious use of sub-sections and references of the form 'see section XX' to avoid repetition and multiple dependency. A fully normalized data

Analytical Network and System Administration. Managing Human–Computer Networks   Mark Burgess
© 2004 John Wiley & Sons, Ltd   ISBN 0-470-86100-2

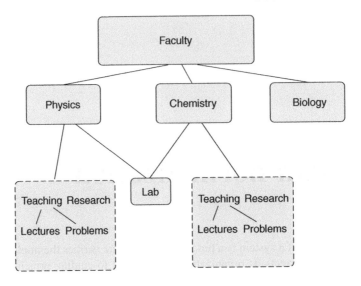

Figure 14.1: An excerpt of a functional structure diagram for a university faculty. This shows that within the different departmental groups, there is a common body of functionality. This repetition suggests that one might rationalize the system by separating this aspect of the groups from the departmental groups' area of responsibility. This process of rationalizing a structure diagram is called *normalization*.

structure does not make for easy reading, but it does allow a great precision and economy of representation, with potentially no inconsistency (see fig. 14.1).

A functional dependency is expressed by the mapping of several parameters k_i, (usually called *keys* in database theory), into a single object O of some information space:

$$O = f(k_1, k_2, \ldots, k_n), \tag{14.1}$$

that is, we have a relationship between the result f and the combination of keys or parameters that produce that result. We say that f depends on $k_1, k_2, \ldots$, since each combination can yield a different result for f. The function f, whatever it is, defines a *relation*. For example, see table 14.1.

In continuous functions, parameters can vary along simple lines of numbers with an infinite variety, for example, a position or a time (see chapter 7). In a discrete structure, such as a database, the parameters are discrete lists, or sets (see chapter 5).

Table 14.1: Comparison of numerical and associative values, through a functional dependency.

Numerical	Abstract (associative) keys
$y = f(k_1, k_2, k_3)$	lecture $=$ f (teacher, time, subject)
$12 = f(14, 3, 6)$	addition $=$ f (Mark, Tuesday, Algebra)

14.2   The database model

A relational database (or diagram) is a functional mapping from a set of parameters (lookup keys) to data objects (usually tables or vectors) of the form

$$\text{Table} = \vec{D}(k). \tag{14.2}$$

The vector arrow denotes the fact that the value returned by the association is assumed to be a table or a vector in the general case, that is, a bundle of values that are related in some way.

Example 131 *Many systems have the form of a rudimentary relational database. For example, a building can be thought of as a set of rooms associated with room numbers. A table of information for each room number (key) might consist of an inventory of the room, who sits there and so on. A database of computers, with a primary key that is the serial number of the computer, might contain tables of information describing the location, operating system type, name, address and so on. In this latter example, it might make sense to have a sub-database (a set of tables within a set of tables), documenting information about each type of operating system. The key would be the operating system name, and the sub-table might contain the version number, the manufacturer, their service telephone number and so on (see fig. 14.2).*

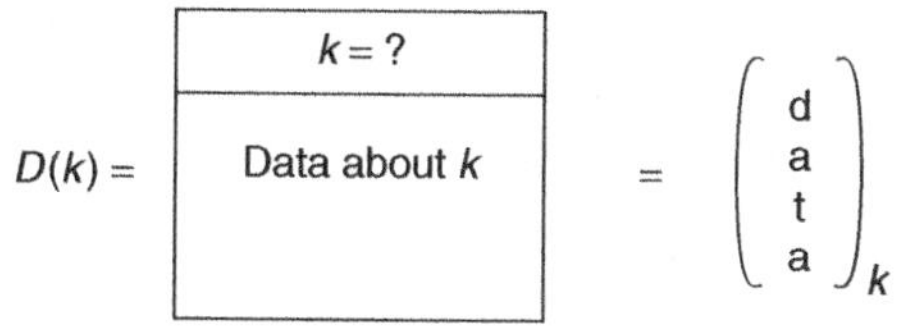

Figure 14.2: Notations for a functional mapping from k to a table, represented as a table with a key and as a vector with subscript.

If we look at this from the viewpoint of a database, then we envisage the system as a set of entities that are related to one another by functional dependencies. To decompose a system, we ask:

- What are the *entities*?

  These include computers, services, humans, departments, job positions, and so on.

- What are the *primary keys* that label and classify data in system administration?

  These include things like host names, group names, project names, departments, and so on.

- What are the *data* and other attributes?

  These are specific instances of computers, disks, data, persons and other resources, and so on.

- Is the system database structure *hierarchical*?

  We often impose a hierarchical structure on systems, out of fear for loss of control. It is a common myth that a hierarchy is the only reliable model of control.

14.3   Normalized forms

Database normalization seeks to ensure consistency of data, no hidden dependencies, and the avoidance of internal conflict. These are desirable properties for any system, and we shall see how the guidelines can be used to learn something about the organization of general systems, by mapping the general systems onto relational databases.

A database is a set of tables (also called vectors), organized into a list, so that each distinct element is labelled by a unique key or a combination of keys. The keys are the 'coordinates' of the objects in the database. For example, a single key database is simply a list of tables:

$$\text{Database} = \{T(k)\} = \{T(1), T(2), \dots T(n)\} \tag{14.3}$$

$$= \left\{ \begin{pmatrix} \vdots \end{pmatrix}_1, \begin{pmatrix} \vdots \end{pmatrix}_2, \dots \begin{pmatrix} \vdots \end{pmatrix}_n \right\}. \tag{14.4}$$

The point of the normal form is to extract structurally similar components, and repeating patterns, and place them in separate abstractions with their own labels. This means that the association is one-to-one, and that data or entities are not repeated unnecessarily. There are both practical and aesthetic reasons for this. A practical reason is that one should not duplicate information or effort in competing locations, since this would lead to contention, competition and thus inconsistency. The aesthetic reason is the same as the one used in programming: repeatedly useful subroutines or data are best separated and called up by reference to a new abstract entity, rather than copying the same code or information in different locations. This makes maintenance easier and it makes explicit the logical structure of the task.

First normal form

The first normal form is really a definition about the type of objects one chooses to call a database. It restricts tables to being simple vectors of a fixed size and shape. The purpose of this definition is to ensure that all of the objects in the database are comparable; that is, in order to be able to compare any two tables in a database, one must be able to compare their contents meaningfully. This is only possible if the tables are constructed in a similar fashion.

For example, the simplest case that each vector or table instance consists only of scalar values is

$$T(k) = \begin{pmatrix} s_1(k) \\ s_2(k) \\ s_3(k) \end{pmatrix} = \begin{pmatrix} \text{Algebra} \\ \text{Tuesday} \\ \text{Mark} \end{pmatrix}_k \tag{14.5}$$

Sub-vectors (sub-tables) are allowed as long as they have a fixed, predictable size; for instance, the following is acceptable because it can easily be rewritten as a larger vector of scalars:

$$T(k) = \begin{pmatrix} s(k) \\ v(k)[6] \\ v(k)[2] \end{pmatrix}. \tag{14.6}$$

However, one could not have the following objects in the same database, because they are not comparable objects:

$$\begin{pmatrix} s(k) \\ v(k)[6] \\ v(k)[2] \end{pmatrix} \neq \begin{pmatrix} s(k) \\ v(k)[5] \\ v(k)[9] \end{pmatrix} \tag{14.7}$$

The comparison of these objects is meaningless.

What does this have to do with systems in general? We understand that consistency of form is important for stacking data in rows, or for stacking boxes in a warehouse, but what does this mean for system administration? The main thing it tells us is that objects that have a similar structure can be handled by the same system, but objects that are structurally dissimilar should be dealt with by separate systems.

Example 132 *In spite of the fact that computers are produced by very different manufacturers and use a variety of software, the information about computers within an enterprise is structurally similar: each computer has a serial number, a name, a location and an operating system. This means that a system that deals with these aspects of computers can be handled by a single system. If we also go deeper and consider the details of the different software (operating system, for example), we find that Windows and Unix are structurally dissimilar in a number of ways. The normalization rule tells us that it is therefore unnatural to try to combine these into a single system (see fig 14.3).*

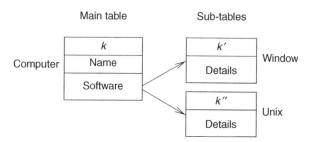

Figure 14.3: How normalization by the first normal form suggests that a system should be organized. The common parts belong to a single system of tables, while the structurally dissimilar items are separated into independent sub-systems.

Second normal form

The second normal form says that, given any 1NF association $\vec{T}(k)$, which has a common repeating pattern, one should extract that pattern and parameterize it with a new label (or key). This is the introduction of 'database subroutines' in the data structure. For example, suppose one has the set of tables as follows:

$$\left\{\vec{T}(k)\right\} = \left\{ \left(\begin{pmatrix} s_1 \\ 1 \\ 1 \end{pmatrix} \right), \left(\begin{pmatrix} s_2 \\ 1 \\ 1 \end{pmatrix} \right), \left(\begin{pmatrix} s_3 \\ 1 \\ 1 \end{pmatrix} \right), \cdots \right.$$

$$\left. \left(\begin{pmatrix} s_4 \\ 1 \\ 0 \end{pmatrix} \right), \left(\begin{pmatrix} s_5 \\ 1 \\ 0 \end{pmatrix} \right), \left(\begin{pmatrix} s_6 \\ 1 \\ 0 \end{pmatrix} \right), \cdots \right\}. \tag{14.8}$$

Here we see a repeating pattern. The vectors

$$\begin{pmatrix} 1 \\ 1 \\ 1 \end{pmatrix}, \begin{pmatrix} 1 \\ 0 \end{pmatrix} \tag{14.9}$$

are common to several of the elements. In geometry, this is called an invariant sub-space. The second normal form demands that we recognize the importance of this structure, and parameterize it as a vector $v[k_s]$ with a new sub-key k_s:

$$\vec{v}[1] = \begin{pmatrix} 1 \\ 1 \end{pmatrix}, \vec{v}[2] = \begin{pmatrix} 1 \\ 0 \end{pmatrix}. \tag{14.10}$$

The normal form thus transforms elements

$$\vec{T}(k) \longrightarrow \left\{ \vec{T}(k; \vec{k}_s), \vec{v}(k_s) \right\}$$

$$\begin{pmatrix} s_k \\ s'_k \\ s''_k \end{pmatrix} \rightarrow \left\{ \begin{pmatrix} s_k \\ \vec{v}[k_s] \end{pmatrix}, \begin{pmatrix} v_{k_s} \\ v'_{k_s} \end{pmatrix} \right\}. \tag{14.11}$$

In the example above, we note that the pattern is simple, and k_s is varying a third as fast as k. This means that the keys are simply related. One could therefore argue that there is really only one key. We could, in principle, repeat this procedure again for sub-tables in sub-tables, and so on, until all of the parameterizations are separated. This is just like the problem of recursively breaking up functions into sub-functions in programming.

The implication of this rule for general systems is that wherever we see repeated substructures in formally separate systems that are associated with one another, these should be removed from the separate systems and be replaced by a single, independent instance that can serve both the systems instead.

Example 133 *The science, engineering and arts faculties of a university all have separate student registration, printing and accounting departments. The task of registering students is the same in all three faculties; the same is true of the financial accounting and printing services. The second normal form therefore suggests that these functions should be removed from the three faculties and be replaced by three independent services: a common registration department, an accounting department and a printer.*

Example 134 *The science and engineering faculties of the same university all have laboratory engineers who maintain and manage the laboratories. Although these engineers have analogous functions in their respective faculties, their tasks are quite dissimilar, so the normalization rule does not apply here. It would not make sense to group dissimilar laboratory engineers into a common service, because they do not have enough in common.*

Third normal form

The third normal form is about parameterizing a structure in a non-redundant way. In geometry, it helps to avoid ambiguity and inefficiency when traversing vector spaces by marking out labels using *linearly independent* or *orthogonal* coordinates. Similarly, the third normal form is about securing this kind of independence in discrete, tabular structures.

Interdependence of the elements within a vector can occur in two ways. The first is a simple linear dependence, in which two of the scalar values contain common information, or depend on a common value. The second is to avoid making convoluted structures that can feed information back through a chain of relationships that lead to a cyclic dependency. Consider the table

$$\vec{T}(k) = \begin{pmatrix} g_1(k) \\ g_2(k) \\ g_3(k) \end{pmatrix}. \tag{14.12}$$

The third normal form seeks to avoid *transitive dependence*, that is, a relation of two non-key elements through a third party.

$$g_1(k) = g_2(g_3(k)) \tag{14.13}$$

Here, two elements g_1 and g_2 are not related directly—they do not contain common information directly, but rather they are both functions, derived from a common item of data, thus they are related because one of them depends recursively. This is a non-linear relation.

Another way of putting this is that one attempts to eliminate items of data in any type of object that do not depend on the key directly and explicitly. The presence of a 'behind-the-scenes' relationship represents an unnecessary symmetry, or a hidden/covert channel from one part of the system to another.

There are other normal forms for even more complex structures, and these can be applied in kind to the simplification of a relational system, but the main points are covered in these three cases.

Example 135 *In a combined Windows and Unix environment, user data are collected both in the SAM database and in the Unix files* /etc/passwd. *The same data are registered on each and every machine. The user registration data thus form a distributed database, and consist of two different types of data records (tables), containing similar information. One way to normalize this database would be to use an LDAP directory service for both.*

Applications and Further Study 14

- *Rationalizing the entities and players in a system for planning and design. Rational decision criteria for segregating processes and information conduits.*

15

System integrity

Integrity is about the preservation of a system's policy, its resources and its ability to function consistently. System integrity can be couched in terms of the communication of those assets from place to place, or from time to time, using information theory; however, transfer of information is not always reliable, nor is the information transmitted always understood by the receiver.

The theory of communication, and its generalization 'information theory', seek to answer questions such as how rapidly or reliably the information from the source can be transmitted over a channel. Insofar as we can define the code by which the work in a human–computer system can be transmitted, we can use the tools of information theory to describe the possibility of flow or corruption of those assets. Shannon's work is significant because it addresses fundamental limits on the ability to resolve and preserve information. He addressed the problem of coding of symbolic information in such a way as to maximize the rate and the fidelity of transmission, given that the representation might change. He also achieved considerable success with his technique of random coding, in which he showed that a random, rather than a sequential, encoding of data can, with high probability, give essentially optimal performance.

15.1 System administration as communication?

The process of communication is essential in any information system. System administration is no different; we see essential bi-directional communications taking place in a variety of forms as follows:

- Between computer programs and their data

- Between computers and devices

- Between collaborating humans (in teams)

- Between clients and servers

Analytical Network and System Administration. Managing Human–Computer Networks   Mark Burgess
© 2004 John Wiley & Sons, Ltd   ISBN 0-470-86100-2

- Between computer users and computer systems

- Between policy decision makers and policy enforcers

- Between computers and the environment (spilled coffee).

The intent of these communications is constantly being intruded upon by an environmental noise. Errors in this communication process occur in two ways:

- Information is distorted, and symbols are changed, inserted or omitted, by faulty communication or by external interference.

- Information is interpreted incorrectly; symbols are incorrectly identified, owing to imprecision or external interference.

For example, suppose one begins with the simplest case of a stand-alone computer, with no users, executing a program in isolation. The computer is not communicating with any external agents, but *internally* there is a fetch-execute cycle, causing data to be read from and written to the memory, with a CPU performing manipulations along the way. The transmission of data, to and from the memory, is subject to errors that are caused by electrical spikes, cosmic rays, thermal noise and all kinds of other effects. These errors are normally corrected by error-correction mechanisms, originating from Shannon's work. The computer program itself manipulates the data in the memory and rewrites it to the memory with a new coding.

From this point of view, one may think of the memory of the computer itself as being both a transmitter and a receiver, and passing through the 'CPU plus computer program' communication channel. This communication channel does not transmit the data unaltered; rather, it transforms the data according to specific rules laid down in the computer program (see fig. 15.1). In other words, the very operation of a computer fits the paradigm of communication over a coded channel. The fact that the channel is also noisy is a result of the physical environment. Computer operation, at this level, is largely *immune* to environmental noise, because it employs error-correction methods. At higher levels, there are no standardized error-correction mechanisms in common usage.

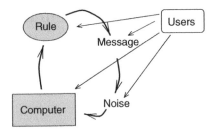

Figure 15.1: An instruction loop, showing the development of a computer system in time, according to a set of rules. The efficacy of the rules may be distorted by users from local and remote domains, who change the conditions under which the message was applicable. This change may be viewed as an intentional change, or as a stochastic error.

Suppose now that an administrator sends a configuration message to a host, or even to a single computer program. Such a message takes place by some agreed form of coding: a protocol of some kind, for example, a user interface, or a message format. Such a configuration message might be distorted by errors in communication, by software errors, by random typing errors. The system itself might change during the implementation of the instructions due to the actions of unknown parties working covertly. These are all issues that contribute uncertainty into the configuration process and, unless corrected, lead to a 'sickness' of the system, that is, a deviation from its intended function.

The idea of *convergence* is introduced in section 10.4, to describe the behaviour of such error-correction. It suggests a process of continual regulation in order to correct deviations from system policy. In other words, it does not simply ensure that a configuration message is transmitted correctly once, it also views the entire time development of the system as the ongoing transmission of the message and seeks to correct it at every stage. This is also our definition of a reliable system (see section 4.11).

At each level of computer operation, one finds messages being communicated between different entities. System administration is a meta program executed by a mixture of humans and machines that concerns the evolution and maintenance of distributed computer systems. It involves the following:

- Configuring systems within policy guidelines

- Keeping machines running within policy guidelines

- Keeping user activity within policy guidelines.

System administration requires computer–computer interaction, human–computer interaction, and human–human interaction (fig. 15.2). Each of these communication channels is subject to error, or misinterpretation.

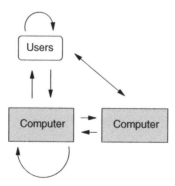

Figure 15.2: The human–computer interaction is also a form of communication. Collaboration between system administrators or users is the least reliable form of communication, since one cannot be sure that the two parties even use the same symbolic alphabets. Thus, there is the danger not only of noise but also of misunderstanding, that is, perversion of meaning.

- *Instruction*: System administration is about developing a policy (instruction manual for machine and humans) for use of the system, including the choice of programs that are executed by the system. A complete policy can therefore be identified with the sum of all programs and *ad hoc* rules that lead to any planned change of the system. This includes normal usage of the system.

- *Propagation*: As the system evolves in time, according to policy (or deviating), it propagates information from the past into the future, modifying the system. If the system is stable, this iterative mapping will not lead to any divergent behaviour; if it is unstable, then even a small error might cause a runaway breakdown of the system.

- *Collaboration*: Programs and humans exchange information in the course of this loop of instruction and propagation. If collaboration is interrupted, or errors occur, then the enactment of policy-correct behaviour is spoiled, with possibly dangerous consequences for the system. Humans also frequently misunderstand each another's commands.

- *Automation*: Automatic processes that monitor systems, even perform routine maintenance, are not immune to errors, because they depend, for input, on data that are influenced by the external environment.

- *Repair*: If some influence causes an error, then the error must be corrected in order to uphold policy, else the correct propagation of policy over time is corrupted. The same thing applies to human policy and purely automated policy. This process of maintenance, reparation or regulation is central to the stability of information systems.

To summarize, one may view system administration as communication over a communications channel at several levels (see fig. 15.3).

- *Input*: Policy, instruction.

- *Noise*: Stochastic user activity, illegal behaviour, random error, systematic error.

- *Output*: The system as we see it.

Having made this identification, the question is does this help us build computer systems that behave in a stable and predictable fashion.

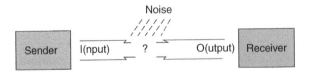

Figure 15.3: The transmission of information over a channel takes place from a sender (channel input) to a receiver (channel output). Along the way, the signal might become distorted by noise.

15.2 Extensive or strategic instruction

The transmission of configuration information requires a language which, in turn, requires an abstract *alphabet* or a set of codes to encode it. This 'alphabet' might be one of strings or of shortest length symbols. The information content will tell us how compressible the actual transmitted configuration is and therefore how concisely we can express system policy.

Each word of symbol in the language must represent a basic operation of the system like 'create file' or 'insert string' and so on. These operations can be represented as single-letter codes, with accompanying data (like opcodes in machine language), or they could be packed into lengthy XML wrappers to name two extremes.

Example 136 *Suppose the operation to change the permissions of a Unix file (chmod) is coded in a data stream by the letter 'A', and the operation to change the owner of a file (chown) is 'B'. The operations need both data parameters and operands. A configuration policy can be written in any language that the system understands, for example,*

Human symbol	Compressed code
chmod	A
chown	B
700	a
770	b
755	c
644	d
600	e
555	f

So, if we number files according to their file system entries (e.g. index node number), the command to set permissions on file 12 would be Ad(12) or A(12)d, where (12) represents a suitable shortest representation of a number 12. This set of symbols will suffice for a limited number of operations.

The amount of information that has to be specified in order to express a configuration depends on the expert knowledge of the system receiving the instruction. There are two approaches to this as follows:

- If we build *expert operators* that know how to do their jobs and only require a simple signal to activate them, then configuration policy can be written in a very short and compact form, for example, as in the single-letter codes above. This reflects the fact that the detailed procedures are coded into operators and therefore do not need to be reiterated in every configuration message. (This strategy allows maximal compression and optimal normalization of information, since there is only one copy of the expertise in the operators. Thus, a single symbol can represent an expert operation.)

- If the operations contain no internal expertise, then each precise sequence of primitive operations must be expressed in the policy message. This, in principle, involves a precise specification with redundant information.

Example 137 *A compressed message to an expert operator providing maintenance might take the form*

```
CheckAndRepair(routine_1)
```

An extensive form of the instruction could take the form

```
Locate panel screws
Rotate screws anti-clockwise
Remove screws
Lift lid
Locate memory slot 12
Insert new memory into slot 12
```

In the latter form, the detailed procedure is described in the message; in the form case the procedure is coded into the operators themselves, and only a short message needs to be passed on to start the operators executing their policy instruction.

An example of the former is found in (Burgess (1995); Couch and Daniels (2001)) and examples of the latter include the Simple Network Management Protocol (SNMP), Arusha (Holgate and Partain (2001)) and so on.

Example 138 *The human representation symbol alphabet used by cfengine, which describes information in a plain text file, consists of all of the printable ASCII symbols; the set of symbols required to implement a policy decision about file permissions is the set of all rules (one for each item), which for Unix might look like this:*

```
files:

 # symbol 1

 file1 mode=0644 owner=mark group=users action=fix

 # symbol 2

 directory1 mode=a+rX owner=root recurse=true action=fix
```

In this case, each entire rule can be a single symbol of the higher- level policy alphabet, and when it is coded in this fashion, since the number of variations is finite. This short symbolic coding of policy, is robust to accidental or random error, and is easy to reapply (retransmit), should external factors alter its result.

The symbol objects represent new effective entities of the system. They form new and higher alphabets of preferably non-overlapping objects. To perform configuration management, we need to reiterate this configuration message over time, correcting for any random error. We now examine how to characterize the error in a stream of these symbols.

Generative configuration

Conventional wisdom suggests that when an instruction for building configuration state is specified, the order of operations is important to the outcome of the configuration (see arguments in Traugott (2002), for instance). However, an alternative prescription for configuration is based on the idea of expert operators and convergence (see section 5.8). We refer to these two alternative forms of instruction as extensive and strategic respectively, and keep the property of convergence towards an ideal state separate, since it can, in principle, be implemented by either approach.

- *Extensive configuration instruction*

  In the extensive approach, each individual decision in the configuration state is represented as part of a tree with exponential complexity. If the number of symbols is N and the alphabet size is m, then the amount of information that must be maintained is of the order m^N.

- *Strategic configuration instruction*

  In a strategic approach, the decision trees are built into the properties of the operators that carry out the maintenance (see Burgess (2004); Couch and Daniels (2001); Couch and Sun (2003)). Here, the complexity of the configuration is at most of order N^2, and the information represented is no larger than N. The ordering of these operators is not essential, provided the configuration message is repeated over and over again as regular maintenance, since the operators measure their activities relative to both desired policy and current environment. If any operation depends on another having preceded it, it simply waits until the necessary conditions exist for it to proceed. In order for this to work, the operations must be quite primitive.

Ordering of dependencies

There is a connection between the ordering of operations and the uniqueness of the task completed by a schedule (see section 9.8, for instance). The information required to perform a schedule depends on whether task precedence matters (i.e. whether the graph is directed or not). This, in turn, depends on whether the task alphabet commutes or not.

Definition 56 (Commuting operations) *If two system operation codes commute then*

$$[\hat{O}_1, \hat{O}_2] \equiv \hat{O}_1 \hat{O}_2 - \hat{O}_2 \hat{O}_1 = 0. \qquad (15.1)$$

If operators commute, it implies that the order of their execution does not matter to the system. This requires special properties and is seldom true in every instance, since ordering reflects the structure that distinguishes a system from a random assembly of components. However, there is a possible solution to this using the concept of orthogonality.

Contrary to many expectations, most simple configuration tasks can be performed convergently, by randomly or cyclically scheduling non-ordered, commuting operations. This can be done by making operation *orthogonal* (see Burgess (2004); Couch and Sun (2003)).

Definition 57 (Orthogonal operators) *An operation is said to be orthogonal to all other operations if it is inequivalent to any combination of other operations, that is, any representation of the operation is linearly independent of all others:*

$$\hat{O}_i \neq \sum_{j \neq i} c_j \, \hat{O}_j, \qquad (15.2)$$

for some constants c_j. Orthogonal operations are automatically commutative and ordering does not matter.[a]

---

[a] This definition, as given, describes strictly linear independence. Orthogonality implies that an inner product of representative basis vectors would vanish. It is possible to find a representation in which this is the case, but we shall not go into it here. See ref. Burgess (2004) for a discussion of an explicit matrix representation.

Consider how configurations are built up. Let the state $|0\rangle$ denote the base state of a host after installation. This is a reference state to which any host may be returned by reinstallation. From this state, one may build up an arbitrary new state $|a, b, c, \ldots\rangle$ through the action of sequences of the configuration operations. The set a, b, c may be regarded as the system policy specification. Once a desirable state has been reached, one may re-normalize these definitions to allow $|0\rangle$ to be the new base-state. Using this representation, one can now define the meaning of convergence in terms of these operations.

Definition 58 (Convergent and idempotent operators) *Let $|s\rangle$ be an arbitrary state of the system. An operator $\hat{O}$ is idempotent if $\hat{O}^2 = \hat{O}$, that is, its potency to operate is always the same. A convergent operator $\hat{C}_\alpha$ has the more general property*

$$(\hat{C}_\alpha)^n |s\rangle = |0\rangle$$
$$\hat{C}_\alpha |0\rangle = |0\rangle, \qquad (15.3)$$

for some integer $n > 1$, that is, the n-th power of the operation is null-potent, and the base state is a fixed point of the operator.

In other words, a convergent operator has the property that its repeated application will eventually lead to the base state, and no further activity will be registered thereafter.

The use of orthogonal, convergent operations implies that only one type of prerequisite dependency can occur. For example, let $\hat{C}_C$ mean 'create object' and let $\hat{C}_A$ mean 'set object attribute'. The following operations do not commute because the setting of an attribute on an object requires the object to exist. On an arbitrary state $|s\rangle$, we have

$$[\hat{C}_C, \hat{C}_A]|s\rangle \neq 0. \qquad (15.4)$$

Thus, the ordering does indeed matter for the first iteration of the configuration tool. This error will, however, be automatically corrected on the next iteration, owing to the property of convergence. To see that the ordering will be resolved, one simply squares any ordering of the above operations.

Theorem 15.2.1 *The square of a create-modify pair is order independent.*

$$([\hat{C}_C, \hat{C}_A])^2 |s\rangle = 0. \qquad (15.5)$$

This result is true because the square of these two operators will automatically result in one term with the correct ordering. Orderings of the operators in the incorrect order are ignored because of the convergent semantics.

To prove this, suppose that the correct ordering (create then set attribute) leads to the desired state $|0\rangle$:

$$\hat{C}_A \hat{C}_C |s\rangle = |0\rangle; \tag{15.6}$$

performing the incorrect ordering twice yields the following sequence:

$$\hat{C}_C \underbrace{\hat{C}_A \hat{C}_C} \hat{C}_A |s\rangle = |0\rangle. \tag{15.7}$$

The action of $\hat{C}_A$ has no effect, since the object does not exist. The under-brace is the correct sequence, leading to the correct state, and the final $\hat{C}_C$ acting on the final state has no effect, because the system has already converged.

The same property is true of sequences of any length, as shown in ref. Couch and Daniels (2001); in that case, convergence of n operations is assured by a number of iterations less than or equal to n.

Theorem 15.2.2 *A sequence of n self-ordering operations is convergent in n iterations, that is, is of order n^2 in the primitive processes.*

The proof may be found by extending the example above, by induction (see Couch and Daniels (2001)).

The important thing about this construction is its predictability. We might not know the exact path required to bring a system into a policy conformant state; indeed, a given specification might meet obstacles and fail to work. However, any policy expressed entirely in terms of convergent, commuting operators is guaranteed to do something, indeed it will always have the same result[1]. Commutation becomes not only a desirable property but also an essential one in ensuring predictability. An extensive approach is not guaranteed to be implementable or stable once implemented, but a convergent strategic approach is (see Burgess (2004); Couch and Daniels (2001); Couch and Sun (2003) for a proof).

One of Shannon's conclusions was that if one splits up a signal into a coding scheme that is randomized in ordering, then many of this is a good defence against random noise, because any systematic error will be reduced to maximum entropy random error, rather than concentrated in one area of greater damage. The advantage of commuting operations is that they do not have to rely on a particular sequence being fulfilled in order to produce a result. A string of commuting operators can thus often be compressed in communication, because the intelligence lies in the operator rather than in the sequence of codes: a short code for a complex task, rather than a detailing of the internals of the task.

15.3 Stochastic semi-groups and martingales

The discussion of convergent operations has an analogue in the mathematics of semi-groups, or transformations that make transitions in only one parameter direction. If there

---

[1] This does not mean that every possible state is necessarily reachable by convergent, commuting operators, but we are suggesting that any states not reachable by this approach represent inappropriate policies.

are n states $\{q\}$, then a transition $T(t; q, q')$ from a state q to a state q', at time t, is an $n \times n$ matrix that acts on a state vector $\vec{q}(t)$.

Definition 59 (Semi-group of transformations) *The set of transformations* $\hat{T} \in \{T(t; q, q')\}$ *is said to form a semi-group (for $t \geq 0$) if:*

1. *$\hat{T}_0 = I$ is the identity transformation that leaves a state vector invariant: $\hat{T}_0 \vec{q} = \vec{q}$.*

2. *$\hat{T}_t$ satisfies the Chapman—Kolmogorov equation*

$$\hat{T}_{s+t} = \hat{T}_s \hat{T}_t, \quad s, t \geq 0. \tag{15.8}$$

A stochastic semi-group is a group of stochastic (probabilistic) transitions (see Grimmett and Stirzaker (2001)). These groups are clearly of central interest to configuration management. For a discussion 'couched' in these terms see (Couch and Sun (2003)).

Another probabilistically convergent process is a *martingale*. The study of martingales is the study of sequences that converge in the sense of their total value. The term martingale comes from gambling.

Suppose a gambler has a fortune. She wagers one euro on an evens bet. If she loses, she wagers two euros on the next play, and 2^n on the nth play. Each amount is calculated to cover her previous losses. This strategy is called a martingale.

Definition 60 (Martingale) *A sequence S_n with $n \geq 1$, is a martingale with respect to another (reference) sequence t_n, $n \geq 1$ if, for all $n \geq 1$:*

1. $\langle |S_n| \rangle < \infty$,

2. $\langle S_{n+1} | t_1, t_2, \ldots, t_n \rangle = S_n$.

Note that we define a reference sequence that is usually the progression of time (ticks of a clock), but any synchronizing pulse sequence would do.

Martingales always converge, in the following sense. If S_n is a martingale with $\langle S_n \rangle < M < \infty$, for some M, n, then there exists a random variable S such that $S_n \to S$, as $n \to \infty$. This is a form of statistical convergence, as discussed in section 10.9. The convergence time of a martingale is a topic of particular interest that is beyond the current text.

15.4 Characterizing probable or average error

The measures of entropy introduced in chapter 9 provide a precise characterization of how much uncertainty there is in the process of propagating any message over any channel. We now have a configuration alphabet that needs to be propagated into the future, or reproduced for repair or backup purposes. We apply the measures of informational uncertainty from chapter 15, to measure the amount of information that must be transmitted.

Example 139 *You decide that, in addition to backing up your user data, you should also back-up your users by copying their DNA. This is rather straightforward, since DNA is written with only a four-symbol alphabet and can be performed by almost any cell. On observing* *the result of cell division, you find the following data for the probability of successful copying. There seems to be a problem with the fidelity of the copying:*

Trans/Recv	A	C	T	G	Marginal
A	$\frac{1}{8}$	$\frac{1}{16}$	$\frac{1}{32}$	$\frac{1}{32}$	$\frac{1}{4}$
C	$\frac{1}{16}$	$\frac{1}{8}$	$\frac{1}{32}$	$\frac{1}{32}$	$\frac{1}{4}$
T	$\frac{1}{16}$	$\frac{1}{16}$	$\frac{1}{16}$	$\frac{1}{16}$	$\frac{1}{4}$
G	$\frac{1}{4}$	0	0	0	$\frac{1}{4}$
Marginal	$\frac{1}{2}$	$\frac{1}{4}$	$\frac{1}{8}$	$\frac{1}{8}$	

Note that we also calculate the marginal distributions here. The marginal distributions are found by summing the rows or columns. The dual distribution is thus

$$p(T, R) = \begin{pmatrix} \frac{1}{8} & \frac{1}{16} & \frac{1}{32} & \frac{1}{32} \\ \frac{1}{16} & \frac{1}{8} & \frac{1}{32} & \frac{1}{32} \\ \frac{1}{16} & \frac{1}{16} & \frac{1}{16} & \frac{1}{16} \\ \frac{1}{4} & 0 & 0 & 0 \end{pmatrix} \tag{15.9}$$

Had the communication been perfect, this would have been

$$p(T, R)_{\text{perfect}} = \begin{pmatrix} \frac{1}{4} & 0 & 0 & 0 \\ 0 & \frac{1}{4} & 0 & 0 \\ 0 & 0 & \frac{1}{4} & 0 \\ 0 & 0 & 0 & \frac{1}{4} \end{pmatrix} \tag{15.10}$$

As it is, however, this diagonal perfection is smeared out in both directions. We note that there is a persistent error in the channel that causes a 'G' to be received as an 'A'.

From the marginal distributions, we have

$$H(R) = -\frac{1}{2}\log_2\frac{1}{2} - \frac{1}{4}\log_2\frac{1}{4} - \frac{1}{8}\log_2\frac{1}{8} - \frac{1}{8}\log_2\frac{1}{8}$$
$$= \frac{1}{2} + \frac{1}{2} + \frac{3}{8} + \frac{3}{8}$$
$$= \frac{7}{4} \text{ bits.} \tag{15.11}$$
$$H(T) = 2 \text{ bits.} \tag{15.12}$$

The conditional distributions are

$$H(R|T) = \sum_{i=1}^{4} p(T = i)H(R|T = i)$$

$$= \frac{\frac{1}{4}H\left(\frac{1}{8}, \frac{1}{16}, \frac{1}{32}, \frac{1}{32}\right)}{\frac{1}{4}} + \frac{\frac{1}{4}H\left(\frac{1}{8}, \frac{1}{16}, \frac{1}{32}, \frac{1}{32}\right)}{\frac{1}{4}}$$

$$+ \frac{\frac{1}{4}H\left(\frac{1}{16}, \frac{1}{16}, \frac{1}{16}, \frac{1}{16}\right)}{\frac{1}{4}} + \frac{\frac{1}{4}H\left(\frac{1}{4}, 0, 0, 0\right)}{\frac{1}{4}}$$

$$= \frac{11}{8} \text{ bits.} \tag{15.13}$$

$$H(T|R) = \frac{13}{8} \text{ bits.} \tag{15.14}$$

$$H(R, T) = \frac{27}{8} \text{ bits.} \tag{15.15}$$

The interpretation of the numbers is as follows. $H(T)$ is the uncertainty per symbol in our ability to describe the message that is transmitted and made available for copying, as what the next symbol is likely to be. $H(R)$ is the uncertainty per symbol in our ability to describe the message that is received, that is, to predict the occurrence of each symbol in a stream. $H(R|T)$ is the uncertainty per symbol in the received message's integrity, given that the transmitted message is known. This is the main quantity of interest to us: it characterizes the likely integrity of the copy, given that the original is completely known.

To get some idea of how high this uncertainty is, there are $\log_2 4$ bits per symbol, that is, 2 bits per symbol $\{A, C, G, T\}$. Thus, the uncertainty in the original transmitted message $H(T)$ is maximal; no one symbol occurs more frequently than any other, so we cannot say anything about the original copy to compress it or to simplify it. We have to know the exact message in order to copy it.

The uncertainty in the received message is actually less than this. Why? Because the copying is imperfect and it biases the message in a systematic error (see $p(R, T)$). The uncertainty in the copy, given that the original is known exactly, $H(R|T)$ is about half a symbol per symbol! This is a very high probability of error, far from appropriate for backing-up our users[2].

15.5 Correcting errors of propagation

One of Shannon's accomplishments was to prove that any communications channel has a limited capacity, and that it is possible to find a coding scheme which achieves reliable transmission of symbols with an efficiency that is arbitrarily close to that channel capacity. This is known as Shannon's theorem. Shannon imagined a generic model for transmission over a noisy channel, shown in fig. 15.4. This model can also be applied to the transmission of *policy*, or system *integrity*.

Errors creep into the transmission of rules and actions, even with digital channels. One of the reasons for introducing graphical user interfaces to computers, for instance, was to reduce the possibility of error, by condensing difficult operations into a few simple symbols

---

[2] It has come to the author's attention that not all readers possess a sense of humour. For the record, it should be pointed out that backing-up users is not an ethical procedure. Please do not try this at home.

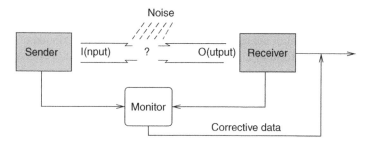

Figure 15.4: Shannon's view of the correction of data on a coded channel can also be applied to the correction of policy propagation or system 'health'. It makes clear the need for a correctional system.

for inexperienced users. In spite of the effort, users still hit the wrong icon, or menu item, and the simplifications that are assumed do not always apply, because external conditions invalidate the assumptions that were made in selecting the symbols.

Shannon's theorem, however, tells us that a suitable coding scheme (protocol) can assure the transmission of a system policy within arbitrary accuracy. Windowing systems use error-correction protocols, such as 'Are you sure?' dialogue boxes, in order to catch random errors—not usually because they believe that users will change their minds very often.

The signal we wish to propagate is this:

$$\text{Ideal signal} = p(\text{usage}|\text{policy}) \tag{15.16}$$

Policy can be communicated by declaration (see, for instance, Burgess (1995); Damianou et al. (2000)), or by simulated dialogue (Couch (2000); Libes (1990)). The latter is becoming more common, with technologies like SNMP (Omari et al. (1999)) and XML-based grid services (XML-RPC (n.d.)). In either case, its transmission or implementation through fallible humans and technologies is subject to the incursion of 'noise'.

The issue now, at this high level, is subtler than for simple bits, however. Do we really want to achieve such a level of precision as to lead to no errors at all? As long as humans are part of the equation, there is the question of user comfort and human welfare.

Thus, in order to apply error correction to larger social and ecological systems, we must choose the symbols rather carefully. Unlike Shannon, we must pay attention to the semantic content of symbols when formulating policies that one wishes to have enforced (see for instance Zadeh (1973)).

Proactive error correction is one way of dealing with this issue; by requiring confirmation of a policy breach, one can avoid spurious overreactions to acceptable transgressions. For example, the double keys used in nuclear missile launches, or the double signals used in co-stimulation of the immune response, are simple security features that prevent potentially damaging responses to innocent errors. Hamming codes and checksum confirmation are other examples of this type of protocol coding.

15.6   Gaussian continuum approximation formula

It is not always convenient or appropriate to provide a complete description of transmission joint probabilities in the form of a matrix. If the number of symbols is effectively infinite, that is, if the signal varies as an arbitrary real number, rather than as a digital signal, then the characterization of a probable error must be performed in terms of functions or distributions rather than matrices. If we believe that a transmission channel is basically reliable, but with a quantifiable source of random error, then it is useful to use a simple continuum approximation model for the expected error. The expression for the capacity of a channel with Gaussian noise is one of the classic results in information theory, and has many applications. Consider the probability distribution

$$p(q) = \frac{1}{\sqrt{2\pi\sigma^2}} \exp\left(-\frac{q^2}{2\sigma^2}\right). \tag{15.17}$$

We have

$$-\ln p(q) = \ln\sqrt{2\pi\sigma^2} + \frac{q^2}{2\sigma^2}, \tag{15.18}$$

and thus entropy

$$
\begin{aligned}
H(q) &= -\int p(q)\ln p(q)\,\mathrm{d}q \\
&= \ln(\sqrt{2\pi\sigma^2})\int p(q)\,\mathrm{d}q + \int p(q)\frac{q^2}{2\sigma^2}\,\mathrm{d}q, \\
&= \ln(\sqrt{2\pi\sigma^2}) + \frac{\sigma^2}{2\sigma^2} \\
&= \frac{1}{2}\ln(2\pi e\sigma^2). \tag{15.19}
\end{aligned}
$$

Now consider a time series $q(t)$ to be a series of real numbers measured at arbitrary times, and let us consider the total system to be a mixture of an average signal $\langle q(t)\rangle_s$ and a noise term $\delta q(t)$; that is,

$$q(t) = \langle q(t)\rangle_s + \delta q(t). \tag{15.20}$$

and $\langle \delta q \rangle = 0$. We shall assume Gaussian noise so that equation (15.17) applies, and obtains the classic result due to Shannon (Cover and Thomas (1991); Shannon and Weaver (1949)). The information that can be transmitted by this continuous channel is infinite in principle, since we can transmit any real number at any time with no error. However, this is not a realizable situation, since there is a physical limit to the information that can be inserted electrically or optically into a physical channel. It is thus normal to calculate the mutual information that can be transmitted, given that there is an upper bound on the average *signal power* $P = S + N$, where S is the signal power and $N = \sigma^2$ is the noise power. The power varies like the signal squared, so we apply the constraint

$$\frac{1}{\Delta T}\int_0^{\Delta T} q^2(t)\,\mathrm{d}t \le P. \tag{15.21}$$

If we exceed the maximum power, the channel could melt or be destroyed.

The channel capacity is defined to be the maximum value of the mutual information in the average signal, given this power constraint:

$$C(1) = \max_{p(q)} H(\langle q \rangle; q). \tag{15.22}$$

This tells us how many digits per sample we are certain are being transmitted, since it is the fractional number of digits of information required to distinguish the average string that was transmitted over the communications channel. In this case, we feed the locally averaged signal $\langle q \rangle$ (the smooth part of the noisy signal that we are responsible for), and we extract the full noisy signal q at the output.

$$\begin{aligned} H(\langle q \rangle; q) &= H(q) - H(q|\langle q \rangle) \\ &= H(q) - H(\langle q \rangle + \delta q | \langle q \rangle) \\ &= H(q) - H(\delta q | \langle q \rangle) \\ &= H(q) - H(\delta q). \end{aligned} \tag{15.23}$$

The last line follows from the independence of $\langle q \rangle$ and δq. Now, for the Gaussian channel, we have

$$H(\delta q) = \frac{1}{2} \ln(2\pi e\sigma^2) = \frac{1}{2} \ln(2\pi eN). \tag{15.24}$$

Thus,

$$\begin{aligned} H(\langle q \rangle; q) &= H(q) - H(\delta q), \\ &\leq \frac{1}{2} \ln(2\pi e(S+N)) - \frac{1}{2} \ln(2\pi eN). \\ &= \frac{1}{2} \ln\left(1 + \frac{S}{N}\right). \end{aligned} \tag{15.25}$$

To reconstruct the signal, we must sample it at twice its maximum frequency, by the Shannon–Nyquist sampling theorem; thus, the channel capacity of a channel of parallel bands of width B cycles per second is

$$C(B) = \sum_{n=1}^{2B} C(1)$$

$$C(B) = B \log_2\left(1 + \frac{P_S}{P_N}\right) \tag{15.26}$$

where P_S and P_N are the total power in all the band frequencies.

Example 140 *A Bangalore Internet Service Provider has a critical copper Internet cable with bandwidth 2 GHz to share amongst customers. This cable is carried on pylons through the city. Calculate the reduction in usable channel capacity C due to thermal noise (power proportional to kT), if the temperature changes from 15 to 35°C, given that the signal to noise ratio is 50 dB at 15°C.*

The signal to noise ratio is defined by

$$10 \log_{10} \left(\frac{P_S}{P_N} \right) = 50, \tag{15.27}$$

hence

$$\left. \frac{P_S}{P_N} \right|_{T=(273+15)} = 10^5. \tag{15.28}$$

Given that the noise power is proportional to absolute temperature, at 35°C the signal to noise ratio is thus

$$\left. \frac{P_S}{P_N} \right|_{T=(273+35)} = 10^5 \times \frac{273 + 15}{273 + 35}. \tag{15.29}$$

The capacity of the cable is thus

$$C(T = 15) = 2 \times 10^9 \log \left(1 + 10^5 \right) \tag{15.30}$$

$$C(T = 35) = 2 \times 10^9 \log \left(1 + 10^5 \frac{(273 + 15)}{(273 + 35)} \right) \tag{15.31}$$

$$C(T = 15) - C(T = 35) = 2.9 \times 10^7 \tag{15.32}$$

Thus, there is a loss of channel capacity of about 30 Megabits per second.

Applications and Further Study 15

- *Relating expectation about a system to observed behaviour.*

- *A method of describing errors (incompatible events) that occur in system processes and information flows.*

- *A method of quantifying losses.*

16

Policy and maintenance

When we speak of policy, we really mean a way of defining and controlling the behaviour of a system. A policy description must include the configuration of a system and offer guidelines as to how it should be allowed to evolve. Policy and configuration management are thus aspects of the same thing. We know that random fluctuations will always lead to some changes that do not agree with policy, so fluctuations themselves cannot be made policy conformant. This implies that we need something to keep fluctuations from permanently altering a system.

16.1 What is maintenance?

When does a process become a maintenance process? The notion of system administration is closely allied with that of maintenance. We need some general notion of maintenance that can be described quantitatively. Maintenance is a process that tends to oppose fluctuation—that is, minimize short-term change and provide medium-term stability. We are not interested in managing systems that cannot achieve a basic level of stability, since these cannot perform any reliable function.

In this chapter, we shall think of maintenance as a response to a stochastic process. There is a parallel here to Shannon's discussion of communication theory (Shannon and Weaver (1949)) discussed in chapter 9. To overlay the language of stochastic systems onto the maintenance process, we need to make a separation into what is policy conformant and what is anomalous. The meaning of conformant and anomalous is not automatically clear, but if fluctuations have finite variance, then it is self-consistent to associate these concepts with slowly and rapidly varying changes, respectively, measured in relation to user behaviour (see Burgess et al. (2001)).

16.2 Average changes in configuration

The separation of slow and rapid changes to configurations can be made precise by observing the system through a *local averaging procedure*. We shall refer to the schematic diagram in fig. 16.1.

Analytical Network and System Administration. Managing Human–Computer Networks   Mark Burgess
© 2004 John Wiley & Sons, Ltd   ISBN 0-470-86100-2

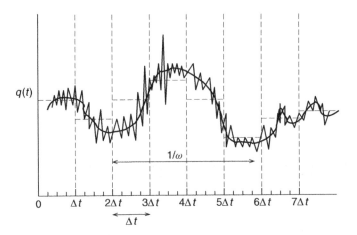

Figure 16.1: A schematic picture of the separation of scales in an open dynamical system, which satisfies eqn. 16.1. The jagged line represents the highest-resolution view of what $q(t)$ is doing. The solid curve is a short-interval local average value of this behaviour, and the solid blocks (dotted line) are a much coarser local average. The order of magnitude of the system's approximate oscillations is ω^{-1}.

Suppose that the total configuration $C = \{q(\vec{x}, t)\}, \forall \vec{x}$, that is the sum of all objects where $\vec{x}$ is the address of each high-level object representation and t is time, is written as a sum of two parts:

$$C(t) \equiv \langle C(t) \rangle + \delta C(t), \tag{16.1}$$

where $\langle C \rangle$ refers to a slowly varying, local average value of C, and δC refers to a rapidly fluctuating, stochastic remainder. This decomposition isolates those parts of the environment that lead to a stable (smooth) average configuration and those parts that tend to be rough and unpredictable. In systems that are *manageable* of interest, one expects $|\delta C| \ll |\langle C \rangle|$, else we are doomed to unpredictability.

Note also that, by definition, $\langle \delta C \rangle = 0$, thus the fluctuations are evenly (though not necessarily symmetrically) distributed about the local mean value. This means that if fluctuations tend in one particular direction, they will drag the mean value with them, preserving their zero mean. If one wishes to avoid a change in the mean value, then one must either offer dynamical resistance to this kind of monotonic drift, or respond to it with a counter-change that balances it on average. This concept of preserving the mean behaviour provides us with a notion of maintenance.

Tasks

The concept of a task is needed to discuss a part of a system that operates autonomously for some purpose, such as maintenance, (Burgess (2003)).

Definition 61 (Task) *Let a* task $\tau(t)$ *be a system contained within a subset s of the total system S:*

$$\tau(\vec{x}, t) = q(\vec{x}, t) : x \in s, \tag{16.2}$$

where the restricted coordinates x ranges only over the sub-system.

Example 141 *A task is an autonomous sub-part of a system, like a computer program or sequence of external changes made by a user. If a task is closed, it does not affect anything outside of its own resources; if it is open, it can affect the state of the rest of the system also. In a distributed environment, a program on one host can affect the state of a program on another host. The actions of a human interacting with the system can also lead to a task.*

We now have a representation of programs running on the system as well as processes carried out by external agents (other computers and humans). One can now define maintenance in terms of the effect of sub-systems on the total system.

Definition 62 (Maintenance Task) *Let* $\tau_M(\vec{x}, \bar{t})$ *be a task in a system S with configuration spanning s, and* $\tau_{M_c}(\vec{x}, \bar{t})$ *be the complement to the subset, that is, the remainder of the configuration of S spanning* s_c*; then* $\tau_M(\vec{x}, \bar{t})$ *is said to be a* maintenance *task if* $\{\tau_M(x \in s, \bar{t})\}$ *is an open system and*

$$\frac{d}{d\bar{t}}\left\langle \sum_{x \in s} \log \tau_M(\vec{x}, \bar{t}) + \sum_{y \in s_c} \log \tau_{M_c}(y, \bar{t}) \right\rangle < \frac{d}{d\bar{t}}\left\langle \sum_{y \in s_c} \log \tau_{M_c}(y, \bar{t}) \right\rangle. \tag{16.3}$$

In other words, the presence of a maintenance task τ_M reduces the total rate of change of the average configuration state $q(\vec{x}, t)$ in S; that is, it counterbalances the information in the fluctuations δq within any smoothed time interval Δt. If the rate of maintenance is less than the rate of fluctuation, it will lead to a window of uncertainty in the value of $\langle q \rangle$, which can result in a real change of average state. The logarithms in these formulae make the ordering and overall scale of the changes unimportant. This is a characterization of the change of information in the configuration, where the spatial ordering is unimportant.

Example 142 *In terms of discrete information coding, a fluctuation composed of operators*

$$\hat{O}_1\,\hat{O}_2\,\hat{O}_6\,\hat{O}_1\,\hat{O}_3$$

can be countered by the string of inverse maintenance operations

$$\hat{O}_3^{-1}\hat{O}_1^{-1}\hat{O}_6^{-1}\hat{O}_2^{-1}\hat{O}_1^{-1}.$$

As the string with entropy H grows longer, the likelihood 2^{-H} *of being able to find the precise counter-string becomes exponentially smaller, if the exact sequence is required. If the operations commute, then there is an average chance* $1/NH(c)$ *of being able to counter the string, since order no longer matters.*

Table 16.1: The separable timescales for changes in a computer system interacting with an environment.

Stochastic open system	Timescale
Fluctuations, system operations δq environmental changes	$T_c \sim T_e < T$
Cycles of persistent behaviour	$T \equiv 2\pi\omega^{-1}$
A coarse grain of N cycles	$\Delta t = NT \gg T$, that is, $(N \gg 1)$
User/policy timescale	$T_p \gg T$
Long-term behavioural trends	$T_b \gg T_p$

The definition of maintenance allows for gradual evolution of the idealized persistent state (e.g. a slow variation in the average length of a queue), since the average value can be slowly modified by persistent fluctuations. This change of the persistent state is said to be *adiabatic* in statistical mechanics, meaning slow compared to the fluctuations themselves. A summary of timescales is shown in table 16.1.

In order to describe and implement a *system policy* for managing the behaviour of a computer system, it must be possible to relate the notion of policy to rules and constraints for time evolution, which are programmed into $q(\vec{x}, t)$. Such rules and constraints are coded as software in $q(\vec{x}, t)$, or are issued verbally to users in the environment of the system. The behaviour of the configuration state is not completely deterministic and is therefore unpredictable. By separating slowly and rapidly varying parts, using a local averaging procedure, we find an average part that is approximately predictable.

We note, as a commentary, that while this shows that the rate of change in the system can be arranged to maintain a particular state over a consistent set of time scales, it does not specify a unique route to such a state through the state space (including space and time scheduling) of the Human–Computer system (see Cheung and Kramer (1996); Traugott (2002)). The existence of inequivalent different routes must be handled by a framework in which they can be compared in some system of returned value. The theory of games, as presented in the final sections of the paper, is suitable for selecting such a route. The existence of a unique path has been addressed in (Burgess (2004)).

16.3 The reason for random fluctuations

In the study of dynamical systems, the environment is not normally modelled as a detailed entity owing to its complexity; rather one considers the projected image of the environment in the main system of interest. The essence of the definition is that the environment leads to a projected component in S, which appears to be partially random (stochastic), because the information about cause and effect is not available. This causes S to behave as an *open dynamical system*.

Definition 63 *An* **open dynamical system** *is the projection of an ensemble of interacting systems* $E = \{S_1, S_2, \ldots, S_N\}$, *onto* S_1. *The time development,* $\hat{D}$, *of the open system, may be considered as operating over a noisy channel, since information from the rest of the ensemble affects the total configuration of the host connected to the ensemble* $C_1(q_1(\vec{x}, t))$. *The closed rule for development of all system is intertwined:*

$$
C_1(t + dt) = (1, 0, \ldots, 0) \begin{pmatrix} \hat{D}_{11} & \hat{D}_{12} & \cdots & \hat{D}_{1N} \\ \hat{D}_{21} & \hat{D}_{22} & & \\ \vdots & & & \\ \hat{D}_{N1} & & & \hat{D}_{NN} \end{pmatrix} \begin{pmatrix} C_1(t) \\ C_2(t) \\ \vdots \\ C_N(t) \end{pmatrix} \tag{16.4}
$$

This definition is an admission of unpredictability in a system that is open to outside influence. Indeed, this unpredictability can be stated more precisely:

Lemma 16.3.1 *The configuration state of an open system S is unpredictable over any interval* $dt \sim T_e$. *(See table 16.1)*

Proof. This follows trivially from eqn. (16.4). There is no equation for the evolution of any part of the system in isolation from the others:

$$
C_1(t + dt) \neq \hat{D} C_1(t), \tag{16.5}
$$

for any $\hat{D}$, since $C_1(t + dt)$ is determined by information unavailable within all of S, if $D_{ij} \neq 0$ for $i \neq j$, which defines the open system.

The only way that a system can become exactly predictable is by isolating itself and becoming closed.

Example 143 *The idea of closed systems can be turned around and made into a requirement. See for instance the approach advocated in Couch et al. (2003). By forcing the closure of a sub-system and placing restrictions or constraints on channels of communication, one maximizes consistency and predictability.*

16.4 Huge fluctuations

In section 10.10, it is remarked that there exist stable stochastic distributions that have large fluctuations with formally infinite variance. These have been observed in network traffic, for instance (see Leland et al. (1994); Willinger and Paxson (1998)). How do these distributions fit into the picture of maintenance described here? In short, these statistical states cannot be maintained without infinite resources. This is not an acceptable maintenance cost.

Statistical stability is thus not enough to ensure maintainability. Fluctuations must be finite in order for us to have a chance. There are two ways to regard this: we can say that the fluctuations are beyond our ability to maintain and resign ourselves to the fact that systems with such behaviour are not maintainable, or we must reinterpret policy to incorporate this environment: stable distributions supersede maintenance—they are already stable. Policy should be based on the appropriate definition of stability for the system. If that includes allowing for power-law fluctuations, then that is the best we can do.

16.5   Equivalent configurations and policy

A high-level partitioning of the configuration, which evolves according to rules for time development at the same level, leads to the appearance of symmetries, with respect to the dynamical evolution of a computer system. A symmetry may be identified whenever a change in a configuration does not affect the further evolution of the system except for the order of its elements. The configurations of the system, which are symmetrical, in this sense, form a group.

Definition 64 *A group $\mathcal{G}$ of transformations is a* symmetry *of the high-level configuration* $q(\vec{x}, t)$, *if for some x and time t, the transformation of the configuration domain*

$$q(\vec{x}, t) = q(g(x), t), \tag{16.6}$$

is an identity, and $g \in \mathcal{G}$.

Thus, a re-labelling of process addresses is unimportant to the configuration, as is any change in $\hat{D}_t$, which leads to a re-labelling in the future. These are just arbitrary labels, unimportant to the functioning of the system. Since the deterministic part of the mapping $\hat{D}_t$ is coded in $q(\vec{x}, t)$, this includes changes in the way the system evolves with time.

Definition 65 *A group Γ of transformations*

$$\gamma : Q_\ell \to Q_\ell. \tag{16.7}$$

is a symmetry *of the state space Q_ℓ, if*

$$q(\vec{x}, t) = \gamma(q(\vec{x}, t)), \tag{16.8}$$

is an identity, and $\gamma \in \Gamma$.

Thus, if two states are equivalent by association, the system is unchanged if we substitute one for the other.

Symmetries are hard to describe formally (they include issues such as the presence of comments in computer code, irrelevant orderings of objects, and so on), but they have a well-defined meaning in real systems.

Example 144 *Renaming every reference to a given file would have no effect on the behaviour of the system. Another example would be to intersperse instructions with comments, which have no systemic function. An important symmetry of systems is the independence of the system to changes in parts of the configuration space R^ℓ, which are unused by any of the programs running on the system.*

The presence of symmetries is of mainly formal interest to a mathematical description of systems, but their inclusion is necessary for completeness. In particular, the notion of equivalence motivates the definition of a factor set of inequivalent configurations

$$P(t) \equiv \frac{C(t)}{\mathcal{G} \otimes \Gamma}, \tag{16.9}$$

which allows us to use one representative configuration from the set of all equivalent configurations. In just a moment, we shall claim that this quantity is intimately associated with the idea of policy. This factored system is now uniquely prescribed by an initial configuration, rules for time development and the environment. It is scarcely practical to construct this factor set, but its existence is clear in a pedantic sense.

16.6 Policy

Up to stochastic noise, the development of the open system is completely described by a configuration of the form of eqn. (16.9), which includes the programs and data which drive it. Conversely, the behaviour at level ℓ is completely determined by the specification of a $P(t)$. With a bottom-up argument about dynamically stable configurations, we have therefore found a set of objects, one for each inequivalent configuration chain, which can be deemed stable and has the potential to be unique in some sense, yet to be clarified. This is therefore a natural object to identify with *system policy* (Burgess (2003)).

In practice, only a part of the configuration will directly impact on the evolution of the system at any time. If a constant part of $P(t)$ can be identified, or if $P(t)$ is sufficiently slowly varying, then this quantity plays the role of a *stable policy* for the system. If no such stability arises, then the policy and configuration must be deemed unstable.

How does this definition of policy fit in with conventional, heuristic notions of policy? A heuristic definition is (i) a system configuration, (ii) rules for behaviour of the system (programmed), (iii) rules for human users (requested), and (iv) a schedule of operations. Of these, the first and the second may be coded into the configuration space without obstacle. The third needs to be coded into the environment, however, the environment is not a reliable channel, and can be expected to obey policy only partially, thus there will be an unpredictable component. The fourth is also programmed into the computer, but there is also a schedule of random events, which belongs to the environment; this also leads to an unpredictability. The resulting 'error' or tendency towards deviation from steady behaviour must be one of two things: a slow drift $\Delta P = P(t) - P(t')$ (systematic error) or a rapid random error $\delta P(t)$ (noise). In order to use a definition of policy such as that above, one is therefore motivated to identify the systematic part of system change.

16.7 Convergent maintenance

The notion of convergence was introduced conceptually in (Burgess (1998a)) and explicitly in (Burgess (1998b)). Some authors later seized upon the word homeostasis to describe this, appealing to a biological analogy (Somayaji and Forrest (2000)). It is related to the idea of the fixed point of a mapping (see Myerson (1991) for an introduction). If $q' = U(q)$ is any mapping, then a fixed point q^* is defined by,

$$q^* = U(q^*). \tag{16.10}$$

This definition is too strict in a dynamical system, rather we need a limiting process that allows for some fuzziness:

$$q^* - U(q^*) < \epsilon. \tag{16.11}$$

As defined, a policy is neither a force for good nor for evil, neither for stability nor for chaos; it is simply an average specification of equivalent system behaviours. Clearly, only a certain class of policies has a practical value in real systems. This refers to policies that lead to short-term stability, thus allowing a stable function or purpose to be identified with the system. A system that modifies itself more rapidly than a characteristic human timescale T_p will not have a stable utility for humans.

The notion of *convergence* is especially useful (Burgess (1998b); Couch and Daniels (2001); Couch and Gilfix (1999)) for regulating systems. A system that possesses a cycle that persists over a given interval of time can be defined as having predictable behaviour over that interval.

Definition 66 (Convergent policy) *A convergent policy $P(t)$, of order n, is one whose chain of time transitions ends in a fixed point configuration $q(\vec{x}, t_f)$, for all values x and times $t_i > t_f$, $f \le n$. That is,*

$$(\hat{D}_t)^n q(\vec{x}, t_i) = q(\vec{x}, t_f), \text{ for some } n \ge 0, t_i < t_f. \tag{16.12}$$

The fixed configuration on which the time development ends is sometimes said to be 'absorbing', since once the system has entered that state, it does not change again. In the language of system administration, one says that the system has converged. In a stochastic, interacting system, this finality cannot be guaranteed precisely. Within a short time period, a change away from the final state can occur at random, thus it is useful to define the notion of average convergence.

Definition 67 (Convergent average policy) *A convergent average policy $P(t)$, of order n, is one whose average behaviour in time ends in an average state $\langle q(\vec{x}, t_f) \rangle$ between any two times t_i and t_f, such that $t_f - t_i > \Delta t$.*

$$\left\langle (\hat{D}_t)^n q(\vec{x}, t_i) \right\rangle = \langle q(\vec{x}, t_f) \rangle, \text{ for some } n \ge 0, t_i < t_f, \tag{16.13}$$

where $\langle \ldots \rangle$ is any local averaging procedure.

This condition is weaker because it allows the final state of exhibit fluctuations that are balanced within the time of the averaging interval.

A discrete chain interpretation of periodicity may be found in Grimmett and Stirzaker (2001); it is convenient here to use the continuum approximation. Over the time interval, it can thus have the general form

$$\langle q(\vec{x}, t) \rangle = \left\langle Q_0(x) + A(t) \, \text{Re} \, \exp\left(i \frac{\omega}{n} t\right) \right\rangle$$
$$= Q_0(x), \tag{16.14}$$

that is, it has an average value and oscillations whose average effect is zero. Since Q is positive, $A < Q_0/2$. Notice that a process that has converged becomes memory-less, that is, its dependence on previous states becomes irrelevant.

A policy in which the average resource usage is constant over the policy timescale T_p is a convergent average policy; for example, a policy of deleting all old temporary files, killing old processes and so on, or by adding new resources, so that that fraction of used resources is constant on an average of a few cycles.

Another example of convergence would be one in which human errors in a configuration file were corrected by an automatic process within a short time interval, by regular checkups, thus preserving the average condition. This has already become a common practice by many system administrators, so convergence is a commonly used strategy for achieving stability.

Persistence

Implicit in the foregoing discussion of averages are two notions of stability, which now crave definition, at the level of the continuum description. These notions form the basis for a self-consistent definition of convergent system policy, which show that system administration is a soluble problem, within clear limits.

The coarse-graining procedure is a redigitization of the timeline. Local averaging procedures are used to separate structures in the time evolution of systems at different levels. One begins by digitizing a details function of time into coarser blocks (like a pixelized image). As one zooms out, the behaviour of a local average looks smooth and continuous again.

Definition 68 (Persistent state) *A persistent state* $\Psi(\vec{x}, t) = q(\vec{x}, t)$ *is a configuration for which the probability of returning to a configuration* $\Psi(\vec{x}, t_0)$ *at a later time* $\Psi(\vec{x}, t_0 + \Delta t)$, *for* $\Delta t > 0$ *is 1. In the continuum description, persistence is reflected in the property that the rate of change of the average state* $\langle \Psi \rangle$ *be much slower than the rate* ω *of* $\delta\Psi$:

$$\left| \frac{1}{\langle \Psi \rangle} \frac{d\langle \Psi \rangle}{dt} \right| = \left| \frac{d}{dt} \log \langle \Psi \rangle \right| \ll \omega \qquad (16.15)$$

that is, the fast variation extends over several complete cycles, of frequency ω *(see table 16.1), before any appreciable variation in the average is seen.*

Example 145 *A system job queue has a fluctuating queue size, whose average length can be determined as a matter of policy, on the basis of observed behaviour, by choice of a scheduling. Since the arrival of jobs in the queue cannot be accurately predicted, the average length will vary slowly, as long as jobs are expedited at only approximately the same rate as they arrive. There is thus a short-term cycle; add job, expedite job that increases and then decreases the queue size. A persistent state is much larger than this cycle. It means that the cycle is locally stable. If the system is characterized by a convergent policy (incoming jobs are indeed expedited at an appropriate rate), then any fluctuations occurring at the rate* ω *will be counteracted at the same rate, leading to a persistent (slowly varying average) state. See fig. 16.2.*

Thus, the meaning of a convergent policy is its resulting persistence. Thus, policy itself must be identified with that average behaviour; this is the only self-consistent, sustainable

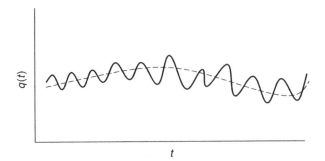

Figure 16.2: A persistent state is one in which the cycle does not vary appreciably over many cycles. Here one sees small variations repeated many times, on a slowly varying background.

definition as long as there are stochastic variables in the system, due to environmental interaction.

The development of an open system is stochastic and this indicates the need for a local averaging procedure to describe it. The split one makes in eqn. (16.1), therefore ensures that the fluctuations are zero on average, distributed about the average behaviour, so by blurring out these fluctuations, one is left with a unique description of the average behaviour. The normalized, coarse-grained policy may now be written:

$$\langle P(\vec{x}, \bar{t})\rangle = \frac{\int_{\bar{t}-\bar{t}/2}^{\bar{t}+\bar{t}/2} dt\, P(t)\, \rho_E(t)}{\int_{t-\bar{t}/2}^{\bar{t}+\bar{t}/2} dt\, \rho_E(t)}$$
$$= \langle C(\bar{t})/(\mathcal{G} \otimes \Gamma)\rangle, \qquad (16.16)$$

In other words, we have shown that the short-term evolution of policy can only be identified with a local average configuration in time; that is, a set of locally average variables, at an appropriate coding level for the system.

16.8   The maintenance theorem

With the meaning of the local averaged mean-field established, it is now a straightforward step to show that local averaging leads to persistence, and hence that this measure of stability applies only to locally averaged states. We thus approach the end of the lengthy argument of this section, which shows that policy can only be an agent for average system state. The theorem suggests that a strategy for maintaining stability in computer systems is to strive for convergence.

Theorem 16.8.1 *In any open system S, a policy $P(\bar{t})$ specifies a class of persistent, locally average states $\langle q(\bar{t})\rangle$ equivalent under symmetry groups $\mathcal{G}$ and Γ, if and only if $P(\bar{t})$ exhibits average convergence.*

The proof of this is found in (Burgess (2003)), and follows basically from the fluctuation rates. From lemma 16.3.1, in an open system S, a configuration is unpredictable over a timescale $T_e \sim \omega^{-1}$ (see table 16.1), hence a configuration can only be guaranteed persistent on average. We thus need only to show that a convergent average policy $\langle P(t) \rangle$, of order n, is persistent for a time $\Delta t \gg T$, since, by definition, this implies a set of equivalent persistent average configurations, under the available symmetries. From the definition of the maintainable fluctuations, one has:

$$\left| \frac{d\langle P \rangle}{d\bar{t}} \right| \ll \omega \left| \langle P \rangle(\bar{t}) \right|, \tag{16.17}$$

hence

$$\left| \frac{1}{\langle P \rangle} \frac{d\langle P \rangle}{d\bar{t}} \right| \ll \omega, \tag{16.18}$$

and $\langle P \rangle$ is persistent. $P(t)$ is associated with a class of states, equivalent under a symmetry group $\mathcal{G}$, which can vary no faster than policy, since it is a part of the policy. Hence a locally average state, resulting from a non-divergent policy specification, is persistent. This completes the proof.

The maintenance theorem provides a self-consistent definition of what a stable state is, and hence what a stable policy is, for a computer interacting with external agents (users, clients etc.). The implication is thus that system administration can be pursed as a regulation technique (see Diao et al. (2002); Goudarzi and Kramer (1996); Hellerstein et al. (1999); Hoogenboom and Lepreau (1993); Seltzer and Small (1997)), for maintaining the integrity of policy, provided one can find a convergent average policy. It sets limits on what can be expected from a policy in a dynamical environment. Finally, the argument makes no reference to the semantic content of policy; it is based purely on information and timing.

It is interesting to note another theorem that is better known but also applicable (and very similar) to the stochastic and semantic views of policy as a propagating influence: it is simply a transcription of Shannon's channel capacity theorem for a noisy channel (Shannon and Weaver (1949)).

Theorem 16.8.2 *There exists a policy, $P(t)$, that can evolve in time with arbitrarily few errors; that is, the system can be forced to obey policy to within arbitrary accuracy.*

Shannon's original theorem stated that 'there exists a message coding which can be transmitted with arbitrarily few errors'; that is, by creating a policy that is so strictly enforced as to police the activities of users in every detail, one could prevent users from doing anything that might influence the strict, predictable development of the system. Such a policy is possible if the average configuration of the host that it represents has sufficiently low entropy that it can be compressed into a part of the system dedicated to maintenance (error correction). It exists because of the finiteness of the digital representation.

16.9   Theory of back-up and error correction

As an application of many ideas thus far in this book, we consider now a basic example of system maintenance: a theory of system back-up. This is both an issue of great practical importance and it is an interesting analytical problem. Unfortunately, this single topic could easily fill a half a book of this size, so we can only outline the analysis and refer to details in (Burgess and Reitan (2004)).

The theory of back-up is a theory of random change in systems and the effort to catch up by making a response to each change. It is the study of trying to hit a moving target. This situation is not unique to back-up, of course; here are some analogous examples:

- Random change of files leads to the need for renewed back-up,

- Configuration error leads to the need for renewed maintenance,

- Accumulation of garbage leads to the need for garbage collection (tidying),

- Arrival of tasks in a queue leads to the need for server or human action.

There are several parts to this model, all of which address the competing random processes in the problem. We must address:

- Change detection (digital or continuous; i.e. symbolic or probabilistic),

- Rates of change or event arrivals (clustered or independent arrivals),

- Rate of measurement (including scheduling intervals or detection 'dead times'),

- Rate of transmission of response (capacity of communication channel).

The arrival of events can be modelled in a number of ways. The events might be faults, intrusions, accidents, arrival of users, cleanup after departure of users and so on.

In traditional statistical theory, arrival events are always assumed to follow the pattern of a Poisson arrival process, that is, as a stream of random, independent arrivals. This assumption is made because it is simple and has special analytical properties. The study of this kind of system models well the detection of particles from a radioactive source, for example, by a Geiger counter. These are truly independent events. Such processes are called *renewal processes* (see Grimmett and Stirzaker (2001)). However, it is known from observation that many arrival processes are not Poisson processes: arrivals of events are often clustered or come in 'bursts'. The only way to determine this is to observe actual systems and collect data.

The most basic observation we can make about user behaviour is that it follows a basic daily rhythm. Taking data from the measurements at the author's site, shown in fig. 2.1, we see a distinct maximum in user processes around midday and a lull at around 5:00 or 6:00 in the morning. Clearly this user behaviour must be correlated with changes in disk files, since there is a direct causation. It is a necessary but not sufficient condition for change of disk data. We thus expect that most changes will occur during the daytime, and few changes during the night. This is all very well, but what about the fluctuation distribution of the arrival process? There are various ways of characterizing this: by deviation from mean or by inter-arrival time.

Arrival fluctuation process

Suppose we consider inter-arrival time distributions. What is the likelihood of being able to predict the next file system change? By taking direct measurements of user disks at the author's College, one finds a highly clustered and noisy distribution of data, with a long tail. There is clearly no theoretical maximum inter-arrival time for file changes, but there is a practical maximum. During holiday periods, inter-arrival times are longest because no one is using the system.

Most changes occur closely together, within seconds, between related files (we are unable to measure multiple changes to the same file simply by scanning the current state as a back-up system does). This short time part of the spectrum dominates by several orders of magnitude over the longer times. However, there is a non-zero probability of measuring inter-arrival times that are minutes and hours and even days. Clearly the breadth of scales is large, and this tells us that a simple independent Poisson type of distribution, such as we used in queueing analysis, is unlikely to be an adequate explanation of this total process.

In the measurements made while writing this book, a rather dirty power-law type distribution or low order was found to approximate the arrival process:

$$p(\Delta t) \sim \Delta t^{-\alpha}, \tag{16.19}$$

where $1 < \alpha < 2$[1]. We can note that many phenomena that are driven by social networks are described by power laws (see for instance Barabási (2002) for a discussion of this). The fit is not very precise, but it is adequate to sketch a functional form[2].

What about the magnitude of each fluctuation? This is harder to gauge, since it would require a memory of what has gone on during the entire history of the system, and since most back-up systems have to transfer an entire file at a time, this is not immediately interesting (though note that the rsync program can transfer differential content, see Tridgell and Mackerras (1996)).

How should we deal with these changes? There are several questions:

1. When should a detector start scanning the filesystem for change, over the daily cycle?

2. What is the risk associated with waiting once a change is made, or having a dead time between back-up processes?

3. What is the cost associated with the back-up process?

Detection process

Clearly, in an ideal world one could make an immediate back-up of every file as soon as it were changed.

Example 146 *This method is used in copy-on-write mirroring technologies, such as the Distributed Computing Environment filesystem DFS. But this is expensive, both in terms of machine resources and in terms of storage. Moreover, this choice involves a policy decision. By simply mirroring the file, we choose not to keep multiple archive versions. Thus if a mistake is added to a file, it will immediately be propagated to the back-up also. This causes us to lose one reason for making a back-up.*

---

[1] One could try to model this process by splitting up the process into a superposition of well-defined processes, and thus take account of causation.

[2] With a noisy distribution, the value of α is uncertain, so we choose it to be a normalizable distribution $\alpha > 1$ to define a probability.

Thus the decision is no longer just one about efficiency, but policy. Let us ignore the copy-on-write policy here and focus on an intermittent back-up.

In order to detect changes without notification by the filesystem directly, we need a detection process. This is a process that must scan through the system, either using its hierarchical structure, or using its disk block structure to detect change. There are two strategies one might use here:

- Back-ups parse file tree as quickly as possible (spans shortest possible time),

- Back-ups parse file tree slowly (spans a large interval of time, several runs do not overlap).

In the first case, the back-up process presents the shortest interruption to system resource availability but with high load (back-up can be a disk and CPU intensive process that disrupts system performance for users). In the latter case, one presents a low load to the system over a longer interval of time. The extreme case here would be to have a continuous scan of the file tree, picking up modified files continuously and back-up them up as a low priority process.

In the first case, one takes a rapid sharply focused snapshot of the filesystem. In the latter case, we take a blurred snapshot capturing more changes over a longer time. It is not obvious which is these strategies can capture most changes over a shortest 'risk interval'. We would like to minimize this risk interval, that is, the time between a file change and its back-up.

In either case, we have a process of file change arrivals overlapping with a detection process. If both processes are random in time, these are two superposed random processes. As we know from section 10.9, the superposition of random processes does not lead to the same kind of random process except in a few special cases. If the detection process is regular in time, we might suppose that detection is not a random process, but this is not the case. In detection, one must parse the file structures in some regular manner in order to cover the file space evenly; one cannot predict when the actual change will be registered during this parsing. With partially ordered events, for instance, a dependency structure that can be represented graphically, results of experiments indicate that a random sampling can be an efficient (see Sandnes (2001)) way of parsing the task list. However, random access makes it difficult to cover the file tree in an even-handed manner, and could demand large disk head movements. We must always remember the physical limitations of disk back-up: it is the mechanical movement of the disk head that retards the system performance the most, and the thus increases the performance cost. Since, interleaving a continuous scan of the disk with random access must lead to large movements at least some of the time. This suggests that copy on write would be a most efficient way of making an immediate copy, since it requires no additional disk head movement.

In the maintenance theorem, we avoided mentioning the form of fluctuation distributions explicitly, dealing only with rates and finiteness since computer systems are finite. Here, observation directs us towards an approximate power-law distribution.

Policy choices

The mechanisms behind the optimization of back-up are rather complex and a full understanding of them goes beyond the scope of this book; however, suppose we have satisfied

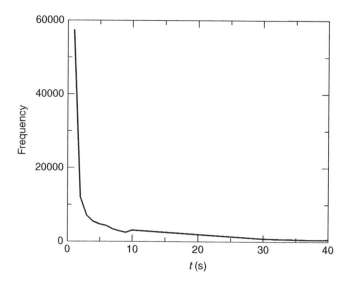

Figure 16.3: A sketch of the distribution of inter-arrival times for file changes. A log–log plot has a gradient of about 3/2, hence the power law in eqn. (16.20)

ourselves as to the causes and have a satisfactory model of file arrivals that describes a file time distribution of the form

$$P(\Delta t) = A \; \Delta t^{-\frac{3}{2}} e^{-\gamma \Delta t},$$ (16.20)

where A is a constant amplitude and γ is the decay time of the distribution that truncates the infinite power-law tail to no longer than a day (since the process is pseudo-periodic, with daily period, there is no point in looking to times longer than this to describe a maintenance procedure that will also periodic). We now consider what this means for the best approach to capturing change with minimal risk of loss.

Most changes are for small $\Delta t < 10$ seconds (see fig. 16.3), so a frequent search is likely to reduce the risk time between change and back-up more effectively than a regular daily back-up. A social-network model should also be used to look at the locations of these files, and relate it to the scanning policy used by the back-up program to see whether changes are likely to be clustered in the file tree as well as in time.

How is risk evaluated for a back-up strategy? The pathways for failure are various:

- Accidental file deletion

- Malicious file deletion

- Disk failure

- Natural disaster.

In defence against these 'gremlins', we have to choose:

- Copy on write

- Back-up daily (over short bursts)

- Back-up daily (over long times)

- Continuous search and copy.

We can set these up in opposition to one another, by creating a matrix of cases, either knowing the relative probabilities of the loss mechanisms, or starting from the worst-case scenario. In chapter 19, a method for calculating and optimizing these strategies is described. The risk or payoff associated with each pair of strategies can then be modelled, for instance, by the probability that the copy is up to date and divided or subtracted by the relative disruption cost of the back-up method to system productivity. We can make various assumptions about systems by observing their nature.

Example 147 *Taking a back-up at night is not obviously the best strategy. It reduces the overhead, but if one accepts that most accidents happen while the users are busy, then this also maximizes the chance of changes being lost because of some catastrophe.*

Example 148 *Using a program that backs up data very quickly provides only a momentary snapshot of one time of day. From the data here, we should can minimize risk by choosing the back-up time to be somewhere close to the maximum of the graph of $R(t)$.*

This vagueness underlines the point that, in science, there is virtually never only one right answer to a question. In system administration, it is about making an arbitrary compromise.

Let us examine slightly more closely how we might evaluate these strategies in terms of data capture rates. If a back-up is to success, then its average performance must be such that its average capture rate is equal to the average rate at which changes are made to the file system. The cumulative change to the file system at rate $R(t)$, over an interval Δt is (see fig. 16.4):

$$C(\Delta t) = \int_0^{\Delta t} R(t)\,dt. \tag{16.21}$$

Note the regions of differing rates during on and off-peak times, m_1 and m_2, where $m_1 > m_2$. In region m_1, there is high risk of losing changes: there is m_1/m_2 times the risk in region m_2. The likelihood of disruption by taking back-up, on the other hand is greater in region m_1, so we might try to even out this disruption by capturing at a rate not exceeding m_2/m_1 of the total rate of change. The simplest solution, when these are in balance, is thus when $m_1 = m_2$ and we have the constant rate of capture with dotted line, in fig. 16.4.

Now, if we want to perform a quick non-disruptive back-up, then this diagram tells us to perform the back-up at the positions marked by the dark blobs, since these lie at the end of the period of maximum change. This assumes that the rate of failure leading to need for back-up is much slower than the rate at which data change on the disk, and can therefore happen with more or less equal likelihood at any moment. By placing the back-up after the maximum rate period, we assume capture of the largest number of changes since the last back-up, while at the same time minimizing the disruption.

It is almost obvious from the diagram that, in this simplified picture, there is no advantage to choosing any special time for a short-burst daily back-up except to minimize disruption. We do not capture any greater number of changes in this way, on average.

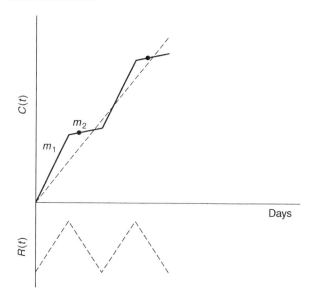

Figure 16.4: A simplified schematic plot of a daily data rate $R(t)$ and cumulative change $C(t)$ of data on a user disk. The straight dotted line shows the rate of continuous detection and transfer required to back-up data change as a continuous process. Gradients m_1 and m_2 of the cumulative graphs indicate the most basic approximation to data rates during peak and off-peak hours (usually day and night). The dark blobs indicate short-burst back-ups. See also fig. 19.5 for a related picture.

However, if we consider making more than one back-up per day, or using a continuous scan technique, there are advantages to picking the time more carefully and a more detailed analysis is required (see Burgess and Reitan (2004)).

If we change any of these simplifying assumptions to make the model more realistic, some of the details will change, but the essence will most likely remain somewhere between continuous transfer and short-burst rates.

Applications and Further Study 16

- *Determining whether a system design is maintainable (sustainable) or not.*

- *Basis for investigating generalized renewable processes.*

- *Foundation for the very existence of systems.*

17

Knowledge, learning and training

The great tragedy of science–
the slaying of a beautiful hypothesis by an ugly fact.

—Thomas Huxley

Human–computer systems are often called *knowledge-based systems* because they rely on a 'database' of procedural or factual expertise to carry out their function. Knowledge has to be acquired somehow, and then summarized and stored. There are two forms of stored knowledge:

- Statistically acquired data that document experience,

- Rules or relationships matched to observation, then summarized in algebraic form.

In the first case, we observe and check, and let experience be our guide; in the second case, we believe that we have understood and so summarize what the data seem to show as a formula, in the context of a theory. Similarly, knowledge is acquired in one of two ways:

- Supervised learning or training: We are certain of the value of information (e.g. is a fact true or false) and all that remains is to incorporate this classification into decision-making procedures of the system.

- Unsupervised learning or just learning: We are not certain of the precise classification of information, and have to rely on the statistical significance of information to formulate a hypothesis.

The scientific process itself is about continually questioning and verifying our beliefs about such knowledge until we are satisfied with their accuracy. It is unquestionably a process of unsupervised learning: no one knows the right answers; we have to make do with dribs and drabs of evidence to make sense of the world.

Analytical Network and System Administration. Managing Human–Computer Networks   Mark Burgess
© 2004 John Wiley & Sons, Ltd   ISBN 0-470-86100-2

Experts tend to overestimate their competence in making judgements; so it is important to have impartial methods that can mitigate this overconfidence with a reality check. Probability theory is a key tool in estimating the uncertainties in the learning process, but even absolute probabilistic descriptions cannot always be made with sufficient accuracy to make knowledge reliable. We therefore need to consider how sufficient knowledge is built up and refined in a quantitative fashion[1]. Note that, while ordinary statistics have no 'direction' (the correlation of two variables implies no ordering or asymmetry between them), the Bayesian viewpoint is directional and can therefore be used to discuss cause and effect, or the arrow of development.

17.1 Information and knowledge

A system or organization's knowledge is built up gradually from experience, by revising facts and procedures until the process converges on some corpus of expertise that enables it to carry out its function. Uncertainty remains throughout this process, but can be quantified and allowances can be made for the imperfection. This chapter brings us back to the importance of empirical observation.

Rather than giving up on a model, if we do not have sufficient data to reasonably estimate the parameters, we can use hints and guesses to 'bootstrap' the investigative procedure, and then revise estimates on the basis of testing out the early assumptions. Bayesian statistics are widely used as a tool for modelling learning (machine learning, human learning, behavioural adaptation etc.). The idea is to gradually refine knowledge and move from a situation with *imperfect information* to one with *perfect information*. When we learn, by gathering new evidence of phenomena, it is equally important to forget outdated knowledge, so that contradictions do not arise.

The idea of expertise and knowledge really brings us back to the philosophical issues about science and its interpretation of the world (see section 2.3). Without getting embroiled in this tar pit, we can refer to the problem of expertise as being a human one, which relates to policy. The roles of humans in making such interpretations are central.

17.2 Knowledge as classification

In chapter 9, we defined information by the number of symbols that must be transmitted in order to exactly reproduce an object or procedure independently of the original. The uncertainty about the original was measured by the informational entropy. Both of these definitions hinged on the need to classify observations into *symbol classes*. Even time series seek to identify from a list of basic features. The more focused a classification, the lower the uncertainty (entropy). We say that the acquisition of information about something decreases out uncertainty about it, so we need to receive classified symbols to achieve this. This tells us that classification is central to identifying data with knowledge.

---

[1] Much is written on the subject of learning and statistical inference that goes way beyond the scope of this book. Readers are referred to books on causal statistics, for example, (Breipohl (1970); Pearl (2000)), or on pattern recognition, for example, (Duda et al. (2001)), for more information on this vast and subtle topic. Bayesian methods are frequently cited, and often used to throw a veil of philosophical subtlety over the subject. Here be dragons!

Definition 69 (Knowledge) *Knowledge is a systematic classification of facts and algorithms, that is, it is about identifying events with a correct hypothesis (class) for their cause.*

We shall now return to the idea of classification to define what we mean by learning. Learning is closely related to the acquisition of information, but it is not identical because information might tell us all kinds of contradictory things about a system. It is our ability to classify those pieces of information into a consistent picture of *probable cause* that is knowledge.

Pattern recognition is central to the ability to classify (see Duda et al. (2001)). As with the classifications of knowledge above, this falls largely into two types: deterministic matching and probabilistic matching of patterns. Biological pattern recognition, as exercised by the immune system, is believed to be a statistical hybrid of a symbolic (deterministic) model of string matching (Perelson and Weisbuch (1997)). This is an approach that has been used to look at integrity checking and anomaly detection (see P.D'haeseleer et al. (n.d.)).

What is impressive about the biological immune system is that it recognizes patterns (antigens) that the body has never even seen before. It does not have to know about a threat in order to manufacture antibody to counter it. Recognition works by jigsaw pattern-identification of cell-surface molecules out of a generic library of possibilities. A similar mechanism in a computer would have to recognize the 'shapes' of unhealthy code or behaviour (Cowan (n.d.); Sun Microsystems (n.d.)). If we think of each situation as being designated by strings of bytes, then it might be necessary to identify patterns over many hundreds of bytes in order to identify a threat. A scaled approach is more useful. Code can be analysed on the small scale of a few bytes in order to find sequences of machine instructions (analogous to dangerous DNA) that are recognizable programming blunders or methods of attack. One could also analyse on the larger scale of linker connectivity or procedural entities in order to find out the topology of a program.

Example 149 *To see why a single scale of patterns is not practical, we can gauge an order of magnitude estimate as follows (Perelson and Weisbuch (1997)). Suppose the sum of all dangerous patterns of code is S bytes and that all the patterns have the same average size. Next, suppose that a single defensive spot check has the ability to recognize a subset of the patterns in some fuzzy region* ΔS*; that is, a given agent recognizes more than one pattern, but some more strongly than others and each with a certain probability. Assume the agents are made to recognize random shapes (epitopes) that are dangerous, then a large number of such recognition agents will completely cover the possible patterns. The worst case is that in which the patterns are randomly occurring (a Poisson distribution). This is the case in biology since molecular complexes cannot process complex algorithms, they can only identify affinities. With this scenario, a single receptor or identifier would have a probability of* $\Delta S/S$ *of making an identification, and there would be a probability* $1 - \Delta S/S$ *of not making an identification, so that a dangerous item could slip through the defences. If we have a large number n of such pattern-detectors, then the probability that we fail to make an identification can be simply written*

$$P_n = \left(1 - \frac{\Delta S}{S}\right)^n \sim e^{-n\frac{\Delta S}{S}}. \tag{17.1}$$

Suppose we would like 50% of threats to be identified with n pattern fragments, then we require

$$-n\frac{\Delta S}{S} \sim -\ln P_n \sim 0.7. \qquad (17.2)$$

Suppose that the totality of patterns is of the order of thousands of average-sized identifier patterns, then $\Delta S/S \sim 0.001$ and $n \sim 7000$. This means that we would need several thousand tests per suspicious object in order to obtain a 50% chance of identifying it as malignant. Obviously this is a very large number, and it is derived using a standard argument for biological immune systems, but the estimate is too simplistic.

Testing for patterns at random places in random ways does not seem efficient, and while it might work with huge numbers in a three-dimensional environment in the body, it is not likely to be a useful idea in the one-dimensional world of computer memory (though see Sandnes (2001)). Computers cannot play the numbers game with the same odds as biological systems. Even the smallest functioning immune system (in young tadpoles) consists of 10^6 lymphocytes, which is several orders of magnitude greater than any computer system. What one lacks in numbers must therefore be made up in specificity or intelligence. The search problem is made more efficient by making identifications at many scales. Indeed, even in the body, proteins are complicated folded structures with a hierarchy of folds, which exhibit a structure at several different scales. These make a lock and key fit with receptors, which amount to keys with sub-keys and sub-sub-keys and so on. By breaking up a program structurally over the scale of procedure calls, loops and high-level statements, one stands a much greater chance of finding a pattern combination that signals danger.

17.3   Bayes' theorem

The basic formula normally used in learning is Bayes' theorem for conditional probability. This prescribes a well-defined method for a learning procedure, but it is not the only method (see section 17.6). We have already seen how conditional probability allows us to attach a causal arrow to the development of system information (see section 9.7). We now take advantage of this to develop a method of increasing certainty, or refined approximation by including the effect of observed evidence.

Bayes' formula is an expression of conditional probability. The probability of two separate events, A and B, occurring together may be written

$$P(A \textbf{ AND } B) = P(A \cap B) = P(A|B)P(B) = P(B|A)P(A). \qquad (17.3)$$

If the events are independent of the order in which they occur, that is, they occur simultaneously by coincidence, then this simplifies to

$$P(A \textbf{ AND } B) \to P(A)P(B). \qquad (17.4)$$

The symmetry between A and B in eqn. (17.3) tells us that

$$P(A|B) = \frac{P(B|A)P(A)}{P(B)}. \qquad (17.5)$$

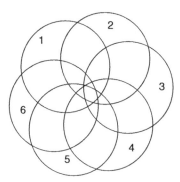

Figure 17.1: When several classes of events can occur simultaneously, the classes overlap and this must be taken into account in approximating the cause of the events. The classes represent different hypotheses for observed events.

This trivial re-expression is the basis for system learning. If we rewrite it for more than two kinds of events (see fig. 17.1), using fixed classes c_i, for $i = 1 \ldots C$, and an unknown event E that could be in any of them, we have

$$P(c_i|E) = \frac{P(E|c_i)P(c_i)}{P(E)} = \frac{P(E|c_i)P(c_i)}{\sum_{i=1}^{C} P(c_i, E)}$$
$$= \frac{P(E|c_i)P(c_i)}{\sum_{i=1}^{C} P(c_i|E)P(c_i)}. \tag{17.6}$$

This is Bayes' formula. The usefulness of this much-adored result lies in a special interpretation of the learning process. We assume that the 'true' classification of observed information is represented by an event of type c_i, and that this is a fixed classification. Our aim in learning is to determine the most probable classification of information, given new information.

The Bayesian philosophy distinguishes between *a priori* probabilities (i.e. our initial estimate or guess, which is based on imperfect information), and *a postiori* probabilities (i.e. our revised estimates of likely classification after receiving new information). The formula is then interpreted as follows.

- The uncorrelated probabilities of seeing different events $P(c_i)$ and $P(E)$ are assumed to be known. We can guess these initially, or simply admit that our previous experience is always limited and that they are estimates of the truth, and might not be very accurate. They are based on prior knowledge, that is, the experience we have before making any new observations.

- The conditional probability $P(E|c_i)$ is interpreted as the likelihood that we are able to classify an event E, given that we know the classes. This is a *likelihood function*. As this probability increases, our guesses become closer and closer to the truth. This is treated as though it were a function of E, and is sometimes written $L(E|c_i)$ to mark this special interpretation.

- The derived (*a postiori*) probability $P(c_i|E)$ is the best estimate of the probability of seeing the classification c_i, given the evidence E is $P(c_i|E)$. It can be thought of as $P(H_i|O)$, the probability of hypothesis being the correct explanation, given the observation O. This result can be used directly, and it can also be used, with a little subtlety, to replace our initial estimate of $P(c_i)$ to reiterate the procedure once more.

Example 150 *Consider a single cause–effect relationship $C = 1$, and let c_1 be computer failure by disk error, and let E be the probability of a computer being down by any cause. The probability of a disk error, on any computer, is found over time to be $P(c_1) = 0.02$.*

The probability that a computer will be down because of disk error is initially unknown, so the system administrator sucks a finger and pokes it into the air, declaring that $P(E|c_1) = L(E|c_1) = 0.6$, that is, there is a 60% chance that disk error will be the cause of computer failure. The probability that disk failure is not the cause is thus $P(E|\neg c_1) = 0.4$, since probabilities sum to one. We can now combine these uncertainties to find the probability that disk error will be the cause, given that a computer is observed to be down:

$$
\begin{aligned}
P(c_1|E) = P(H|O) &= \frac{L(E|c_1)P(c_1)}{P(c_1)L(E|c_1) + P(\neg c_1)L(E|\neg c_1)} \\
&= \frac{0.6 \times 0.02}{0.6 \times 0.02 + 0.4 \times 0.98} \\
&= 0.03.
\end{aligned}
\tag{17.7}
$$

The probability that the true cause is a disk failure is really only as low as 0.02. The uncertainty flies in the face of the system administrators finger estimate, and reflects the fact that 98% of computers do not show the symptoms of disk failure and yet 40% of computers will be down anyway, due to other causes. Note that the result is larger than the independent estimate, but only slightly: 0.03 is now our belief of the probability of disk error, rather than 0.02 that was measured. This tells us that our initial finger-in-the-air estimate was very wrong and it has been adjusted almost all the way down to the independent measurement. We should therefore replace this new value with the old $P(E|c_1) = L(E|c_1) = 0.03$ and use the formula again, if we obtain new data.

17.4   Belief versus truth

It is a philosophical conundrum for science that, in spite of a search for absolute truth, one is forced to settle for making a value judgement about belief. This is an inevitable consequence of unsupervised learning; the world does not give up its secrets easily, and never according to a classification that is preordained. It is therefore up to a process of *inference* to determine what is a reasonable belief about the truth and falsity of our hypotheses. This realization is both liberating and complicating for the administration of human–computer systems, since so much of human involvement is based on belief.

This should not be received as a signal to abandon method however. As always, we are in the business of reducing uncertainty by proper observation and analysis. Rather, it opens up an alternative viewpoint in the form of Bayesian statistical methods. There is only space to mention these briefly here. Readers are referred to (Breipohl (1970); Pearl (2000)) and Duda et al. (2001) for more details.

Example 151 *Network Intrusion Detection Systems (NIDS) examine arriving packets of data on the network using pattern-matching rules for possible intrusion attempts by crackers. Suppose the IDS signals an alarm. What is the likelihood that an intrusion has taken place? Clearly, we cannot be certain about this; indeed, does the question even have any meaning? We can talk about our degree of belief in the matter.*

Suppose that there is a 95% chance that an attempted intrusion will trigger an alarm:

$$P(\text{alarm}|\text{intrusion}) = 0.95. \tag{17.8}$$

On the basis of the possibility of false alarms, there is a 6% chance that an alarm will be false:

$$P(\text{alarm}|\neg\text{intrusion}) = 0.06. \tag{17.9}$$

Figures from security watchdog organizations indicate that there is a 0.01% chance of being attacked on a given day, so $P(\text{intrusion}) = 0.0001$. What is our basis for belief that the alarm is a true signal of attack? Let A stand for alarm, and I for intrusion.

$$
\begin{aligned}
P(I|A) &= \frac{P(A|I)P(I)}{P(A|I)P(I) + P(A|\neg I)P(\neg I)} \\
&= \frac{0.95 \times 0.0001}{0.95 \times 0.0001 + 0.06 \times (1 - 0.0001)} \\
&= 0.00158.
\end{aligned}
\tag{17.10}
$$

Thus, this additional information about the behaviour of the alarm in response to data has increased our belief in the validity of the alarm from 0.0001 to 0.0.00158, some 16-fold. Although the likelihood of truth is still tiny, there is a significant improvement due to the constraints on uncertainty implied by the conditional probabilities.

17.5 Decisions based on expert knowledge

Decision-making is central to system administration. A system that claims to perform an expert function cannot afford to be as badly wrong as in the finger-sucking example 150 above, so it is crucial to make reasoned estimates of the truth of our hypotheses. Bayes' formula tells us how close our estimate of knowledge is to independently measured values. The Bayesian method allows iteration of the formula, and it identifies this iterative revision of probabilities with learning, or the refinement of probable hypothesis fitting (data recognition). The likelihood function $L(H|O)$ tells us the latest state of our knowledge about the reasonableness of a hypothesis H, given the observed data O. The programme is

1. Formulate C hypotheses to classify data.

2. Formulate a discriminant criterion for deciding when a hypothesis is actually true. This is needed to train the likelihood function; otherwise, we can never make decisions.

3. Train the likelihood function by working out probabilities from independent measurements, as far as possible. Even if these are imperfect, they allow us to make a start.

4. Work out the revised estimates probability for probable hypothesis, or update the likelihood function with new data.

Consider the following example of using Bayes' theorem to make a decision on the basis of training, or supervised learning, in which a human is able to pass on expert judgement to a computer program in order to manage a flow of data.

Example 152 *One of the classic examples of Bayesian hypothesis discrimination in system management is the filtering of junk mail by mail servers. There are various approaches to codifying the hypotheses about what constitutes junk mail, some are rule-based and others are probabilistic.*

Suppose that we initially separate junk mail by hand and collect statistics about the percentage of junk mail arriving. Using the mail history, we train the likelihood function $P(\text{data}|\text{junk}) = L(\text{junk}|\text{data})$, giving

$$L(\text{junk}|\text{data}) = 0.95$$
$$L(\neg\text{junk}|\text{data}) = 0.1 \tag{17.11}$$

This tells us that when we feed junk into the likelihood function, and update the probabilities on the basis of certain knowledge (discrimination of junk from non-junk) we obtain a probability estimate of seeing the discriminating features in the data. In reality, there will be many triggers that cause us to classify mail as junk mail (spam), but they all result in a final classification of either junk or NOT junk (in set notation '$\neg$ junk'). Note that some mail is not junk, even though it passes all the tests that we make for it; thus, the separation is not a clean one.

The data we use to train our likelihood function are the test results of expert probes over thousands of e-mails. The data that we use to discriminate each incoming mail then comes from each individual e-mail, after training has ceased. We thus compare the likelihood that each individual message is spam, on the basis of what has been learned from all the others.

On examining a new message, we test the hypothesis that it is junk. We find that about 60% of the junk tests on this mail are found to be positive: $P(c_1) = P(\text{junk}) = 0.6$, $P(\neg\text{junk}) = 0.4$, where the $\neg$ symbol means NOT or complement.

$$\begin{aligned} P(\text{junk}|\text{training}) &= \frac{L(\text{junk}|\text{data})\,P(\text{junk})}{L(\text{junk}|\text{data})\,P(\text{junk}) + L(\neg\text{junk}|\text{data})\,P(\neg\text{junk})} \\ &= \frac{0.95 \times 0.6}{0.95 \times 0.6 + 0.4 \times 0.1} \\ &= 0.93. \end{aligned} \tag{17.12}$$

The probability that this is actually junk is slightly less than the training estimate, that is, we have an almost maximally high degree of likelihood that the message is spam. In this example, our belief is amplified by the high proportion of e-mail messages that are correctly identified by the data sample tests. The main effect of the Bayesian formula is to go from a simple true/false picture of whether an e-mail is junk or not to a more refined threshold-based decision that can be dealt with by making a policy decision about the correct threshold for discarding a message.

The previous example is rather simplistic. In general, effects have many causes, each of which is a potential hypothesis for failure. Bayesian networks are a way of modelling the dependencies in more involved cause trees. Here, we shall offer only a simple example.

Example 153 *Consider a situation in which a worker is trying to establish a Virtual Private Network connection with their company from home over an Internet Service Provider (ISP) line. The probability of a fault F (not being able to establish a connection) depends on two factors: either the server is busy, or the network in between is too slow. Let us denote these two causes by B and S respectively. From monitoring software, the probabilities of these independent events can be estimated:*

$$P(B) = P(\text{busy}) = 0.1, P(\neg B) = 0.9$$
$$P(S) = P(\text{slow}) = 0.4, P(\neg S) = 0.6. \tag{17.13}$$

Moreover, we might believe the following probabilities about how these probabilities link up the independent events in a dependency network.

$$P(F|BS) = 0.8 \qquad P(F|B\neg S) = 0.6$$
$$P(F|\neg BS) = 0.5 \qquad P(F|\neg B\neg S) = 0.5$$
$$P(\neg F|BS) = 0.2 \qquad P(\neg F|B\neg S) = 0.4$$
$$P(\neg F|\neg BS) = 0.5 \quad P(\neg F|\neg B\neg S) = 0.5$$

Note that when the server is busy, we do not really have any prior knowledge of the cause of the failure: it is 50/50.

We begin by calculating the likelihood of a fault given these rough estimates of our beliefs. It is surprisingly complicated to take account of these biases fairly.

$$P(F) = P(FBS) + P(F\neg BS) + P(FB\neg S) + P(F\neg B\neg S)$$
$$= P(F|BS)P(BS) + P(F|\neg BS)P(\neg BS)$$
$$+ P(F|B\neg S)P(B\neg S) + P(F|\neg B\neg S)P(\neg B\neg S). \tag{17.14}$$

Now, since B and S are independent, $P(BS) = P(B)P(S)$.

$$P(F) = P(F|BS)P(B)P(S) + P(F|\neg BS)P(\neg B)P(S)$$
$$+ P(F|B\neg S)P(B)P(\neg S) + P(F|\neg B\neg S)P(\neg B)P(\neg S)$$
$$= (0.8 \times 0.1 \times 0.4) + (0.5 \times 0.9 \times 0.4) + (0.6 \times 0.1 \times 0.6) + (0.5 \times 0.9 \times 0.6)$$
$$= 0.518 \tag{17.15}$$

Thus, the best we can conclude is that the probability that there will be a fault is a little over even odds. The reason for this is our lack of knowledge about the network hypothesis.

Suppose now we make more observations. This would be expected to reduce the uncertainty and we have the opportunity to learn from the added evidence. We use the Bayes' formula for each branch of the tree in fig. 17.2.

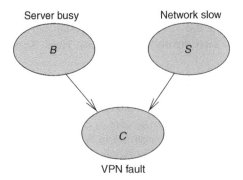

Figure 17.2: A simple Bayesian network with two possible cause hypotheses.

If we learn that B caused the time out, that is, the server is busy, then we apply the Bayes' formula for propagating that fact into the probability for B being the correct hypothesis.

$$P(B|F) = \frac{P(F|B)P(B)}{P(F)}$$
$$= \frac{[P(F|BS)P(S) + P(F|B\neg S)P(\neg S)]\,P(B)}{P(F)}$$
$$= \frac{(0.8 \times 0.4 + 0.6 \times 0.6)}{0.518}$$
$$= 0.131. \tag{17.16}$$

This is the probability that it was B that caused the fault F. It should be compared with P(B)—it is slightly larger, that is, the new evidence makes the likelihood of this hypothesis more likely (though still not very likely). If we learn, on the other hand, that the slow network S is the cause of the error, we use

$$P(S|F) = \frac{P(F|S)P(S)}{P(F)}$$
$$= \frac{[P(F|BS)P(B) + P(F|B\neg S)P(\neg S)]\,P(S)}{P(F)}$$
$$= \frac{(0.8 \times 0.1 + 0.5 \times 0.9)}{0.518}$$
$$= 0.409. \tag{17.17}$$

This can be compared with P(S). Again, it is slightly larger, adding more support to that hypothesis. To take account of this learned experience, we might consider replacing P(B), P(S) with the new estimates 0.131, 0.409 and using the Bayesian rule again. This process of iteration can sometimes result in the determination of optimum estimates, in which case one speaks of Bayesian learning.

Taking the new estimate and feeding it back into the formula to obtain a new one forms the basis of what is known as *Bayesian learning*. It is a gradual refinement of certainty about estimated values. To make this idea precise, we must stray into more technical discussions about parameter estimation that are beyond the scope of this book, so we leave this as an open problem for the reader to investigate further. The implications for measurement are that even an imperfect experiment is better than no experiment. We can make precise use of imperfect data, in a way that can be revised later if more observations can be made.

17.6 Knowledge out of date

The process of learning is not only advantageous; an excess of knowledge amounts to a prejudice. Some kinds of knowledge have a sell-by date, after which the knowledge no longer applies. One example of this is in anomaly detection, in which the distant past history is of little interest to the recent past. If knowledge accumulates for longer than the period over which one expects policy to be constant, then it becomes anomalous itself according to the new policy.

There are two approaches to retaining finite window knowledge: a fixed-width sliding window can be used to eliminate any data points that were accumulated before a certain fixed-width interval, measured backwards in time from the present; alternatively, old knowledge can gradually be degraded by assigning newer knowledge a higher weight (see Burgess (2002b)).

Example 154 *Anomaly detection is usually performed using one of two techniques: off-line time series analysis, or real-time event threshold processing. Time series data consume a lot of space and the subsequent calculation of local averages costs a considerable amount of CPU time as the window of measurement increases. In (Burgess (2002b)), it is shown how compression of the data can be achieved, and computation time can be spared by the use of iterative updating, by geometrical series convergence. The key to such a compression is to update a sample of data iteratively rather than using an off-line analysis based on a complete record.*

The approximate periodicity observed in computer resources allows one to parameterize time in topological slices of period P, using the relation

$$t = nP + \tau. \tag{17.18}$$

This means that time becomes cylindrical, parameterized by two interleaved coordinates (τ, n), both of which are discrete in practice. This parameterization of time means that measured values are multi-valued on over the period $0 \geq \tau < P$, and thus one can average the values at each point τ, leading to a mean and standard deviation of points. Both the mean and standard deviations are thus functions of τ, and the latter plays the role of a scale for fluctuations at τ, which can be used to grade their significance.

The cylindrical parameterization also enables one to invoke a compression algorithm on the data, so that one never needs to record more data points than exist within a single period. It thus becomes a far less resource-intensive proposition to monitor system normalcy.

An iteration of the update procedure may be defined by the combination of a new data point q with the old estimate of the average $\overline{q}$.

$$\overline{q} \to \overline{q}' = (q|\overline{q}) \tag{17.19}$$

where

$$(q|\overline{q}) = \frac{w\,q_1 + \overline{w}\,\overline{q}}{w + \overline{w}}. \qquad (17.20)$$

This is somewhat analogous to a Bayesian probability flow. The repeated iteration of this expression leads to a geometric progression in the parameter $\lambda = \overline{w}/(w + \overline{w})$:

$$(q_1|(q_2|(\ldots(q_r|(\ldots|\overline{q}_n)))) = \frac{w}{w + \overline{w}}\,q_1 + \frac{\overline{w}w}{(w + \overline{w})^2}\,q_2 + \cdots$$
$$+ \frac{w\,\overline{w}^{\,r-1}}{(w + \overline{w})^r}\,q_r + \cdots \frac{\overline{w}^{\,n}}{(w + \overline{w})^n}\,\overline{q}_n. \qquad (17.21)$$

Thus on each iteration, the importance of previous contributions is degraded by λ. *If we require a fixed window of size N iterations, then* λ *can be chosen in such a way that, after N iterations, the initial estimate* q_N *is so demoted as to be insignificant, at the level of accuracy required. For instance, an order of magnitude drop within N steps means that* $\lambda \sim |10^{-N}|$. *Using the definition of the pseudo-fixed-window average* $\langle \ldots \rangle_N$, *we may now define the average standard deviation, or error, by*

$$\langle \sigma(\tau) \rangle \equiv \sqrt{\langle (\delta q(\tau))^2 \rangle_N}$$
$$\delta q(\tau) = q(t) - \langle q(\tau) \rangle_N \qquad (17.22)$$

This has similar properties to the degrading average itself, though the square root makes the accuracy more sensitive to change.

In order to satisfy the requirements of a decaying window average, with determined sensitivity $\alpha \sim 1/N$, *we require,*

1. $\frac{w}{w+\overline{w}} \sim \alpha$, *or* $w \sim \overline{w}/N$.

2. $\left(\frac{\overline{w}}{w+\overline{w}} \right)^N \ll \frac{1}{N}$, *or* $\overline{w}N \ll w$.

Consider the ansatz $w = 1 - r$, $\overline{w} = r$, *and the accuracy* α. *We wish to solve*

$$r^N = \alpha \qquad (17.23)$$

for N. With $r = 0.6$, $\alpha = 0.01$, *we have* $N = 5.5$. *Thus, if we consider the weekly update over five weeks (a month), then the importance of month-old data will have fallen to one-hundredth. This is a little too quick, since a month of fairly constant data is required to find a stable average. Taking* $r = 0.7$, $\alpha = 0.01$, *gives* $N = 13$. *On the basis of experience with off-line analysis, this is a reasonable arbitrary value to choose.*

17.7   Convergence of the learning process

The process of learning should converge with time; that is, we would like the amount of information on which a system depends to be non-increasing. If the information required to perform a function diverges, we have no hope of managing the process. Then, we must be contented with watching the system in either despair or wonder as it continues to operate.

If, however, the information base of the system is constant or even diminishing, then there is a possibility to learn sufficient information to manage system expertise.

Feeding the *a postiori* estimates back into the input of Bayes' formula provides a method (with some reservations) of iteratively determining the 'true' cause of the data, as one of the available hypotheses. It is when the prior probabilities (in their nth iteration) tend towards a very specific answer (low uncertainty or low entropy) that one says the true value has been *learned*. In section 9.11, we noted that the principle of maximum informational entropy is a way of modelling the effect of maximum uncertainty, under given conditions. The maximum entropy principle is a model for forces that tend to increase uncertainty, such as random fluctuations and errors. Bayesian learning is the logical opposite of this; thus we can think of learning and entropy as being competing forces of uncertainty and certainty in a system.

Finally, note that this brings us back, once again, to the idea of *convergence* and *policy* (see sections 5.8 and 10.4). In a well-posed system, learning will tend towards a unique result and stay there in the limit of many observations. This is a stability criterion for the algorithmic flow of cause and effect in the system. Creating such systems predictably is still an open problem in many areas, but note the approaches of convergent configuration management in (Burgess (2004); Couch and Sun (2003)).

Applications and Further Study 17

- *Discussion of knowledge and expertise in a quantitative fashion.*

- *A strategy for collecting data as a basis for expert knowledge.*

- *Automation of data acquisition (machine learning).*

18

Policy transgressions and fault modelling

'And now remains
That we find out the cause of this effect
Or rather say, the cause of this defect,
For this effect defective comes by cause.'

—Shakespeare, *(Hamlet II. ii.100-4)*.

Non-deterministic systems are usually affected by the arrival of events or random occurrences, some of which are acceptable to system policy and some of which are not. Events that are not acceptable may be called faults. The occurrence of system faults is an extensive and involved topic that is the subject of whole texts (see, for instance NRC (1981) and Steinder and Sethi (2003)). This chapter could not cover the breadth of this subject in any detail, so we attempt to distill one aspect of its essence in relatively simple terms and extract some conclusions from a management perspective for general synthesis in this work. Readers may consult (Steinder and Sethi (2003)) for a survey of technological methods of fault localization in systems.

18.1 Faults and failures

System faults fall into three main categories:

- *Random faults*: Unpredictable occurrences or freaks of nature.

- *Emergent faults*: Faults that occur because of properties of the system which it was not designed for. These usually come about once a system is in contact with an environment.

- *Systemic faults*: Faults that are caused by logical errors of design, or insufficient specification.

Analytical Network and System Administration. Managing Human–Computer Networks   Mark Burgess
© 2004 John Wiley & Sons, Ltd   ISBN 0-470-86100-2

The IEEE classification of computer software anomalies (IEEE (n.d.)) includes the following issues: operating system crash, program hang-up, program crash, input problem, output problem, failed required performance, perceived total failure, system error message, service degraded, wrong output and no output. This classification touches on a variety of themes, all of which might plague the interaction between users and an operating system. Some of these issues encroach on the area of performance tuning, for example, service degraded. Performance tuning is certainly related to the issue of availability of network services and thus, this is a part of system administration. However, performance tuning is of only peripheral importance compared to the matter of possible complete failure.

Many of the problems associated with system administration can be attributed to input problems (incorrect or inappropriate configuration) and failed performance through loss of resources. Unlike many software situations, these are not problems that can be eliminated by re-evaluating individual software components. In system administration, the problems are partly social and partly owing to the cooperative nature of the many interaction software components. The unpredictability of operating systems is dominated by these issues.

Another source of error is found at the human end of the system:

- Management errors

- Forgetfulness/carelessness

- Misunderstanding/miscommunication

- Confusion/stress/intoxication

- Ignorance

- Personal conflict

- Slowness of response

- Random or systematics procedural errors

- Inability to deal with complexity

- Inability to cooperate with others.

In system administration, the problems are partly social and partly due to the cooperative nature of the many interaction software components. The unpredictability of operating systems is dominated by these issues.

Humans filter all communication through their own view of the world. We respond to things that make sense to us, and we tend to reject things that do not. This can lead to misunderstanding, or only partial understanding of a communicated message (see chapter 15). It can be modelled as the projection of a signal into a digital alphabet that describes our limited domain of understanding. We match input to the closest concept we already know.

Unlike machines, humans do not generally use reliable protocols for making themselves understood (except perhaps in military operations). A system administrator or a user can easily misunderstand an instruction, or misdiagnose a problem.

When a problem arises, it means that an undesirable change has occurred in the system. Debugging is a meta process—a process about the system, but not generally within the system itself. It involves gathering evidence and tracing cause-relationships. A fault might be

- a conflict with policy

- a logical error

- a random error

- an emergent fault.

We can adopt different strategies for solving a problem, as follows:

- Mitigate the damage by relieving symptoms,

- Fix the cause of the problem at source.

In many complex systems, it is profitable to employ both of these.

Example 155 *The immune system that protects higher animals from infections uses short-term counter measures to prevent the spread of infection (chemicals that retard cell replication processes, for instance), while at the same time synthesizing a counter-agent to the specific threat (antibodies and killer cells).*

18.2 Deterministic system approximation

A simplistic, but effective, mode of analysis of systems is to treat them as deterministic Boolean directed graphs. These are system representations that classify systems into those that either work or do not work. It is essentially a dependency analysis of systems, composed of networks of components.

Consider a system of order n components, labelled by a vector or coordinates $\vec{x} = (x_1, x_2, \ldots, x_n)$, joined together by links that transmit information. It is normal to have one or more entry points to this network, and one or more exit points. Information flows along the links from input to output.

A deterministic analysis assumes a two state model, in which

$$x_i = \begin{cases} 1 & \text{– if component works} \\ 0 & \text{– if component is broken} \end{cases} \qquad (18.1)$$

This appears, at first, to be rather simplistic; however, its chief value is in showing that the concept of 'not working' can be broken down into a more detailed view in which the source of failure is determined to be a single component. An alternative interpretation of this model is to view the connections between components to mean 'depends on' rather than 'results in'.

Let us define a system configuration in terms of the so-called *structure function* $\phi(\vec{x})$ that summarizes the dependencies of the system on its components. (When we extend this analysis to include stochastic (random) failure events, this will become a macrostate function.)

Definition 70 (Structure function) *The system's state of repair is described uniquely by* $\phi(\vec{x})$:

$$\phi = \begin{cases} 1 & \text{– if the system works} \\ 0 & \text{– if the system is broken} \end{cases} \qquad (18.2)$$

Figure 18.1: System components in series, implies dependency.

Example 156 *Let x_i, ($i = 1, 2, 3, 4$) describe the CPU, disks and memory, and kernel of a computer database system. The structure function for this system is*

$$\phi(\vec{x}) = x_1 \cdot x_2 \cdot x_3 \cdot x_4. \tag{18.3}$$

According to this deterministic model, if a single component is not working, then the entire system is not working.

The example above shows the simple view taken by this analysis. Clearly, there are various degrees by which the memory or disks of a computer system might not work, but we are not able to describe that yet. Nonetheless, there is something to be learned from this simple approach, before we extend it to cope with partial or probabilistic failures.

A serial dependency structure (fig. 18.1) works if and only if each of the components works. The combination is by the Boolean AND operation:

$$\phi(\vec{x})_{\text{serial}} = x_1 \text{ AND } x_2 \text{ AND } \ldots x_n = \prod_{i=1}^{n} x_i$$

$$= \min_{i} x_i. \tag{18.4}$$

A parallel dependency structure (fig. 18.2) works if at least one of its components works. The combination is by the Boolean OR operation:

$$\phi(\vec{x})_{\text{serial}} = x_1 \text{ OR } x_2 \text{ OR } \ldots x_n = 1 - \prod_{i=1}^{n} (1 - x_i)$$

$$= \coprod_{i=1}^{n} x_i. \tag{18.5}$$

The 'voting' gate or k of n requires k out of the n components to work. If $k = 1$, this is a parallel connection; if $k = n$, it is a serial connection. Clearly, this interpolates between these two cases.

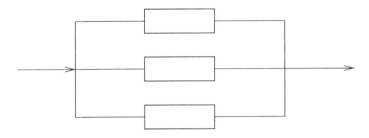

Figure 18.2: System components in parallel, implies redundancy.

So far we have looked at the system from the viewpoint of a random failure of a whole component. The dual description of the system describes the viewpoint of an attacker or a saboteur.

Definition 71 (Dual structure) *Given a structure function* $\phi(\vec{x})$, *we can define the dual function* ϕ^D, *or the dual vector* $\vec{x}^D$.

$$\phi^D(\vec{x}^D) = 1 - \phi(\vec{x}), \tag{18.6}$$

and

$$\vec{x}^D = (1 - x_1, 1 - x_2, \ldots, x_n) = \vec{1} - \vec{x}. \tag{18.7}$$

From the dual viewpoint, we see the vulnerabilities of the system more explicitly. If a system works when only a single component functions (parallel system), then it does not work if $(n - 1)$ additional components are destroyed. If a system works only when all components function, then we have only to destroy 0 additional components to destroy the system.

Normalization criteria

The concept of irrelevant components is a way of identifying and eliminating redundant parameterization in the structure function. It is a way of pruning the graph of irrelevant nodes. We are interested only in the relevant components.

Definition 72 (Relevant components) *The* ith *component of a system is relevant to its structure if* $\phi(\vec{x})$ *is a non-constant function of* x_i, *that is,*

$$\frac{\partial \phi}{\partial x_i} \neq 0, \tag{18.8}$$

that is, the function depends non-trivially on x_i.

Another criterion for discussing only rationally constructed system is to consider only those systems in which the repair of a component never makes the system worse[1]. Such systems are said to be *coherent* or *monotonic*, in reliability analysis.

Definition 73 (Coherent and monotonic systems) *A system of components is said to be coherent if and only if* $\phi(\vec{x})$ *is a non-decreasing function of* x_i, *and all the components are relevant, that is, iff*

$$\frac{\partial \phi}{\partial x_i} > 0. \tag{18.9}$$

It is additionally monotonic if we have $\phi(\vec{0}) = 0$ *and* $\phi(\vec{1}) = 1$, *which is equivalent to requiring at least one relevant component.*

The requirement of coherence might seem superficially obvious, but if there are mutually exclusive events in a system, parameterized by separate coordinates x_i, the positivity of the structure dependence is not guaranteed.

---

[1] Note the similarity of this concept to that of convergence in section 5.8

Example 157 *Consider a system for providing fault-free access to a network server, using a fail-over server. Let x_1 be non-zero if server 1 is active and x_2 be non-zero if server 2 is active. Since the events are mutually exclusive, the structure function is a convex mixture of these:*

$$\phi(\vec{x}) = x_1(1 - x_2) + x_2(1 - x_1). \tag{18.10}$$

Clearly,

$$\frac{\partial \phi}{\partial x_i} < 0, \ i = 1, 2. \tag{18.11}$$

The same example could be applied to different shifts of human workers, in providing a 'round-the-clock service. The fact that the system is non-coherent means that the failure of one server does not leave the system in a non-vulnerable state.

Redundancy folk theorem

A folk theorem about redundancy that follows from this simple deterministic model concerns where to arrange for redundancy in a system. Roughly speaking, it says that a parallel coupling of components (i.e. a low-level parallelism) is never worse than a high-level parallelism. In formal terms, this follows from two inequalities. Using the notation,

$$x_i \bigsqcup y_j \equiv 1 - (1 - a_i)(1 - b_j), \tag{18.12}$$

and

$$\vec{x} \bigsqcup \vec{y} = (x_1 \bigsqcup y_1, x_2 \bigsqcup y_2, \ldots, x_n \bigsqcup x_n), \tag{18.13}$$

we have the inequalities for parallelization

$$\phi(\vec{x} \bigsqcup \vec{y}) \geq \phi(\vec{x}) \bigsqcup \phi(\vec{y}). \tag{18.14}$$

This tells us that the working condition of a system with redundant components is never worse than a redundant combination systems of non-redundant components. For serialization, the opposite is true:

$$\phi(\vec{x} \cdot \vec{y}) \leq \phi(\vec{x}) \cdot \phi(\vec{y}), \tag{18.15}$$

that is, the working condition of a system with strong serial dependencies is never better than a series of dependent systems. The message in both the cases is that the lower the level at which we can introduce redundancy, the better.

Example 158 *Keeping a server reliable using RAID disks arrays, multiple CPUs and error correcting memory, is never worse than keeping two independent systems with single disks, single CPUs and so on.*

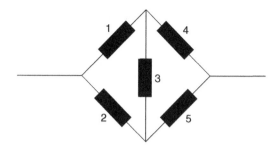

Figure 18.3: A Wheatstone bridge configuration of components provides a good example for demonstrating the concept of minimal path sets and minimal cut sets.

Pathways and cut sets

Two concepts about flow graphs that illuminate their vulnerabilities are pathways and cut sets. We can partition the components in a system into two sets: those that work and those that do not. If $\phi(\vec{x})$ be a monotonic system, then a vector $\vec{x}$: $\phi(\vec{x}) = 1$, that is, there is a pathway through the system that works. A minimal path vector is one in which all the components work along the path. A vector is a cut vector if $\phi(\vec{x}) = 0$, that is, if the vector leads to a broken system. The minimal cut set can then be defined in relation to this as a set of components such that, if all components in the set are broken, the system is broken.

Example 159 *Consider the network in fig. 18.3, taking an input on the left of the graph to an output on the right-hand side. The minimal path sets are seen by referring to the figure.*

$$P_1 = \{1, 4\}, \quad P_2 = \{2, 5\}, \quad P_3 = \{1, 3, 5\}, \quad P_4 = \{2, 3, 4\}. \tag{18.16}$$

The minimal cut sets are

$$C_1 = \{1, 2\}, \quad C_2 = \{4, 5\}, \quad C_3 = \{1, 3, 5\}, \quad C_4 = \{2, 3, 4\}. \tag{18.17}$$

18.3 Stochastic system models

The deterministic analysis above is a useful point of reference that can sometimes be applied directly to human–computer systems. The weakness of the deterministic view of systems is that one cannot ask questions like the following: If only 0.1 percent of my disk has an error, what is the likelihood that this will prevent the whole system from working at any given time? Or, if the probability of reaching the customer help-desk service on the telephone within 30 minutes is 0.3, what is the likelihood that the customer's enterprise will lose business?

A more flexible analysis of the system treats component states as probabilistic variables. Such a stochastic analysis can also be used in a predictive way, to develop architectural strategies (see chapter 13). It is important for continuum approximations of systems, as only probabilities or averages can vary smoothly and characterize the changing external conditions of a system.

Using the notation of the previous section, we define the probability of a component's working (i.e. the reliability of the component) as the expectation value of the component's state:

$$p_i = P(x_i = 1) = \langle x_i \rangle. \tag{18.18}$$

The reliability of the whole system is then given by

$$\rho \equiv P(\phi(\vec{x}) = 1) = \langle \phi(\vec{x}) \rangle. \tag{18.19}$$

ρ is called the reliability function; it is a generalization of the structure function from Boolean estimates to real-valued probabilities.

Relationship between ϕ and ρ

We can see the relationship between these by taking the deterministic limit of the stochastic model, that is, the limit in which probabilities are either zero or one. Then, it follows that they are functionally the same:

$$\rho(\vec{p}) = \phi(\vec{p}). \tag{18.20}$$

This follows since, if $p_i = \{0, 1\}$, $x_i = p_i$

$$\rho(\vec{p}) = \langle \phi(\vec{x}) \rangle = \phi(\vec{x}) = \phi(\vec{p}). \tag{18.21}$$

This tells us how to compute the probabilities of system reliability, given component reliabilities.

Example 160 *For a serially coupled system:*

$$\rho(\vec{p}) = \langle x_1 \text{ AND } x_2 \text{ AND } \ldots \text{ AND } x_n \rangle$$

$$= \langle \prod_{i=1}^{n} x_i \rangle$$

$$= \prod_{i=1}^{n} p_i, \tag{18.22}$$

that is, the combined probability is the product of the component reliabilities.

Example 161 *For a system with a parallel coupling of components:*

$$\rho(\vec{p}) = \langle \coprod x_i \rangle$$

$$= \langle 1 - \prod_{i=1}^{n} (1 - x_i) \rangle$$

$$= 1 - \prod_{i=1}^{n} (1 - \langle x_i \rangle)$$

$$= \coprod_{i=1}^{n} p_i. \tag{18.23}$$

These rules allow us to generalize the folk theorems about system redundancy in a straight-forward manner:

$$\rho\left(\vec{p} \coprod \vec{p}'\right) \geq \rho(\vec{p}) \coprod \rho(\vec{p}')$$
$$\rho\left(\vec{p} \cdot \vec{p}'\right) \leq \rho(\vec{p}) \cdot \rho(\vec{p}').$$
(18.24)

In other words, the probability that a single parallelized system will be working is greater than or equal to the probability that parallel components will both be working. Conversely, the probability that a system that depends on serialization is working is always less than the probability that the components are working. Note that there is no impediment to making probabilities into functions of time, in order to track changing conditions.

Birnbaum measure of structural importance

One measure of component importance that is related to the minimal pathways and cut sets of the deterministic analysis is the Birnbaum measure, defined as follows:

Definition 74 (Birnbaum importance) *The partial rate of change of the system reliability ρ with respect to a given component reliability p_i indicates its structural dependence on the component.*

$$I_{\mathrm{B}}^{(i)} \equiv \frac{\partial}{\partial p_i}\rho(\vec{p}).$$
(18.25)

This measure is easily calculated from a knowledge of the structure function of the system, and it describes the probability that the ith component lies in a critical path vector of the system, that is, that a failure of component i would lead to a failure of the system. To see this, we note that, for a monotonic system,

$$\rho(\vec{p}) = p_i\rho(1_i, p_{-i}) + (1 - p_i)\rho(0_i, p_{-i}),$$
(18.26)

that is ,the reliability of the system is equal to the probability that i is working multiplied by the reliability of the system, given that i is working, plus (OR) the probability that component i is not working and that the rest of the components' states are unknown. Using this expansion that gives special prominence to i, we can now examine the Birnbaum measure for component i:

$$\begin{aligned}
I_{\mathrm{B}}^{(i)} &= \frac{\partial}{\partial p_i}\rho(\vec{p}) \\
&= \rho(1_i, p_{-i}) - \rho(0_i, p_{-i}).
\end{aligned}$$
(18.27)

This is an expression of the conditional probability that the system is working, given that i is working, minus the conditional probability that the system is working given that i is not working; that is, it is the change in reliability as a result of i being repaired.

We can rewrite this as follows:

$$\begin{aligned}
I_{\mathrm{B}}^{(i)} &= \Delta_i\rho(\vec{p}) \\
&= \langle\phi(1_i, x_{-i})\rangle - \langle\phi(0_i, x_{-i})\rangle \\
&= P\left(\left[\phi(1_i, x_{-i}) - \phi(0_i, x_{-i})\right] = 1\right) \\
&= P\left((1_i, x_{-i}) \text{ is a critical pathway}\right).
\end{aligned}$$
(18.28)

Put another way, if the change in structure $\Delta_i \phi = 1$, then it must mean that the system breaks when i is destroyed. Thus, the Birnbaum measure tells us the probable importance of the ith component to overall reliability, assuming that we know the structural form of the system.

Example 162 *For a system of components in series:*

$$I_{\mathrm{B}}^{(i)} = \frac{\partial}{\partial p_i} \left(\prod_j p_j \right) = \prod i \ne j \, p_j. \tag{18.29}$$

Suppose we order the reliabilities of the components (this results in no loss of generality), so that

$$p_1 \le p_2 \le p_3 \cdots \le p_n. \tag{18.30}$$

This implies that

$$I_{\mathrm{B}}^{(1)} \ge I_{\mathrm{B}}^{(2)} \ge I_{\mathrm{B}}^{(3)} \cdots \ge I_{\mathrm{B}}^{(n)}. \tag{18.31}$$

Or, the component with the lowest reliability has the greatest importance to the reliability of the system, that is, 'a chain is only as strong as its weakest link'.

Example 163 *For a system of components in parallel:*

$$I_{\mathrm{B}}^{(i)} = \frac{\partial}{\partial p_i} \left(\coprod_j p_j \right)$$
$$= \prod_{i \ne j} (1 - p_j). \tag{18.32}$$

Ordering probabilities again,

$$p_1 \le p_2 \le p_3 \cdots \le p_n. \tag{18.33}$$

we have that

$$I_{\mathrm{B}}^{(1)} \le I_{\mathrm{B}}^{(2)} \le I_{\mathrm{B}}^{(3)} \cdots \le I_{\mathrm{B}}^{(n)}. \tag{18.34}$$

Or, the component with the highest reliability has the greatest structural importance to the overall reliability since, if it has failed, it is likely that all the others have failed too.

Correlations and dependencies

The interdependence of components can be important to a system in a number of ways. For instance, if one component fails, others might fail too. Or, related components might experience heavy loads or stresses together. In sub-systems that are used to balance load between incoming information, the failure of one component might lead to an extra load on the others.

The correlation of component reliabilities is an indication of such interdependence. One measures this using the statistical covariance or un-normalized correlation function of the variables. If variables are associated, then

$$\begin{aligned} \text{cov}(x, x') &= \langle (x - \langle x \rangle)(x' - \langle x' \rangle) \rangle \\ &= \langle xx' \rangle - \langle x \rangle \langle x' \rangle \\ &\geq 0. \end{aligned} \tag{18.35}$$

Readers are referred to texts on reliability theory, for example (Natvig (1998); Høyland and Rausand (1994)), for details on this. We note in passing that ignoring correlations can lead to erroneous conclusions about system reliabilities. The assumption of independence of components x_i in a serial structure leads to an *underestimation* of the reliability ρ, in general, whereas the assumption of independence of components in a parallel structure leads to an *overestimation* of ρ, in general.

18.4 Approximate information flow reliability

One of the aims of building a sturdy infrastructure is to cope with the results of failure. Failure can encompass hardware and software. It includes downtime due to physical error (loss of power, communications etc.) and also downtime due to software crashes. The net result of any failure is loss of service.

Our main defences against actual failure are parallelism (*redundancy*) and maintenance. When one component fails, another can be ready to take over. Often, it is possible to *prevent failure* altogether with pro-active maintenance (see the next chapter for more on this issue). For instance, it is possible to vacuum clean hosts, to prevent electrical short-circuits. It is also possible to perform garbage collection that can prevent software error. System monitors (e.g. cfengine) can ensure that crashed processes get restarted, thus minimizing loss. Reliability is clearly a multifaceted topic. We shall return to discuss reliability more quantitatively in section 18.4.

Component failure can be avoided by parallelism, or redundancy. One way to think about this is to think of a computer system as providing a service that is characterized by a flow of information. If we consider fig. 18.2, it is clear that a flow of service can continue, when servers work in parallel, even if one or more of them fails. In fig. 18.1, it is clear that systems that are dependent on other series are coupled in series and a failure prevents the flow of service. Of course, servers do not really work in parallel. The normal citation is to employ a *fail-over* capability. This means that we provide a backup service. If the main service fails, we replace it with a backup server. The backup server is not normally used, however. Only in a few cases can one find examples of load-sharing by switching between (de-multiplexing) services.

Reliability cannot be measured until we define what we mean by it. One common definition uses the *average (mean) time before failure* as a measure of system reliability. This is quite simply the average amount of time we expect to elapse between serious failures of the system. Another way of expressing this is to use the *average uptime*, or the amount of time for which the system is responsive (waiting no more than a fixed length of time for a response). Another complementary figure is then the *average downtime*, which is the average amount of time the system is unavailable for work (a kind of informational entropy). We can define the reliability as the probability that the system is available:

$$\rho = \frac{\text{Mean uptime}}{\text{Total elapsed time}} \tag{18.36}$$

Some like to define this in terms of the Mean Time Before Failure (MTBF) and the Mean Time To Repair (MTTR), that is,

$$\rho = \frac{\text{MTBF}}{\text{MTBF} + \text{MTTR}}. \tag{18.37}$$

This is clearly a number between 0 and 1. Many network device vendors quote these values with the number of 9's it yields, for example, 0.99999.

Flow of services

The effect of parallelism, or redundancy on reliability can be treated as a facsimile of the Ohm's law problem, by noting that service provision is just like a flow of work (see also section B.24 for examples of this).

Rate of service (delivery) = rate of change in information / failure fraction

This is directly analogous to Ohm's law for the flow of current through a resistance:

$$I = V/R \tag{18.38}$$

The analogy is captured in this table:

Potential difference V	Change in information
Current I	Rate of service (flow of information)
Resistance R	Rate of failure or delay

This relation is simplistic. For one thing, it does not take into account variable latencies (although these could be defined as failure to respond). It should be clear that this simplistic equation is full of unwarranted assumptions, and yet its simplicity justifies its use for simple hand-waving. If we consider fig. 18.2, it is clear that a flow of service can continue when servers work in parallel, even if one or more of them fails. In fig. 18.1, it is clear that systems that are dependent on other series are coupled in series and a failure prevents the flow of service. Because of the linear relationship, we can use the usual Ohm's law expressions for combining failure rates:

$$R_{\text{series}} = R_1 + R_2 + R_3 + \cdots \tag{18.39}$$

and

$$\frac{1}{R_{\text{Parallel}}} = \frac{1}{R_1} + \frac{1}{R_2} + \frac{1}{R_3} \cdots \tag{18.40}$$

These simple expressions can be used to hand-wave about the reliability of combinations of hosts. For instance, let us define the rate of failure to be a probability of failure, with a value between 0 and 1. Suppose we find that the rate of failure of a particular kind of server is 0.1. If we couple two in parallel (a double redundancy), then we obtain an effective failure rate of

$$\frac{1}{R} = \frac{1}{0.1} + \frac{1}{0.1} \qquad (18.41)$$

that is, $R = 0.05$, the failure rate is halved. This estimate is clearly naive. It assumes, for instance, that both servers work all the time in parallel. This is seldom the case. If we run parallel servers, normally a default server will be tried first, and, if there is no response, only then will the second backup server be contacted. Thus, in a fail-over model, this is not really applicable. Still, we use this picture for what it is worth, as a crude hand-waving tool.

The Mean Time Before Failure (MTBF) is used by electrical engineers who find that its values for the failures of many similar components (say light bulbs) has an exponential distribution. In other words, over large numbers of similar component failures, it is found that the probability of failure has the exponential form

$$P(t) = \exp(-t/\tau) \qquad (18.42)$$

or that the probability of a component lasting time t is the exponential, where τ is the mean time before failure and t is the failure time of a given component. There are many reasons why a computer system would not be expected to have this simple form. One is *dependency*, which causes events to be correlated rather than be independent.

Thus, the problem with these measures of system reliability is that they are difficult to measure and assigning any real meaning to them is fraught with subtlety. Unless the system fails regularly, the number of points over which it is possible to average is rather small.

18.5 Fault correction by monitoring and instruction

Let us now use the flow approach described above to analyse the likely success of a number of common network topologies. Many systems rely on centralized management, but we know that centralization, while a cheap strategy, is fragile since it results in many points of failure. The efficiency of fault correction models has been estimated under different communication infra-structures to analyse the scalability of the solutions (see Burgess and Canright (2003)). A simple estimate of the scalability of fault correction can be found by using the system model in chapter 15 for time evolution and error correction.

The simplest estimate is made by assuming that the reliability of each component in a system and each channel is independent of all others, so that the probabilities of resource availability are all independent random variables. This suffices to discuss many aspects of reliability and scaling. If a system component or dependency fails or becomes outdated, a 'repair' or update requires a communication with the component from some source of 'correctness' or policy.

Let a set of components or resources in a system be defined by a column vector of probabilities

$$\vec{C} = \begin{pmatrix} p_1 \\ p_2 \\ \vdots \\ p_N \end{pmatrix} \tag{18.43}$$

where $p_i (i = 1 \ldots N)$ is the probability that component i is available. If the probabilities are 1, the hosts are said to be reliable, otherwise they are partially reliable.

The channels of information and flow that link the components are represented in the adjacency matrix of the network. This matrix need not be symmetrical in practice, but we shall not address that issue here.

We define a simple measure of the availability of a service, using the connectivity of the graph χ (see section 6.2). χ has a maximum value of 1, when every node is connected to every other, and a minimum value of zero, when all nodes are disconnected.

For a fixed topology and time-independent node availabilities, χ is a constant characterizing the network. In general χ is time-dependent, as the system evolves; one then obtains a static figure for the network by taking the long-time average:

$$\langle \chi \rangle = \lim_{n \to \infty} \frac{1}{n} \sum_{i=1}^{n} \chi(t_i). \tag{18.44}$$

The purpose of this measure is that it enables us to gauge and compare different network configurations on equal terms. It is also a measure for comparison by which we can map the problem of unreliable components in a fixed network onto a corresponding problem of reliable components in an *ad hoc* network.

A duality: *ad hoc* networks and unreliable components

Ad hoc networks are networks whose adjacency matrices are subject to a strong, apparently random time variation. If we look at the average adjacency matrices, over time, then we can represent the probability of connectivity in the network as an adjacency matrix of probabilities.

Example 164 *In an ad hoc communications network, with a fixed number of components, the links are not independent variables. They are constrained both by the physical geography in which the components move (only nearby components are candidates for links), and by interference effects among the set of components near a given component. Any given component thus may or may not establish a working link with a near component, depending on interference from other near components.*

For our purposes here, these dependencies are not important; the important property of the ad hoc net is the intermittency of the links, towing to the components' mobility.

Definition 75 (Ad hoc adjacency matrix) *An ad hoc network is represented by a symmetric matrix of probabilities for adjacency. Thus, the time average of the adjacency matrix (for e.g., four components) may be written as*

$$\langle A \rangle = \begin{pmatrix} 0 & p_{12} & p_{13} & p_{14} \\ p_{21} & 0 & p_{23} & p_{24} \\ p_{31} & p_{32} & 0 & p_{34} \\ p_{41} & p_{42} & p_{43} & 0 \end{pmatrix} \tag{18.45}$$

An ad hoc network is therefore a partially reliable network.

To motivate our discussion further, we note that:

Theorem 18.5.1 *A fixed network of partially reliable components, C_i, is equivalent to an ad hoc network of reliable components, on average.*

Proof. This is easily seen from the definition of the connectivity, using a matrix component form:

$$\begin{aligned} N(N-1)\langle \chi \rangle &= \vec{C}(p')^{\mathrm{T}} \langle A(1) \rangle \, \vec{C}(p) \\ &= \vec{C}(1)^{\mathrm{T}} \langle A(pp') \rangle \, \vec{C}(1) \\ &= \sum_{ij} C_i(p_i) \langle A_{ij}(1) \rangle \, C_j(p_j) \\ &= \sum_{ij} C_i(1) \langle A_{ij}(p_i p_j) \rangle \, C_j(1). \end{aligned} \tag{18.46}$$

The proof demonstrates the fact that one can move the probabilities (uncertainties) for availability from the host vectors to the connectivity matrix and vice versa; for example,

$$\begin{pmatrix} p_1 \\ p_2 \\ p_3 \end{pmatrix}^{\mathrm{T}} \begin{pmatrix} 0 & 1 & 1 \\ 1 & 0 & 1 \\ 1 & 1 & 0 \end{pmatrix} \begin{pmatrix} p_1 \\ p_2 \\ p_3 \end{pmatrix} =$$

$$\begin{pmatrix} 1 \\ 1 \\ 1 \end{pmatrix}^{\mathrm{T}} \begin{pmatrix} 0 & p_1 p_2 & p_1 p_3 \\ p_2 p_1 & 0 & p_2 p_3 \\ p_3 p_1 & p_3 p_2 & 0 \end{pmatrix} \begin{pmatrix} 1 \\ 1 \\ 1 \end{pmatrix}. \tag{18.47}$$

Thus, an array of system components with reliability probabilities p_i, is equivalent to an array of completely reliable components in an unreliable network, where the probability of communication between them is the product of probabilities (assumed independent) from the reliability vector.

Policy current in a graph

As networks grow, some system structure topologies do not scale well. We are interested in examining the scaling properties of different configuration management schemes, especially in the context of network models that look to the future of configuration management.

Using even the most simplistic analysis, we can consider a number of cases, in order of decreasing centralization, to find the worst-case scaling behaviours. Our discussion follows Burgess and Canright (2003)[2].

We assume a simple linear relationship between the probability of successful maintenance and the rates of communication with the policy and enforcement sources. This need not be an accurate description of reality in order to lead to the correct scaling laws (see section 12.8). Let us suppose that a change of configuration ΔQ is proportional to an average rate of information flow I, over a time Δt; that is,

$$\Delta Q = I \Delta t. \tag{18.48}$$

This equation says that I represents the time-averaged flow over the interval of time for which is acts. As we are interested in the limiting behaviour for long times, this is sufficient for the job.

Now we apply this simple picture to configuration management for dynamic networks. We take the point of view of a 'typical' or 'average' host. It generates error in its configuration at the (average) rate I_{err}, and receives corrections at the rate I_{repair}. Hence, the rate of increase of error for the average node is

$$I_{fail} = (I_{err} - I_{repair}) \, \theta (I_{err} - I_{repair}). \tag{18.49}$$

The Heaviside step-function is only non-zero when its argument exceeds zero:

$$\theta(x) = \begin{cases} 1 & x > 0 \\ 0 & x <= 0 \end{cases} \tag{18.50}$$

and we use it to incorporate the fact that, if the maintenance rate exceeds the error rate, then (on average, over long times) nothing remains outstanding and there is no net rise in configuration error. Thus, this averaged quantity is never negative.

If random errors and changes to configuration occur at a rate I_{err} and the configuration agent is unavailable to correct them, then $I_{fail} = I_{err}$. If this holds during a time Δt, the configuration falls behind by an amount:

$$\underset{(\Delta Q)}{\underset{\text{missing}}{\text{Bytes}}} = \underset{(I_{err})}{\text{bytes/sec}} \times \underset{(\Delta t)}{\underset{\text{unavailable}}{\text{seconds}}} \cdot$$

In the following, we will use p to denote the average (over time, and over all nodes) probability that configuration management information flow (repair current) is not available

---

[2] This ordering also corresponds, roughly, to decreasing predictability. However, this interpretation may be misleading, since centralized control schemes are also prone to noise, and local or even catastrophic system-wide failures. The various cases that we consider are presented in Table 18.1 below.

Table 18.1: Comparison of models from the viewpoint of the different dimensions: policy dissemination, enforcement, freedom of choice, whether hosts can exchange chosen policy ideas with peers and how political control flows. A 'push' model implies a forcible control policy, whereas 'pull' signifies the possibility to choose. Model 3 lies between these two, in having the possibility but not the inclination to choose.

Model	Application Topology	Enforcement	Policy Freedom	Policy Exchange	Control Structure
1	Star	Transmitted	No	No	Radial push
2	Star	Transmitted	No	No	Radial push
3	Mesh	Local	No	No	Radial pull
4	Mesh	Local	Yes	No	Radial pull
5	Mesh	Local	Yes	Yes	Hierarchical pull
6	Mesh	Local	Yes	Yes	P2P pull

to a node. This unavailability may come from either link or node unreliability. We can lump all the unreliability into the links (see above) and write

$$p = (1 - \langle A_{ij} \rangle) , \tag{18.51}$$

where $\langle A_{ij} \rangle$ denotes both time and node-pair average. Each node then can only receive repair current during the fraction $(1 - p)$ of the total elapsed time.

The repair current is generated by two possible sources in our models: i) a remote source and ii) a local source. In each case, the policy can be transmitted and/or enforced at a maximum rate given by the channel capacity of the source. We shall denote the channel capacities by C_R and C_L for remote and local sources for clarity, but we assume that $C_R \sim C_L$, since source and target machines are often comparable, if not identical. If the communication by network acts as a throttle on these rates, then one can further assume that $C_R < C_L$. In any case, the weakest link determines the effective channel capacity. Note that in the case of a confluence of traffic, as in the star models below, the channel capacity will have to be shared by the incoming branches.

We now have a criterion for eventual failure of a configuration strategy. If

$$I_{\text{fail}} = \frac{\Delta Q}{\Delta t} > 0, \tag{18.52}$$

the average configuration error will grow monotonically for all time, and the system will eventually fail in continuous operation. Our strategy is then to look at the scaling behaviour of I_{fail} as the number of nodes, N, grows large (see table 18.1).

18.6 Policy maintenance architectures

Model 1: Star model

The traditional (idealized) model of host configuration is based on the idea of remote management (e.g. using SNMP). Here, one has a central manager who decides and implements policy from a single location, and all networks and hosts are considered to be completely reliable. The manager must monitor the whole network, using bi-directional communication. This leads to an N:1 ratio of clients to manager (see fig 18.4). This first model is an

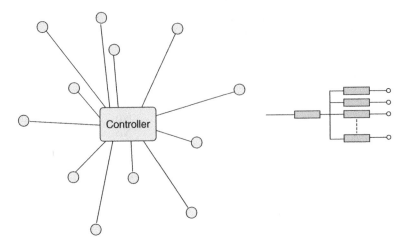

Figure 18.4: Model 1: the star network. A central manager maintains bi-directional communication with all clients. The links are perfectly reliable, and all enforcement responsibility lies with the central controller.

idealized case in which there is no unreliability in any component of the system. It serves as a point of reference.

The topology on the left-hand side of fig 18.4 is equivalent to that on the right-hand side. We can assume a flow conservation of messages on average, since any dropped packets can be absorbed into the probabilities for success that we attribute to the adjacency matrix. Thus, the currents must obey Kirchoff's law:

$$I_{\text{controller}} = I_1 + I_2 + \ldots I_N. \tag{18.53}$$

The controller current cannot exceed its maximum capacity, which we denote by C_S. We assume that the controller puts out a 'repair current' at its full capacity (since the Heaviside function corrects for lower demand), and that all nodes are average nodes. This gives that

$$I_{\text{repair}} = \frac{C_S}{N}. \tag{18.54}$$

The total current is limited only by the bottleneck of queued messages at the controller, thus the throughput per node is only $1/N$ of the total capacity. We can now write down the failure rate in a straightforward manner:

$$I_{\text{fail}} = \left(I_{\text{err}} - \frac{C_S}{N}\right) \theta \left(I_{\text{err}} - \frac{C_S}{N}\right). \tag{18.55}$$

As $N \to \infty$, $I_{\text{fail}} \to I_{\text{err}}$—that is, the controller manages only a vanishing repair current per node. The system fails, however, at a finite $N = N_{\text{thresh}} = C_S/I_{\text{err}}$. This highlights the clear disadvantage of centralized control, namely the bottleneck in communication with the controller.

Model 2: Star model in intermittently connected environment

The previous model was an idealization, and was mainly of interest for its simplicity. Realistic centralized management must take into account the unreliability of the environment (see fig. 18.5).

In an environment with partially reliable links, a remote communication model bears the risk of not reaching every host. If hosts hear policy, they must accept and comply; if not, they fall behind in the schedule of configuration. Monitoring in distributed systems has been discussed in (Abdu et al. (1999)).

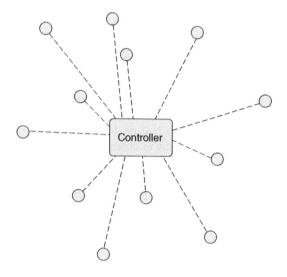

Figure 18.5: Model 2: a star model, with built-in unreliability. Enforcement is central as in Model 1.

The capacity of the central manager C_S is now shared between the average number of hosts $\langle N \rangle$ that is available, thus

$$I_{\text{repair}} = \frac{C_S}{N\langle A_{ij} \rangle} \equiv \frac{C}{\langle N \rangle} \ . \tag{18.56}$$

This repair current can reach the host, and serve to decrease its policy error ΔQ, during the fraction of time $(1 - p)$ that the typical host is reachable. Hence, we look at the net deficit ΔQ accrued over one 'cycle' of time Δt, with no repair current for $p\Delta t$, and a maximal current $C_S/\langle N \rangle$ for a time $(1 - p)\Delta t$. This deficit is then

$$\Delta Q(\Delta t) = I_{\text{err}}p\Delta t + \left(I_{\text{err}} - \frac{C_S}{\langle N \rangle} \right)(1 - p)\Delta t \tag{18.57}$$

(here it is implicit that a negative ΔQ will be set to zero). Thus, the average failure rate is

$$I_{\text{fail}} = I_{\text{err}}p + \left(I_{\text{err}} - \frac{C_S}{\langle N \rangle} \right)(1 - p) = I_{\text{err}} - \frac{C_S}{N} \ . \tag{18.58}$$

(Again there is an implicit θ function to keep the long-time average failure current positive.) This result is the same as for Model 1, the completely reliable star. This is because we assumed the controller was clever enough to find (with negligible overhead) those hosts that are available at any given time, and so to only attempt to communicate with them.

This model then fails (perhaps surprisingly), on average, at the same threshold value for N as does Model 1. If the hunt for available nodes places a non-negligible burden on the controller capacity, then it fails at a lower threshold.

Model 3: Mesh topology with centralized policy and local enforcement

The serialization of tasks in the previous models forces configuration 'requests' to queue up on the central controller. Rather than enforcing policy by issuing every instruction from the central source, it makes sense to download a summary of the policy to each host and empower the host itself to enforce it (see fig. 18.6).

There is still a centrally determined policy for every host, but now each host carries the responsibility of configuring itself. There are thus two issues: (i) the update of the policy and (ii) the enforcement of the policy. A pull model for updating policy is advantageous here, because every host then has the option to obtain updates at a time convenient to itself, avoiding confluence contentions; moreover, if it fails to obtain the update, it can retry until it succeeds. We ask policy to contain a self-referential rule for updating itself.

The distinction made here between communication and enforcement is important, because it implies distinct types of failure, and two distinct failure metrics: (i) distance of the locally understood policy from the latest version, and (ii) distance of host configuration from the ideal policy configuration. In other words: (i) communication failure, and (ii) enforcement failure.

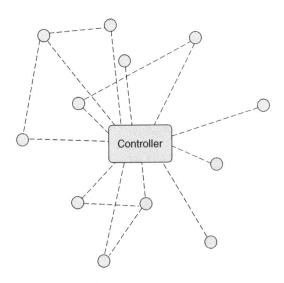

Figure 18.6: Model 3. Mesh topology. Nodes can learn the centrally mandated policy from other nodes as well as from the controller. Since the mesh topology does not assure direct connection to the controller, each node is responsible for its own policy enforcement.

The host no longer has to share any bandwidth with its peers, unless it is updating its copy of the policy, and perhaps not even then, since policy is enforced locally and updates can be scheduled to avoid contention.

Let I_{update} be the rate at which policy must be updated. This current is usually quite small compared to I_{err}. On the basis of the two failure mechanisms present here, we break up the failure current into two pieces:

$$I_{\text{fail}} = I_{\text{fail}}(i) + I_{\text{fail}}(ii) \ . \tag{18.59}$$

The former term is

$$I_{\text{fail}}(i) = (I_{\text{err}} - C_L)\theta(I_{\text{err}} - C_L) \ ; \tag{18.60}$$

this term is independent of N and may be made zero by design. $I_{\text{fail}}(ii)$ is still determined by the ability of the controller to convey policy information to the hosts. However, the load on the controller is much smaller since $I_{\text{update}} \ll I_{\text{err}}$. Also, the topology is a mesh topology. In this case, the nodes can cooperate in diffusing policy updates, via flooding[3], that is, by asking each neighbour to pass on the policy to its neighbours, but never back in the direction it came from.

The worst case—in which the hosts compete for bandwidth, and do not use flooding over the network (graph)—is that, for large N, $I_{\text{fail}} \rightarrow I_{\text{update}}$. This is a great improvement over the two previous models, since $I_{\text{update}} \ll I_{\text{err}}$. However, note that this can be further improved upon by allowing flooding of updates: the authorized policy instruction can be available from any number of redundant sources, even though the copies originate from a central location. In this case, the model truly scales without limit, that is, $I_{\text{fail}} = 0$.

There is one caveat to this result. If the meshed network of hosts is an *ad hoc* network of mobile nodes, employing wireless links, then connections are not feasible beyond a given physical range r. In other words, there are no long-range links: no links whose range can grow with the size of the network. As a result of this, if the ad hoc network grows large (at fixed node density), the path length (in hops) between any node and the controller scales as a constant times $\sqrt{N}$. This growth in path length limits the effective throughput capacity between node and controller, in a way analogous to the internode capacity. The latter scales as $1/\sqrt{N}$ (see Gupta and Kumar (2000); Li et al. (2001)). Hence, for sufficiently large N, the controller and AHN will fail collectively to convey updates to the net. This failure will occur at a threshold value defined by

$$I_{\text{fail}}(ii) = I_{\text{update}} - \frac{C_S}{c\sqrt{N_{\text{thresh}}}} = 0 \ , \tag{18.61}$$

where c is a constant. The maximal network size N_{thresh} is in this case proportional to $\left(\frac{C_S}{I_{\text{update}}}\right)^2$—still considerably larger than for Models 1 and 2.

Model 4: Mesh topology with partial host autonomy and local enforcement

As a variation on the previous model, we can begin to take seriously the idea of distance from a political centre. In this model, hosts can choose not to receive policy from a cen-

---

[3] Note, flooding in the low-level sense of a datagram multicast is not necessarily required, but the effective dissemination of the policy around the network is an application layer flood.

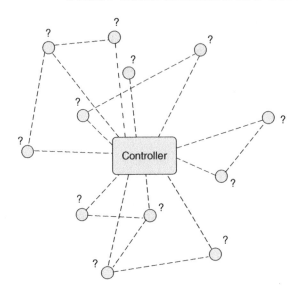

Figure 18.7: Model 4. As in Model 3, except that the hosts can choose to disregard or replace aspects of policy at their option. Question marks indicate a freedom of hosts to choose.

tral authority if it conflicts with local interests. Hosts can make their own policy, which could be in conflict or in concert with neighbours. Communication thus takes the role of conveying 'suggestions' from the central authority, in the form of the latest version of the policy (see fig. 18.7).

For instance, the central authority might suggest a new version of widely used software, but the local authority might delay the upgrade owing to compatibility problems with local hardware. Local enforcement is now employed by each node to hold to its chosen policy P_i. Thus, communication and enforcement use distinct channels (as with Model 3); the difference is that each node has its own target policy P_i, which it must enforce.

Thus, the communications and enforcement challenges faced by Model 4 are the same (in terms of scaling properties) as for Model 3: that is, I_{fail} is the same as that in Model 3. Hence, this model can ,in principle, work to arbitrarily large N.

Model 4 is the model used by cfengine (Burgess (1995, 2004)). The largest current clusters sharing a common policy are known to be of order 10^4 hosts, but this could soon be of order 10^6, with the proliferation of mobile and embedded devices.

Model 5: Mesh, with partial autonomy and hierarchical coalition

An embellishment of Model 4 is to allow local groups of hosts to form policy coalitions that serve to their own advantage. Such groups of hosts might belong to one department of an organization, or to a project team, or even to a group of friends in a mobile network.

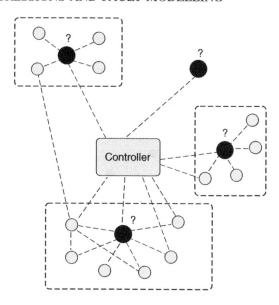

Figure 18.8: Model 5. Communication over a mesh topology, with policy choice made hierarchically. Sub-controllers (dark nodes) edit policy as received from the central controller, and pass the result to members of the local group (as indicated by dashed boxes). Question marks indicate the freedom of the controllers to edit policy from above.

Once groups form, it is natural to allow sub-groups and then a generalized hierarchy of policy refinement through specialized social groups (see fig. 18.8).

If policies are public, then the scaling argument of Model 3 still applies since any host could cache any policy; but now, a complete policy must be assembled from several sources. Once can thus imagine using this model to distribute policy so as to avoid contention in bottlenecks, since load is automatically spread over multiple servers. In effect, by delegating local policy (and keeping a minimal central policy) the central source is protected from maximal loading. Specifically, if there are S sub-controllers (and a single-layer hierarchy), then the effective update capacity is multiplied by S. Hence, the threshold N_{thresh} is multiplied (with respect to that for Model 3) by the same factor.

Model 6: Mesh, with partial autonomy and inter-peer policy exchange

The final step in increasing autonomy is the free exchange of information between arbitrary hosts (peer to peer) (see fig. 18.9). Hosts can now offer one another information, policy or source materials in accordance with an appropriate trust model. In doing so, impromptu coalitions and collaborations wax and wane, driven by both human interests and possibly machine learning. A peer-to-peer policy mechanism of this type invites trepidation amongst those versed in traditional control mechanisms, but it is really no more than a distributed genetic algorithm. With appropriate constraints it could equally be made to lead to sensible convergent behaviour, or to catastrophically unstable behaviour.

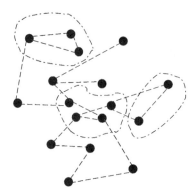

Figure 18.9: Model 6. Free exchange of policies in a peer-to-peer fashion; all nodes have choice (dark). Nodes can form spontaneous, transient coalitions, as indicated by the dashed cells. All nodes can choose; question marks are suppressed.

Before a distributed policy exchange nears a stable stationary point, policy updates could be much more numerous here than for the previous models. This could potentially dominate configuration management behaviour at early times.

Example 165 *A collaborative network that has led to positive results is the Open Source Community. The lesson of Open Source Software is that it leads to a rapid evolution. A similar rapid evolution of policy could also be the result from such exchanges. Probably, policies would need to be weighted according to an appropriate fitness landscape. They could include things like shared security fixes, best practices, code revisions, new software and so on.*

Note that this model has no centre, except for a dynamically formed centre represented by centrality (see section 6.4). Hence it is, by design, scale-free: all significant interactions are local. Therefore, in principle, if the model can be made to work at small system size, then it will also work at any larger size.

In practice, this model is subject to potentially large transients, even when it is on its way to stable, convergent behaviour. These transients would likely grow with the size of the network. Here we have confined ourselves to long-time behaviour for large N—hence, we assume that the system can get beyond such transients, and so find the stable regime.

Finally, we note that we have only assessed the success of the given models according to their ability to provide an integrity preserving, or error correcting stream, as discussed in chapter 15, that communicates and enforces policy.

18.7   Diagnostic cause trees

From the previous sections, we recognize that the causal relationships within a system can form complex networks. Unravelling such networks is difficult. In many cases, we can simplify the causal structure by replacing part of the network with an effective tree that more clearly describes the causal relationships. The price for this simplification is that the events are non-deterministic; by hiding details, we lose complete information about the system, but achieve the illusion of a higher-level understanding.

Charting cause trees is a systematic method used in fault diagnosis. The idea is to begin by building lists of possible causes, then causes of those causes, and so on, until one has covered an appropriate level of detail. Once a cause tree has been constructed for a system, it becomes a road map for fault finding for the future also. The use of cause trees is sometimes called *Root Cause Analysis* (RCA). A related method called *Event Tree Analysis* (ETA) maps out every single eventuality, as a true/false binary tree, where every possibility is documented, but only certain pathways actually occur. The latter is mainly a way of documenting the extent of a system; it has little analytical value.

Many of the techniques described in this chapter were pioneered over the last half century by authorities working with nuclear power, where the risk of accidents takes on a whole different level of importance. The keyword in causal analyses is *dependencies*. All of the immediate causes of a phenomenon or an event are called dependencies, that is, the event depends on them for its existence. The cause tree for the diagnostic example 166 is shown in fig. 18.10. The structure is not completely hierarchical, but it is approximately so.

Example 166 (Network services become unavailable.) *A common scenario is the sudden disappearance of a network service, like, say, the WWW. If a network service fails to respond it can only be due to a few possibilities:*

- *The service has died on the server host,*

- *The line of communication has been broken,*

- *The latency of the connection is so long that the service has timed out.*

A natural first step is to try to send a network ping to server-host:

```
ping www.domain.country
```

to see whether it is alive. A ping signal will normally return with an answer within a couple of seconds, even for a machine halfway across the planet. If the request responds with

```
www.domain.country is alive
```

then we know immediately that there is an active line of communication between our host and the server hosts and we can eliminate the second possibility. If the ping request does not return, then there are two further possibilities:

- *The line of communication is broken,*

- *The DNS lookup service is not responding.*

The DNS service can hang a request for a long period of time if a DNS server is not responding. A simple way to check whether the DNS server is at fault or not is to bypass it, by typing the IP address of the WWW server directly:

```
ping 128.39.74.4
```

If this fails to respond, then we know that the fault was not primarily due to the name service. It tends to suggest a broken line of communication. The `traceroute` *command on Unix-like operating systems, or* `tracert` *on Windows can be used to follow a net connection through various routers to its destination. This often allows us to narrow down the point of failure to a particular group of cables in the network. If a network break has persisted for more than a few minutes, a ping or traceroute will normally respond with the message*

```
ICMP error: No route to host
```

and this tells us immediately that there is a network connectivity problem.

But what if there is no DNS problem and ping tells us that the host is alive? Then the natural next step is to verify that the WWW service is actually running on the server host. On a Unix-like OS, we can simply log onto the server host (assuming it is ours) and check the process table for the `httpd` *daemon that mediates the WWW service.*

```
ps aux | grep httpd    BSD
ps -ef | grep httpd    Sys V
```

On a Windows machine, we would have to go to the host physically and check its status. If the WWW service is not running, then we would like to know why it stopped working. Checking log files to see what the server was doing when it stopped working can provide clues or even an answer. Sometimes a server will die because of a bug in the program. It is a simple matter to start the service again. If it starts and seems to work normally afterwards, then the problem was almost certainly a bug in the program. If the service fails to start, then it will log an error message of some kind, which will tell us more. One possibility is that someone has changed something in the WWW service's configuration file and has left an error behind. The server can no longer make sense of its configuration and it gives up. The error can be rectified and the server can be restarted.

What if the server process has not died? What if we cannot even log onto the server host? The latter would be a clear indication that there was something more fundamentally wrong with the server host. Resisting the temptation to simply reboot it, we could then try to test other services on the server host to see if they respond. We already know that the ping service is responding, so the host is not completely dead. There are therefore several things that could be wrong:

- *The host is unable to respond (e.g. it is overloaded),*

- *The host is unwilling to respond (e.g. a security check denying access to our host).*

We can check that the host is overloaded by looking at the process table, to see what is running. If there is nothing to see there, the host might be undergoing a denial of service attack. A look at `netstat` *will show how many external connections are directed towards the host and their nature. This might show something that would confirm or deny the attack theory. An effective attack would be difficult to prevent, so this could be the end of the line for this particular investigation and the start of a new one, to determine the attacker. If there is no attack, we could check that the DNS name service is working on the server host. This could cause the server to hang for long periods of time. Finally, there are lots of reasons why the kernel itself might prevent the server from working correctly: the TCP connection close time in the kernel might be too long, leading to blocked connections; the kernel itself might have gone amok; a full disk might be causing errors that have a knock-on effect (the log files from the server might have filled up the disk), in which case, the disk problem will have to be solved first. Notice how the DNS and disk problems are problems of dependency: a problem in one service having a knock-on effect in another.*

A cause tree for diagnosing a full disk is shown in fig. 18.11. This is a particularly simple example; it simply becomes a flat list.

Causal analyses can be used at different levels. At the level of human management, it takes on a more heuristic role, for example,

- inadequate procedures

- inadequate training

- quality control

- miscommunication

- poor management

- social/human engineering

- supervision error

- preventative maintenance lacking.

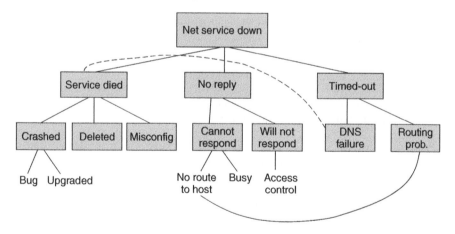

Figure 18.10: Attempt at cause tree for a missing network service.

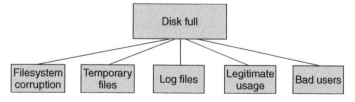

Figure 18.11: Attempt at cause tree for a full disk.

Information is collected about an incident or a phenomenon and this is broken down into cause-effect relationships. Analysts must understand the systems they model thoroughly from the highest levels, down to the component level.

The construction of an event tree is just like the top-down analysis performed in programming. Breaking the event up into component causes is like breaking up a task into subroutines. The benefit is the same: a complex problem is reduced to a structured assembly of lesser problems.

18.8   Probabilistic fault trees

How can we go beyond mapping cause and effect to calculating the likely outcomes of the different pathways through a cause tree, to include some of the stochastic reliability analysis from the start of this chapter? This would give us an approximate way of performing reliability analysis based on a kind of spanning tree approximation to diagnosis. To accomplish this, we must acknowledge that not all of the possible pathways occur all of the time: some occur only infrequently, some are mutually exclusive, some are co-dependent and others are uncorrelated. To make serious headway in estimating likely cause, we thus need to add probabilities and combinatorics to the discussion. This is the value of fault tree analysis. The discussion here follows that of Apthorpe (2001), based on NRC (1981).

18.8.1   Faults

For the purposes of modelling, fault tree analysis distinguishes between:

- *Failures*: Abnormal occurrences,

- *Faults*: Systemic breakdowns within the system.

An important subset of faults is formed by *component faults*.
Component faults fall into three categories:

- *Primary faults*: Occur when a component is working within its design limits, for example, a web server that is rated at 50 transactions per second fails when it reaches 30 transactions per second.

- *Secondary faults*: Occur when a fault is operating outside its design specification. For example, a web server that is rated at 50 transactions per second fails when it reaches 90 transactions per second.

- *Command faults*: Are faults that occur when a system performs its specified function, but at the wrong time or place. For example, a Web server that begins querying a database persistently when no request is being made by an external agent.

Faults occur in response to events. The events are also categorized, this time depending on their position within the tree structure:

- *Top*: This is the top of the tree—the end phenomenon that we are trying to explain. It is analogous to the 'main' function in a computer program.

- *Intermediary*: This is a dependency within the tree, but not a root cause of the phenomenon. It is analogous to a subroutine of the main program; it has deeper dependencies that are subroutines of itself.

- *Primary*: This is an event that is either a root cause, or as deep an explanation as we can manage to determine. In a computer program analogy, it is like a basic library function, that is, the lowest level of control available. Events that we cannot say much about are called *undeveloped events* because although we cannot dig any deeper, we know that there is more going on than we can say. Events that have no further explanation are called *basic events*. These are the primitive atoms of causality: the very root causes.

Events are drawn according to the symbols in fig. 18.12.

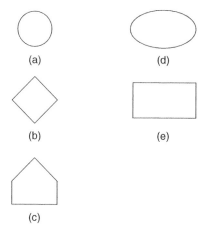

Figure 18.12: Basic symbols for fault trees.

18.8.2 Conditions and set logic

When several smaller causes lead to an intermediate event or phenomenon, there arises a question about how many of the sub-events were needed to trigger the higher-level event—All of them? Any of them? A certain number? Events thus combine in ways that can be represented by simple combinatoric set notation—with 'AND' and 'OR' or other conditions. These are best known to computer scientists in the form of *logic gates*[4]. Figure 18.13 shows the standard symbols for the gates types. Although there are many gate types, for a richness of expression, in practice 'AND' and 'OR' suffice for most cases.

The properties of the gates in combining the probabilities are noted below. Note that it makes a difference whether or not events are independent, in the probabilistic sense: that is, the occurrence of one event does not alter the probability of occurrence of another.

---

[4] One might be forgiven for believing that Boolean logic arrived with digital computers, but this is not the case. Mechanical logic gates may be created, for example, with hydraulics.

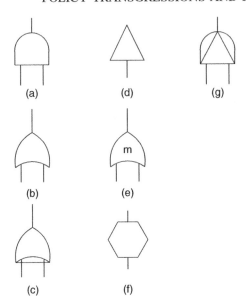

Figure 18.13: Basic gate types: (a) **AND** , (b) **OR** , (c) **XOR** , (d) Transfer partial result to separate sub-tree, (e) Voting gate (m of n), (f) Inhibit conditional of 'if' gate, and (g) Priority AND (inputs ordered from left to right).

- In OR gates, probabilities combine so as to get *larger*.

$$P(A \ \mathbf{OR} \ B) = P(A) + P(B) - P(A \ \mathbf{AND} \ B). \tag{18.62}$$

In general,

$$P(A_1 \ \mathbf{OR} \ A_2 \ \mathbf{OR} \ \dots A_n) = \sum_{i=1}^{n} P(A_i) - \sum_{i=1}^{n-1} n \sum_{j=i+1}^{n} P(A_i)P(A_j) + \dots$$
$$+ (-1)^{n+1} P(A_1)P(A_2) \dots P(A_n). \tag{18.63}$$

- In AND gates, probabilities combine so as to get *smaller*:

$$P(A \ \mathbf{AND} \ B) = P(A)P(B|A), \tag{18.64}$$

or in general:

$$P(A_1 \ \mathbf{AND} \ A_2 \ \mathbf{AND} \ \dots A_n) = \prod_{i=1}^{n} P(A_i). \tag{18.65}$$

If A and B are independent, then

$$P(A)P(B|A) = P(A)P(B), \tag{18.66}$$

which is smaller than $P(A)$ or $P(B)$; but if the events are not independent, the result can be much greater than this.

- XOR gates have no predictable effect on magnitudes.

$$P(A \ \mathbf{OR} \ B) = P(A) + P(B) - 2P(A \ \mathbf{AND} \ B) \qquad (18.67)$$

Thus, if we see many OR pathways, we should be scared. If we see many AND pathways, we should be pleased—the latter means that things are tied down quite tightly with redundancy or protections.

18.8.3   Construction

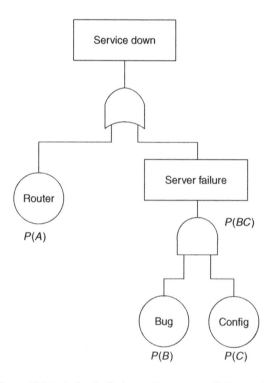

Figure 18.14: A simple fault tree for an unavailable service.

As a simple example, consider how to work out the probability of failure for a system attack, where an attacker tries the obvious pathways of failure: guessing the root password, or exploiting some known loopholes in services that have not been patched (see fig. 18.14).

We split the tree into two main branches: first try the root password of the system, 'OR' try to attack any services that might contain bugs.

- The two main branches are 'independent' in the probabilistic sense, because guessing the root password does not change the sample space for attacking a service and vice versa (it is not like picking a card from a deck).

- On the service arm, we split (for convenience) this probability into two parts and say that hosts are vulnerable if they have a service that could be exploited AND the hosts have not been patched or configured to make them invulnerable.

- Note that these two arms of the AND gate are time-dependent. After a service vulnerability becomes known, the administrator has to try to patch/reconfigure the system. Attackers therefore have a *window of opportunity*. This adds a time dimension to the fault analysis, which we might or might not wish to address.

Since all the events are independent, we have

$$P(\text{break in}) = P(A \text{ OR } (\text{NOT } A \text{ AND } (B \text{ AND } C))) \qquad (18.68)$$
$$= P(A) + (1 - P(A)) \times P(B)P(C) \qquad (18.69)$$

Suppose we have, from experience, that

- Chance of router problem $P(A) = 5/1000 = 0.005$.

- Chance of server problem $P(B) = 50/1000 = 0.05$.

- Chance that server is mis-configured $P(C) = 10\% = 0.1$.

$$
\begin{aligned}
P(T) &= 0.005 + 0.995 \times 0.05 \times 0.1 \\
&= 0.005 + 0.0049 \\
&= 0.01 \\
&= 1\% \qquad (18.70)
\end{aligned}
$$

Notice how, even though the chance of guessing the root password is small, it becomes an equally likely avenue of attack, because of the chance that the host might have been upgraded. Thus, we see that the chance of a break-in is a competition between an attacker and a defender.

A *cutset* is a set of basic events that are essential for a top-level fault to occur. A *minimal cutset* is a cutset in which the removal of a single event no longer guarantees the occurrence of the top-level event. One of the aims of fault tree analysis is to identify these cut sets. They represent the critical dependencies of the system.

Applications and Further Study 18

- *Analysis of structure and its effects on failure modes.*

- *Determining the likely place and time window of fault occurrence.*

- *Quantifying reliability.*

- *Avoidance of flawed structures.*

- *Choosing reliable strategies in policy decisions.*

Queries

19

Decision and strategy

Decisions must be made at many levels in Network and System Administration, from the top level decisions about policy to the low-level choices made during diagnostics. Rational decision-making is therefore central to the human–computer system. However, humans do not always behave rationally, and even computers do not always behave predictably, so one must take into account the possibility that systems will not always be deterministic.

This chapter is about how one can evaluate the rational limits of strategy in a system, and how one sets choosing the best strategies to maximize the result delivered by the system.

19.1 Causal analysis

All human and computer systems satisfy basic laws of physics that tell us how the world works. We cannot escape such laws; they bind us to basic truths about systems, even when the physics of systems seems utterly buried from view. Causality is the term used to express a basic truth about the world: that for every effect there is a cause that precedes it.

Sometimes, authors confuse the necessity of this basic law with the ability to identify the precise root of a cause. It is important to realize that effects need not have one simple cause, but that even complex systems that are practically unpredictable obey the law of causality. Causality says simply that physics is a directed graph, at a low level. However, we know from weather charts that when we combine a lot of arrows pointing in different directions, the result can be far from easy to predict. This is why we are often more interested in systems that are predictable than in necessarily being able to trace the exact sequence of changes that led to the state of the system. Still, we are bound by the basic constraint of causality in every decision we make.

When a causally directed system ends up in a state that is not going anywhere, we say that it has reached an equilibrium, as discussed in chapter 10. Such a state is stable to small perturbations in the conditions under which it formed. This concept returns in this chapter in connection with decision-making. We would like to ask the following question: What is the likelihood of being able to base decisions on causal, rational information that can tell us a stable kind of 'truth' that will not be undone by the smallest change in the

Analytical Network and System Administration. Managing Human–Computer Networks Mark Burgess
© 2004 John Wiley & Sons, Ltd ISBN 0-470-86100-2

environment? The weather forecast suggests to us that this is not going to be easy, but can we at least minimize the uncertainty in decision-making by using all of the available information about a human–computer system?

19.2 Decision-making

The simplest decisions are made by associating an action with the occurrence of a state in a chain of events[1]. States are identified with actions by combining predicates about the world using logical operations such as **AND** , **OR** and **NOT** . These state-based classifiers are set constructions (see chapter 5).

Sets effectively make decisions because they classify the world into regions such as inside the set, outside the set, inside this set *and* that set and so on. Thus, if we give every set a name, then we have labelled all of the objects that lie within the sets also, and this is what we use to sift through members and identify their properties in terms of a state that is given by the label of the set.

Example 167 *The system administration tool cfengine (Burgess (1993)) makes decisions by classifying systems according to their set membership. When the cfengine agent starts executing on any computer, it tries to identify the sets or classifiers to which it belongs:*

```
Defined Classes = ( Saturday Hr12 Min10 Min10_15 Day7 June Yr2003
solaris nexus 32_bit sunos_5_9 sun4u sunos_sun4u_5_9 sparc
myname_domain ipv4_128_39_89_10 )
```

This list of sets identifies the type of operating system, the name of the computer, its Internet address and so on. Notice that even the time is classified into sets that describe the days of the week, the hours of the day and intervals of minutes and so on. Any property can be classified in this way, using sets. That is the essence of logic and reasoning.

Some sets do not overlap, or are mutually exclusive:

$$\texttt{linux AND windows} = \texttt{linux} \cap \texttt{windows} = \emptyset \tag{19.1}$$

Others do overlap and provide nuances of description to decide whether or not to take one course of action or another:

```
linux AND Saturday          ::   action 1

linux AND Sunday            ::   action 2

(linux OR windows) AND Monday ::  action 3
```

Another way in which decisions are made is by statistical confidence: We have processes such as *voting*. If sufficient *support* is given to a certain proposition, we can view it as being statistically 'true', or at least *significant*.

Finally, we have *games*. The point of game theory is to find out the limits of rational choices with varying degrees of information about competing inputs. If one player has a high expectation of pay-off from a particular decision, the game can provide a reality check on those expectations. The results of the game set the baseline for expectation. There will then be noise on top of this, since not all agents in the game are likely to behave rationally.

---

[1] This is like a switch-case or if-then-else construction in programming

19.3   Game theory

The Theory of Games was first significantly developed by Von Neumann and Morgenstern (1944) and later developed by Nash (1996) and many others. It is a method for pitting a set of pre-emptive and defensive strategies against one another, and finding their point of balance in order to see how they fare against one another. By doing this, one tries to maximize gain and minimize loss in a competitive setting.

Games are played in many contexts; in fact, a wide variety of interactive processes can be formulated as some kind of game. The simple pendulum, mentioned in chapter 4, is a game that is played between gravity and motion for winning 'energy'. At a supermarket, customers and merchants play a game with each other in which prices are used to lure customers to a particular seller but also to maximize profits. A balance must be then found between setting the prices as high as possible without losing customers and using a cheaper strategy to sell more items but with a lesser profit per item.

Game theory is applicable in all cases where it is difficult to evaluate the gains generated by following particular policies. This occurs whenever the number of choices is large or the effects are subtle. Contests that are caused by conflicts of interest between system policy and user wishes unfold in this framework as environmental interactions that tend to oppose convergence and stability (see Dresher (1961); Neumann and Morgenstern (1944)). Game theory introduces 'players', with goals and aims, into a scheme of rules and then analyses how much each player can win, according to those restrictions. Each pair of strategies in a game affords the players a characteristic value, often referred to as the 'pay-off'. Game theory has been applied to warfare, to economics (commercial warfare) and to many other situations.

Game theory is a vast subject, with many technical challenges. Here, we shall restrict our examples to games between two players, since this is adequate for many situations and presents a sufficient range of issues for an introductory text.

Who are the players?

The players in a game are any actors or influences that affect the transfer of value to any of the other players; that is, they are entities who exchange some form of system currency. Each player has their own viewpoint of what is best (most valuable) for them but that viewpoint is constrained to work within and share the same system as all of the other players. In some cases, players have opposing interests, in which case we speak of non-cooperative games. In other cases, players share some common interests and can collaborate leading to partially cooperative games.

Players are labelled by Roman indices $i = 1 \ldots n$ and abstract players can be made to represent many opposing viewpoints about a system:

- System users versus system policy (or system administrator),

- System policy versus entropy—chance degradation of the system,

- A rogue user versus the other $n - 1$ users.

In many games, it is not necessary to interpret a player as a person or a rational entity, as one does in classical game theory. As we have seen in chapter 16, the random forces of disorder, measured as entropy, are a sufficient counter player in many situations. The principle of maximizing entropy is sufficient to make even random chance into a 'rational' player in the game theoretical sense: It is an influence that cares nothing for the other players and that is statistically biased against them. In that sense, it can be viewed as seeking to maximize its own gains. We can therefore think of system administrators as playing a two-person game against 'gremlins' in the system, and this will be a profitable way of formulating the strategic problem.

However, we should be cautious with this viewpoint. There is a slight difference between playing a game against a rational user and playing a game against chance. It is not necessarily true that the most likely outcome of chance is an optimal strategy in a game. When we look for the strategies played by the forces of chance, the most reliable guide is to measure them with experimental data to find out what they actually do, rather than necessarily trusting the formalism of the game that would like to assume the worst. In either case, it is instructive to assume the worst and then compare how efficient chance is at maximizing its effect. If actual data are procured later, they can be substituted and the table elements with a known (sub-optimal) strategy can be summed over to reduce the problem to one of optimization with respect to one less variable.

Definition 76 (Worst-case scenario) *We define this to be the mixed-strategy minimax solution of a zero-sum game. Even if our counter player is 'nature' or the forces of chance, the equilibrium solution tells us that chance is playing its most destructive hand.*

Refinement of reasonable belief

In chapter 17, the issue of learning or the refinement of belief was raised for the attainment of expert knowledge. Decisions must clearly be made on the basis of this kind of expert knowledge. This applies to games or to any other form of decision. If one does not have a reasonable observational basis for making decisions, a strategy of confining the limits of possibility can be used. For example, one begins with the worst-case scenario, and then refines it as more data become available. The worst-case scenario is bound to be pessimistic, but it is the only rational starting point in the absence of more data.

Principle 9 (Policy confinement) *In the absence of perfect information about a problem, one adopts a strategy of finding the bounds of reason, and refining these as new information is acquired.*

There is thus a synergy between decision theory and Bayesian learning.

Pay-off or 'utility'

What is it that players win or lose in a system administration game? In classical game theory, the pay-off has often been money, as game theory was employed as a means for studying economic competition. In section 4.9, we looked at the ways of measuring gain in a human–computer system in terms of system resources or even social capital (status and privilege). There is no simple answer, nor recipe, for what the pay-off is in a game within the system. If we can formulate a game in which players compete for something, then that is a sufficient justification for doing so. One can imagine pay-off being formed from a combination of several importance factors: for example, memory share, CPU resources, disk space, money, privilege, time for human recreation and so on.

Pay-off[2] is represented by a function Π_i for each player i. For two players, this function is a matrix with a finite number of pure strategies s_i. Games fall into two distinct types: Constant sum games and non-constant sum games. In a constant sum game, each element of the pay-off sums to a constant value over all players:

$$\sum_{i=1}^{n} \Pi_i = \text{const} \times \mathbf{1}, \tag{19.2}$$

where $\mathbf{1}$ is a matrix or table filled with ones.

What is a strategy?

A *pure strategy* is a single course of action taken by a player throughout a game. We can think of it as a mode of behaviour. There are two interpretations of strategies: in the *extensive* form of a game, a strategy represents a single set of moves by a single player from start to finish; in the *strategic* or 'normal' form of the game, the strategy represents an average mode of play, without specifying the details of individual moves.

Different courses of action lead to different returns or pay-offs, and the point of the game is to compare the results of using different strategies on the final pay-off. The method of solution of a game is to vary each player's strategy, or mixture of strategies so as to optimize the amount they can win.

Suppose that the set of all pure strategies s_i for player i is denoted by S_i, so that $s_i \in S_i$. The set of all players' strategies is denoted by the outer product:

$$S = S_1 \times S_2 \times \ldots \times S_n. \tag{19.3}$$

Sometimes, it is not advisable for players to play a single strategy, but to mix several different approaches to playing a game. For instance, we might discover that it pays more to play one strategy half the time and a different strategy the remainder of the time. This is expressed by defining mixtures of pure strategies.

A *mixed strategy* is a probability distribution over pure strategies, and is denoted σ_i for player i. In other words, if player i plays strategy α with probability $P(\alpha)$ then,

$$\sigma_i(\alpha) = P_i(\alpha). \tag{19.4}$$

Clearly, the sum of probabilities for all alternative strategies is one, for every player, so

$$\sum_{\alpha} \sigma_i(\alpha) = 1, \forall i. \tag{19.5}$$

Mixed-strategy probabilities can be interpreted in various ways as follows:

- The average play over time within a single execution of a game,

- The likelihood of choosing a particular pure strategy on repeated invocations of similar games,

- The average strategies of multiple players of a game, over multiple trials.

---

[2] The pay-off is also called the player's *utility* in many texts.

Mixed strategies are important because they make the theory of games into a tool for statistical inference. A certain randomness of strategy can often compensate for uncertainty by randomly hitting a randomly moving target.

The value for a player

The value of that what is earned or 'won' by a player in a game is given by the scalar product of the pay-off function Π_i with the strategy profiles of the users.

The value of any player is weighted by the choices made by all the players. Thus, no player can win an arbitrary amount, without other players being able to downgrade their potential pay-off by counter-play.

Example 168 *Consider a two-person game with pay-off matrix*

$$\Pi_1 = \begin{pmatrix} 4 & 5 & 6 \\ 2 & 8 & 3 \\ 3 & 9 & 2 \end{pmatrix} \tag{19.6}$$

for Player 1 and pay-off matrix

$$\Pi_2 = \begin{pmatrix} 3 & 1 & 2 \\ 1 & 4 & 6 \\ 0 & 6 & 8 \end{pmatrix} \tag{19.7}$$

for Player 2. These two matrices are often combined as follows:

$$\Pi_{(1,2)} = \begin{pmatrix} (4,3) & (5,1) & (6,2) \\ (2,1) & (8,4) & (3,6) \\ (3,0) & (9,6) & (2,8) \end{pmatrix}. \tag{19.8}$$

Now let σ_1^{T} be the transpose of a general mixed-strategy vector for Player 1, that is,

$$\sigma_1^{\mathrm{T}} = (P(s_1), P(s_2), P(s_3)) = \left(\frac{1}{3}, \frac{1}{3}, \frac{1}{3} \right), \tag{19.9}$$

and let σ_2 be the mixed-strategy vector for Player 2, that is,

$$\sigma_2 = \begin{pmatrix} P(s_1') \\ P(s_2') \\ P(s_3') \end{pmatrix} = \begin{pmatrix} 0 \\ \frac{1}{2} \\ \frac{1}{2} \end{pmatrix}. \tag{19.10}$$

The value of the pay-off to Player 1

$$v_1 = \sigma_1^{\mathrm{T}} \Pi_1 \sigma_2 = \left(\frac{1}{3}, \frac{1}{3}, \frac{1}{3} \right) \begin{pmatrix} 4 & 5 & 6 \\ 2 & 8 & 3 \\ 3 & 9 & 2 \end{pmatrix} \begin{pmatrix} 0 \\ \frac{1}{2} \\ \frac{1}{2} \end{pmatrix} = \frac{11}{2}. \tag{19.11}$$

The value of the game for Player 2 is

$$v_2 = \sigma_1^{\mathrm{T}} \Pi_2 \sigma_2 = \left(\frac{1}{3}, \frac{1}{3}, \frac{1}{3}\right) \begin{pmatrix} 3 & 1 & 2 \\ 1 & 4 & 6 \\ 0 & 6 & 8 \end{pmatrix} \begin{pmatrix} 0 \\ \frac{1}{2} \\ \frac{1}{2} \end{pmatrix} = \frac{27}{6}. \tag{19.12}$$

19.4 The strategic form of a game

The *strategic* or *normal* form of a game consists of a number of players, strategies and rewards or pay-offs that result from the use of the strategies.

1. A set of players $i = 1, 2, \ldots n$.

2. Sets of pure strategies S_i for each player i. The vector $\vec{s} = (s_1, s_2, \ldots, s_n)$, where $s_i \in S_i$ is called a strategy profile for the game, that is, a choice of strategies for each player. Note that each s_i is also a vector whose number of elements is the number of pure (independent) strategies available to player i.

3. A function $\Pi_i(s)$ for player i that describes the player's pay-off when a certain combination of strategies s is played by all the players.

Example 169 *A simplistic formulation of a game to weigh the advantages and disadvantages of upgrading software by various methods. The pay-off can be thought of as the level of 'convenience' to the players. Thus, the row player, who is the system administrator, considers the advantage to the system, while the column player, who represents the users of the system, considers the advantage to themselves.*

	Security hole	Bug in function	Missing function
Upgrade version now	*(10,5)*	*(10,0)*	*(5,−5)*
Test, then upgrade	*(5,5)*	*(3,9)*	*(0,8)*
Keep parallel versions	*(−10,5)*	*(−1,10)*	*(0,10)*

The system administrator believes that the maximum advantage (10) to the system arises from upgrading software immediately when faced with a security hole, while the hostile user is dealt a maximum blow by a quick upgrade so this is also of advantage to the users of the system, who do not have the same level of advantage (5) from the strategy since they are perhaps protected in other ways, with redundancy and back-up. If the users are missing some important functionality that only exists in a newer version of the program they have a high level of pay-off by getting the upgrade quickly (8), however new versions often incorporate new bugs, so parallel versions give the maximum benefit to users (10), but neither of these cases are of any great interest to the system administrator who does not use the software.

If there is a bug in the software, the administrator benefits from upgrade by not having to deal with irate users (3), while users clearly benefit from upgrade (9). Again parallel versions suit users (10) but might disadvantage administrators (−1) since multiple versions often present technical and administrative challenges to administrators.

We can continue in this way, posing values for the pay-offs. The pay-off values here certainly depend on factors other than that we have considered in the primitive example, for example, how reliable the new versions tend to be from the software producer. A better model of this game would take these explicitly into account.

The real challenge in formulating strategic form games is how to model the pay-offs using actual numbers. Inspired guesswork is the most primitive level of approximation. One might imagine that this would not lead to any useful result, but often surprises result when the game is solved. Games automatically take into account all of the competing forces in the game, which often leads to results that are not easily seen or guessed from the individual estimations of pay-offs.

Measurement of the system over time is another way to develop pay-off models. A semi-empirical model can easily be used to gauge the relative advantages of different strategies. This also allows one to introduce dimensions such as time variation into the games. Note however that a game has no concept of causal time in its strategic form, with one strategy leading to another. Time becomes at best an average parameter changing conditions for the whole game.

19.5   The extensive form of a game

How do we find the strategic pay-offs from a game that has many complicated moves and courses of action? In a board game, such as chess for instance, we normally think of a complex interaction of the players' moves and counter-moves. In system administration, games between users and the system might extend over considerable numbers of moves and counter-moves. The *extensive* form of a game is thus based on a complete tree of every detailed possible move or sub-decision between the players (fig. 19.1). It is a way of

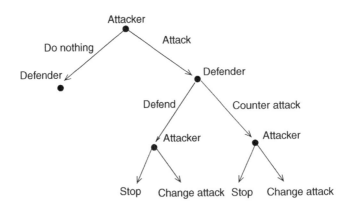

Figure 19.1: The extensive form of a game is a complete history of the state space of the players. By tracing the choices made from the start of the game, a player can examine the pay-off accrued and predict the best course of action for the remaining moves.

mapping out the behaviour of the players causally, though not necessarily deterministically. The decision trees in game theory are related to the fault trees in chapter 18: We can examine the tree for the players' moves in turn (one moves, then the other counter-moves etc), or we can separate the two decision trees of the players, if we do not necessarily know the order in which the moves are made. The two cases are referred to as games with perfect and imperfect information, respectively.

The extensive form consists of a number of players, a game tree and a set of pay-offs. A game tree is a graph (x, Γ) consisting of nodes and edges or arcs (see section 6.2). Each edge represents a move in the game, and each node is a player whose turn it is to make a decision. The graph for the game must be a *tree*, that is, every node must have exactly one path back to a root node, that is, must have a unique parent. This is somewhat like a state diagram, except that the same 'state' in a system can appear several times in the game tree, if it can be reached by a number of different means. The game tree codifies the history of the transitions as well as the actual states of the players.

A player in a game might have complete knowledge of the game tree when it is his or her turn to move, but he might also not be able to remember how he arrived at a given node—thus, if several nodes have the same state, there might be several alternatives to play that seem appropriate. In fact, only one choice is the true choice, but the player might not be able to determine this and could believe that the best course of action forward is from one of the other nodes that look the same. Game play can therefore also describe situations in which the players have perfect recall of the entire game history, and situations in which that information is lost.

Example 170 *In an organization, records are usually kept only for a certain length of time. In a computer system, logs of system events are rotated and deleted after a certain length of time. In each case, information must eventually be forgotten because of limited resources. This might affect the ability of system administrators and managers to determine the best course of action forward in time.*

The extensive form of a game can finally be reduced to a strategic form, by summing the pay-offs of the individual pathways in the game. Two extensive forms of a game are said to be equivalent if they reduce to the same strategic form.

In this book, we shall not look at extensive games in any detail; such a topic could probably be a book in its own right. The strategic form of game theory will be the most immediately useful form for ordinary decision-making, however, the extension to using causal trees as an interpretation of the extensive form of the game allows more detailed analyses of system dependencies by means of game theory.

19.6 Solving zero-sum games

Zero-sum games are games that satisfy conservation of pay-off constraints, that is, we are neither allowed to create nor destroy the currency of the game. This is a familiar idea in the physical world, where energy is conserved for closed systems. Certain simplifications arise for this type of game, as a result of this constraint. The basic approach to solution of zero-sum games begins with the minimax theorem, due to Von Neumann.

Consider a two-person zero-sum game, with pay-off matrices (Π_1, Π_2) and pure strategy sets (S_1, S_2). The pay-off matrices satisfy

$$\Pi_2(s_1, s_2) = -\Pi_1(s_1, s_2), \quad \forall s_1 \in S_1, \ s_2 \in S_2. \tag{19.13}$$

The minimax theorem tells us that all games have a solution that is expressible in terms of mixed strategies, and games that have an immediate saddle point have a solution in terms of pure strategies. The minimax theorem says that it is always possible to find a pair of mixed strategies (σ_1, σ_2) such that there is a unique equilibrium between the players that gives value v_1 to Player 1 and $-v_2$ to Player 2:

$$v_1 = \max_{\sigma_1} \min_{\sigma_2} \ \sigma_1^T \Pi_1 \sigma_2 = \min_{\sigma_2} \max_{\sigma_1} \ \sigma_1^T \Pi_1 \sigma_2. \tag{19.14}$$

Moreover, the limiting case of the theorem occurs when the pay-off matrix alone has a saddle point, that is,

$$\max_{\sigma_1} \min_{\sigma_2} \ \Pi_1 = \min_{\sigma_2} \max_{\sigma_1} \ \Pi_1. \tag{19.15}$$

For the two-player game, this condition is very easy to check, by looking along the rows and the columns of the pay-off matrix for either of the players.

Example 171 *The following game has a pure strategy saddle point. Let the pay-off or utility matrix for player A be given by*

$$\Pi_A = \begin{pmatrix} 1 & -3 & -2 \\ 2 & 5 & 4 \\ 2 & 3 & 2 \end{pmatrix}. \tag{19.16}$$

This game is zero-sum, so $\Pi_B = -\Pi_A$. We look for a saddle point:

$$\max_{\updownarrow} \min_{\leftrightarrow} \Pi_A = \max_{\updownarrow} \begin{pmatrix} -3 \\ 2 \\ 2 \end{pmatrix}$$

$$= 2. \tag{19.17}$$

$$\min_{\leftrightarrow} \max_{\updownarrow} \Pi_A = \min_{\leftrightarrow} (2, 5, 4)$$

$$= 2. \tag{19.18}$$

Since these two values are equal, we have found the value of the game $v = 2$, and see two optimal strategy saddle points, with row-column coordinates: $(r^*, c^*) = (2, 1)$ and $(3, 1)$.

19.7 Dominated strategies

We wish to discuss varying the strategies of a single player i, while holding the opponents' strategies fixed. Let s_i be an element of S_i, the strategy space of player i, and let s_{-i}

denote a strategy selection for every player other than i (i.e. using the set notation from section 5.1, this is the strategy for the set $-i$ or 'not' i). We can now write a complete strategy selection for all players, but singling out i as (s_i, s_{-i}). Better still, we can draw attention to the fact that we are looking at a trial strategy for i by writing $s_i \to t_i$, so that a complete strategy profile is given by

$$\vec{s} = (t_i, s_{-i}). \tag{19.19}$$

Similarly, for mixed strategies, we can write (σ_i, σ_{-i}) or

$$\vec{\sigma} = (\tau_i, \sigma_{-i}). \tag{19.20}$$

We say that a pure strategy t_i is *strictly dominated* for player i if

$$\Pi_i(\sigma_i, s_{-i}) > \Pi_i(t_i, s_{-i}), \forall s_{-i}, \tag{19.21}$$

that is, a player is always better off using some other mixture of strategies than choosing t_i, regardless of what the other players do. If the strict inequality above is replaced by a weak inequality $\geq$, then the strategy is said to be weakly dominated. Notice that this means that the definition in eqn. (19.21) also applies for any opponent mixed strategies, since $\sigma_{-i} = \sum_i p_i s_{-i}$, but all s_i are covered in this relation, and all $p_i \leq 1$, hence $\sigma_{-i} \leq s_{-i}$. Similarly, given any pure strategy that is dominated, a mixed strategy that gives non-zero weight to this strategy is also dominated.

19.8 Nash equilibria

The Nash equilibrium is probably the most important solution concept in game theory. For two-person zero-sum games, it corresponds to the minimax saddle point for mixed strategies; however, it also generalizes this concept, as it is not limited to zero-sum games. Nash proved that all games have at least one equilibrium in terms of mixed strategies.

The concept of a Nash equilibrium is related to the idea of fixed points and equilibria encountered in chapter 10. It is most easily formulated for the strategic form of the game. Suppose a game has n players with pure strategy sets S_i, and pay-off functions $\Pi_i: S \to R^1$ for $i_1, \ldots, n$. The set or space of all random strategy profiles is defined by

$$\begin{aligned}\Sigma &= \sigma(s_1) \times \sigma(s_2) \times \cdots \times \sigma(s_n) \\ &= \times_{i \in n} \sigma(s_i).\end{aligned} \tag{19.22}$$

A Nash equilibrium is a mixed strategy profile σ^* for each and every player, such that each player's strategy is an optimal response to all of the other's strategies, that is,

$$\Pi_i(\tau_i^*, \sigma_{-i}^*) \geq \Pi_i(\sigma_i, \sigma_{-i}^*), \ \forall \sigma_i \in \Sigma_i. \tag{19.23}$$

The Nash equilibrium is related to the Kakutani fixed point theorem, by forming a correspondence between every mixed strategy and its optimal response. Let us define the optimal response mapping as the function R_i that maps a certain combination of opponents strategies σ_{-i} to an optimal strategy σ_i for player i:

$$\sigma_i = R_i(\sigma_{-i}) = \text{argmax}_{\tau \in \Sigma} \ \Pi_i(\tau, \sigma_{-i}), \tag{19.24}$$

that is, it selects the value of the argument that maximizes the pay-off and returns it as its value. Although this function does not need to know the value of σ_i, since it actually selects it, it does no harm to make R_i functionally dependent on it in a trivial way, that is, we can equally well write this for all the players' σ_i:

$$\sigma_i = R_i(\sigma). \tag{19.25}$$

Finally, we can form the product correspondence of all of these functions for all the players:

$$R(\sigma) = R_1(\sigma) \times R_2(\sigma) \times \cdots \times R_n(\sigma). \tag{19.26}$$

The Nash equilibrium is then defined as strategy profile σ^*, which is the fixed point of this correspondence;

$$\sigma^* = R(\sigma^*). \tag{19.27}$$

Nash used this construction to prove that such a fixed point must exist in a finite game, for mixed strategies.

Example 172 (Competition or cooperation for service?) *Consider, for simplicity, just two customers or users A and B, who wish to share a service resource. We shall assume that the service 'market' is a free-for-all; that is, no one player has any* a priori *advantage over the other, and that both parties behave rationally.*

The users could try to cooperate and obtain a 'fair' share of the resource, or they could let selfish interest guide them into a competitive battle for the largest share. The cooperation or collaboration might, in turn, be voluntary or, it might be enforced by a service provider.

These two strategies of competition and collaboration are manifestly reflected in technologies for networking, for instance

- *Voluntary sharing: Ethernet is an example of voluntary sharing, in which any user can grab as much of a share as is available. There is a maximum service rate that can be shared, but it is not necessarily true that what one user loses is automatically gained by the other. It is not a zero-sum game.*

- *Forced sharing: Virtual circuits (like MPLS, ATM or Frame Relay networking technologies) are examples of forced sharing, over parallel circuits. There are thus fixed quotas that enforce users' cooperation. These quotas could be allocated unevenly to prioritize certain users, but for now we shall assume that each user receives an equal share of the resource pot.*

We analyse this situation in a very simple way using a classic game theoretical approach. The customers can 'win' a certain amount of the total service rate R (e.g. bytes per second, in the case of network service), and must choose strategies for maximizing their interests. We can therefore construct a 'pay-off' matrix for each of the two users (see tables 19.1, 19.2, 19.3).

Thus, we assume that each of the users assumes an equal share $R/2$ when they cooperate with one another. The relative sizes of the pay-off are important. We have

$$\delta R \le \frac{R}{2} \tag{19.28}$$

$$\left(\frac{R}{2} - \delta R\right) \le \frac{R}{2} \le \left(\frac{R}{2} + \delta R\right). \tag{19.29}$$

Table 19.1: A's pay-off matrix in two-customer sharing.

A	B Cooperate	B Compete
A Cooperate	$\frac{R}{2}$	$\frac{R}{2} - \delta R$
A Compete	$\frac{R}{2} + \delta R$	R_c

Table 19.2: B's pay-off matrix in two-customer sharing.

B	B Cooperate	B Compete
A Cooperate	$\frac{R}{2}$	$\frac{R}{2} + \delta R$
A Compete	$\frac{R}{2} - \delta R$	R_c

Table 19.3: A, B combined pay-off matrix in two-customer sharing. This is the usual way of writing the pay-off matrices. We see that, when both customers collaborate (either willingly or by forced quota), they obtain equal shares. If one of them competes greedily, it can obtain an extra δR that is then subtracted from the other's share. However, if both users compete, the result is generally worse (R_c) than an equal share.

A, B	B Cooperate	B Compete
A Cooperate	$\frac{R}{2}, \frac{R}{2}$	$\frac{R}{2} - \delta R, \frac{R}{2} + \delta R$
A Compete	$\frac{R}{2} - \delta R, \frac{R}{2} + \delta R$	R_c, R_c

In other words, by competing, a selfish user might be able to gain an additional amount of the service capacity δR to the other's detriment. The sum of both users' shares cannot exceed R. If both users choose to compete, the resulting competition might lead to an amount of waste that goes to neither of the users. This is the case in Ethernet, for instance, where collisions reduce the efficiency of transmission for all parties equally. We model this by assuming that both users then obtain a share of $R_c < R/2$.

This leaves us with two separate cases to analyse:

1. $R_c > R/2 - \delta R$: If the result from competitive 'attacks' against one another is greater than the result that can be obtained by passively accepting the other customer's aggressiveness, then we are inclined to retaliate. This becomes an instance of the Prisoner's Dilemma game. It has a solution in terms of Nash equilibria by dominant strategies.

2. $R_c < R/2 - \delta R$: If the pay-off for mutual competition is less than the penalty for collaborating, then the situation becomes equivalent to another classic game: the

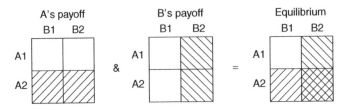

Figure 19.2: With $R_c > R/2 - \delta R$, the game becomes a classic game theoretical problem of 'Prisoner's Dilemma'. The dominant Nash equilibrium is where both players decide to compete with one another. If the customers are altruistic and decide to collaborate (or are forced to collaborate) with one another, they can win the maximum amount. However, if they know nothing about each other's intentions then they realize, rationally, that they can increase their own share by δR by choosing a competitive strategy. However, if both choose to be competitive, they cannot achieve exactly this much: the balance point for mutual competition is R_c. This value is determined by the technology used by the service provider. If either one of the players decided to cooperate with the other, they would lose.

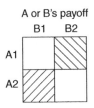

Figure 19.3: With $R_c < R/2 - \delta R$ the game becomes another classic game of Dove-Hawk. If both players are 'hawkish' and attack greedily, they both lose out. The stable equilibria are that one player is greedy and the other is submissive.

Maynard-Smith Dove-Hawk game. Both players see that they can win an important share by being greedy, but if the other player retaliates, they both stand to win less. Thus one player can afford to be aggressive (hawkish) but then the other must be peaceful (dove-like). This is the case with Ethernet, for instance. If there is excessive contention, there is an exponential 'back-off' from collisions leading to significantly worsened performance.

We can ask if there is a mixed strategy of partial cooperation that would succeed at countering the poor result from mutual competition, but which yields slightly more. To show that this is not the case, let us pick B's strategy and then allow A to choose cooperation with probability p:

(a) B cooperates: *compare then the pay-offs for A and B and ask, is there a value of p such that*

$$p\frac{R}{2} + (1-p)\left(\frac{R}{2} + \delta R\right) > p\frac{R}{2} + (1-p)\left(\frac{R}{2} - \delta R\right)? \quad (19.30)$$

Either we must have $\delta R = 0$ or $p = 0$, so the answer is no; there is no way to improve on this strategy as long as there is something to gain from competition.

(b) B *competes: compare then the pay-offs for A and B and ask, is there a value of p such that*

$$p\left(\frac{R}{2} - \delta R\right) + (1 - p)R_c > p\left(\frac{R}{2} + \delta R\right) + (1 - p)R_c? \quad (19.31)$$

Again, consistency forces us to take $p = 0$ or $\delta R = 0$.

These simple games capture the essence of the issues involved in sharing. They reflect both human strategies for competition and technological ones. We see that there is no clear answer as to whether Ethernet (hawkish) or fixed-quota virtual circuit (dove-like) behaviour is preferable; it depends on the level of traffic.

Example 173 *The model in the previous example addresses many situations. We can use it for human scheduling too. The basic result from game theory, although simplistic, tells us that random event competition works well as long as the number of jobs is small, that is, as long as there is plenty of time and no one is taxed to the limit. However, as a human becomes very busy, it becomes better to schedule fixed-quota time slices for jobs, otherwise all jobs tend to suffer and nothing gets done. The overhead of swapping tasks at random can lead to great inefficiency if time is short.*

19.9 A security game

Electronic banking, and other on-line services that require users to identify themselves in a secure way, use a variety of mechanisms to achieve this authentication. Many banks issue 'calculator'-like smart cards that generate one-time passwords on the basis of a personal code and the time of day. Others use a Transmission Layer Security (TLS)[3] certificate mechanism to download a secret key from the bank, using a login and Personal Identification Code that they receive from the bank by postal mail. At the lowest end level, some sites are password-protected, or could use some kind of biometric secret. The attacks that can be levelled against these measures include actually stealing a user's credentials, hijacking the TCP session, spoofing the web site and placing a Trojan horse in the user's browser.

Let us formulate a game based on a simple system versus adversary model. At the most simplistic level, security can be regarded as a zero-sum game. What is secured properly is lost to the potential attacker and what is not secured is gained. However, if we add in the cost of buying or developing and then maintaining the security system, the 'gains' of the defender are somewhat mitigated by the cost of ownership. This would make the game non zero-sum. We can therefore examine the game in stages. We begin by defining the pay-offs in the 'game'.

We estimate the pay-offs on a scale of 0–100%. We shall consider the pay-off to the defender of the system, so that the pay-off is essentially the same as the estimated level of security. Suppose that the security levels are given as in table below for defence and attack (D, A).

---

[3] TLS was formerly known as Secure Socket Layer (SSL).

Security	Steal credentials	Hijack session	Spoof site	Trojan in browser
Smart card	(90,10)	(90,10)	(40,60)	(30,70)
Certificate	(70,30)	(80,20)	(70,30)	(10,90)
Password	(50,50)	(70,30)	(10,90)	(10,90)

These values are posed only for illustration and are not actually measurements or estimates, but we can justify them approximately as follows. Smart cards are difficult to steal, since they can be carried with a person and kept safe; thus, they offer a high level of security against theft. Certificates are a little easier to steal, since they reside on the user's computer and could therefore be eavesdropped on or extracted somehow. Passwords are easier still to steal or tease out of users by social engineering or eavesdropping.

If an attacker can hijack a session, smart card systems require code confirmations of all important operations, so without the smart card, such session hijacking would not be a useful strategy. Certificates are slightly less secure, since they can sometimes be extracted from transactions; again, passwords offer the weakest security.

An attacker who spoofs a web site can trick a user with a smart card to entering a code that could be used by the attacker instead for a short interval of time, so the security level to spoofing is quite low. A certificate is relatively better here, however, since it is used to encipher information for the specific site, and thus would provide only nonsense to a spoofed web site. Passwords would be trivially collected at a spoof site. They offer essentially no security.

Finally, if an attacker can sneak Trojan code into the user's browser or computer, then none of the security mechanisms are really secure, but the smart card is a little more secure than either certificates or passwords, since it is external to a user's computer and can only be abused for a limited time window when a user enters a code from the smart card.

19.9.1 Zero-sum approximation

In order to apply the minimax theorem, we re-normalize the matrix elements so that the sum of attack and defence is not 100 but 0, thus we subtract 50 from all the values, giving

$$\Pi_D = -\Pi_A \begin{pmatrix} 50 & 50 & -10 & -20 \\ 20 & 30 & 40 & -40 \\ 0 & 20 & -40 & -40 \end{pmatrix}. \tag{19.32}$$

We begin by looking for a saddle point in the game.

$$\max_{\updownarrow} \min_{\leftrightarrow} \Pi_D = \max_{\updownarrow} \begin{pmatrix} -20 \\ -40 \\ -40 \end{pmatrix}$$

$$= -20. \tag{19.33}$$

$$\min_{\leftrightarrow} \max_{\updownarrow} \Pi_D = \min_{\leftrightarrow}(50, 50, 40, -20)$$

$$= -20. \tag{19.34}$$

The matrix yields a single saddle point and the value of the game as -20, with the optimal strategies being smart cards versus Trojan horses. This is a surprising result that is not clearly obvious from the original pay-off table. First of all, the value of the game is a negative number, which means that the result is in the attacker's favour. This might not have been expected by looking at the original estimates of security, but the conclusion is the stable equilibrium of the two opposing sides, and therefore reveals the 'rational' conclusion in the data.

We might, however, consider this to be an unfair conclusion, since the likelihood that an attacker will be able to insert a Trojan horse into a user's computer is quite low and might be detected by anti-virus software. We could therefore delete the Trojan strategy and re-compute the equilibrium. Thus, the game is still zero-sum, but we now have

$$\Pi'_D = -\Pi'_A \begin{pmatrix} 50 & 50 & -10 \\ 20 & 30 & 40 \\ 0 & 20 & -40 \end{pmatrix}. \tag{19.35}$$

We now find that there is no saddle-point equilibrium in the matrix, thus there is no optimal pure strategy contest here. This means that the solution must be in terms of a mixture of strategies and is therefore rather harder to solve. We begin by trying to eliminate any obviously weak strategies from the mixture, since these cannot lead to any optimal behaviour. Examining the matrix in eqn. (19.35), we see that rows $i = 1, i = 2$ both strictly dominate the row $i = 3$; hence $i = 3$ is a weak strategy for the defending row player 'α'; we can delete it, giving

$$\Pi''_D = -\Pi''_A \begin{pmatrix} 50 & 50 & -10 \\ 20 & 30 & 40 \end{pmatrix}. \tag{19.36}$$

There is still no saddle point, and there are no more cases of strict dominance. We must now find a way of solving the correct linear combinations. There are many ways one might proceed, but this case is quite simple. The value of the pay-off for the defending row player is

$$v = \vec{\alpha}^T \Pi''_D \vec{\beta}, \tag{19.37}$$

where $\vec{\alpha}$ is the vector of strategies for the defence that now lies between a convex mixture of the smart card strategy and the certificate strategy, and $\vec{\beta}$ is a convex mixture of the attack strategies, all of which are active at present:

$$\vec{\alpha}^T = (\alpha, 1 - \alpha) \; , \; \vec{\beta} = \begin{pmatrix} \beta_1 \\ \beta_2 \\ \beta_3 \end{pmatrix}, \tag{19.38}$$

where $\beta_1 + \beta_2 + \beta_3 = 1$ and $\alpha > 0$, $\beta_i > 0$. We can try to maximize the pay-off for the defending player, by examining the rates of change and looking for any stationary values. This is not a solution method, but it does provide an indication of how to choose the values of the free parameters $(\alpha, \beta_1, \beta_2, \beta_3)$. The partial derivatives are

$$\frac{\partial v}{\partial \beta_1} = 30\alpha + 20$$

$$\frac{\partial v}{\partial \beta_2} = 20\alpha + 30$$

$$\frac{\partial v}{\partial \beta_3} = -50\alpha + 40$$

$$\frac{\partial v}{\partial \alpha} = 30\beta_1 + 20\beta_2 - 50\beta_3 = 0. \tag{19.39}$$

We set the last derivative to zero, since we are looking for a mixture of $\vec{\beta}$ that makes the pay-off stationary for the players. Thus, whatever the attacker would like to play, this condition will limit the success of the attack, by virtue of the defending player's optimal responses. We notice first that the derivative with respect to β_3 has a possibility of going negative. This means that the β-attack player will try to play this strategy to make the defence player's pay-off v less. Using the fact that the β_i sum to one, we have

$$\frac{\partial v}{\partial \alpha} = 30\beta_1 + 20\beta_2 - 50(1 - \beta_1 - \beta_2) = 0$$
$$= 80\beta_1 + 70\beta_2 - 50 = 0. \tag{19.40}$$

Now, let us examine some cases of this:

1. Suppose the attack player decides to avoid session hijack and play β_2, since it looks like a marginally weak strategy (it is a weakly dominant column of the defence player's pay-off—and this is a zero-sum game), then we have

$$\beta_1 = \frac{5}{8}, \beta_3 = \frac{3}{8}. \tag{19.41}$$

If we choose this mixture of strategies for $\vec{\beta}$ in eqn. (19.37), then the value of the game is

$$v = \frac{1}{8}(\alpha, 1 - \alpha) \begin{pmatrix} 220 \\ 220 \end{pmatrix}. \tag{19.42}$$

that is, the pay-off to the defence player is $v = 27.5$ regardless of what mixture of strategies α is chosen. Thus, both smart cards and certificates seem to be equally valid defences to this attack. This conclusion is not particularly intuitive.

2. What if we choose $\beta_1 = 0$? Then we have

$$\frac{\partial v}{\partial \alpha} = 70\beta_2 - 50 = 0. \tag{19.43}$$

That is,

$$\beta_1 = \frac{5}{7}, \beta_3 = \frac{2}{7}. \tag{19.44}$$

and the value of the game for the defence player is

$$v = \frac{1}{7}(\alpha, 1 - \alpha) \begin{pmatrix} 230 \\ 230 \end{pmatrix}. \tag{19.45}$$

Again, surprisingly, the choice of α is irrelevant to the outcome. However, crucially, the value of the game, $v = 230/7$, is now greater for the defence player. This is to the defence player's linking, but, alas, he is not calling the shots and choosing $\vec{\beta}$. The attacker would not choose this strategy, since he is trying to minimize the defence player's pay-off. Remarkably, in both these cases, the defence player is a 'sitting duck', with no way of improving his defence by choosing one strategy or another, in any given mixture.

We have not proved that case 1 above is the optimal strategy, beyond doubt, but with a little work, it is not difficult to see that it is indeed the solution to the game. Indeed, we can see from eqn. (19.40), the value of $80\beta_1 + 70\beta_2$ can never be greater than when $\beta_2 = 0$, thus the value of the denominator in the fraction cannot be made larger and the value of the game cannot be made smaller by choice of β_i.

What is interesting about this example is that the conclusion is not at all obvious from the original security level evaluations. This analytical procedure selects the limits of the tug-of-war contest in an impartial way.

19.9.2   Non-zero sum approximation

The zero-sum approximation does not allow us to take into account other sources of loss and gain. What is lost to the defender is not necessarily gained by the attacker. The cost of implementing a technological solution should be factored into the calculations in considering 'cost of ownership'. It is not *gratis* to implement a certificate system, for example. We can add in costs of this kind by modifying the pay-off:

Pay-off = Security level − cost of strategy.

What should the exact formula be for this cost? This depends on our estimation of the relative importance of these. We need to relate the currencies to one another, using the same scale (see section 11.3).

Let us define for clarity α to be the defending player and β to be the attacking player. There is no rational way to relate security level to cost of implementation, so we must define this relationship as a matter of policy. For the defender,

$$\Pi_1 = \Pi_1^{(0)} - k_1 C_\alpha, \tag{19.46}$$

where $\Pi_1^{(0)}$ is the basic constant sum estimation of pay-off, k_1 is a policy constant, and C_α is the cost of investing in the security technology. Similarly, for the attacker,

$$\Pi_2 = \Pi_2^{(0)} - k_2 C_\beta, \tag{19.47}$$

where $\Pi_2^{(0)}$ is the attacker's basic pay-off, k_2 is the attacker's own estimate of how gain relates to invested time C_β in carrying out the attack.

The addition of the cost of strategy term is a perturbation to the basic pay-off. One can use the solution for the Nash equilibrium of the resulting game to test how much of a perturbation must be added to the constant sum conclusion, before the conclusion about optimal strategies is altered.

Add cost to defender $\sigma_1 = \alpha$

We now try to combine the information about the different strategies to modify the pay-off, as above. What is the cost of implementing the security technologies?

Smart cards cost money and need to be replaced sometimes, so there is an expense; however, we can make customers pay for these, so there is no effect on the pay-offs for the bank. A certificate system, on the other hand, is costly since it must be set up and maintained, depending on the local web services. Also, cryptographic certificates do not work consistently[4] in all browsers, so there is much programming, debugging and maintenance to keep the system working. Passwords are available to everyone with no investment, so these are also unaffected.

We shall thus suppose, as a combination of judgement and policy, that $k_1 C = -20$ for the certificate strategy, and zero for the others. Accordingly -20 is subtracted from the second row of the pay-off table, for the defender,

Security	Steal credentials	Hijack session	Spoof site	Trojan in browser
Smart card	(90,10)	(90,10)	(40,60)	(30,70)
Certificate	(50,30)	(60,20)	(50,30)	(−10,90)
Password	(50,50)	(70,30)	(10,90)	(10,90)

Now that the pay-offs are not constant sum, different solution methods are required than those used for the zero-sum case. The solution of non-constant sum games is beyond the scope of this book, however the open source Gambit software package, Gambit (n.d.) is a useful tool for solving for Nash equilibria. Feeding these data into the computer software, we obtain an answer that is, in fact, the same as that for the zero-sum case: the pure strategy equilibrium is (α_1, β_4), that is, use of smart cards for the defender and Trojan horse for the attacker.

Add cost to attacker $\sigma_2 = \beta$

Now we consider the cost of engagement from the attacker's viewpoint. The bank is not the only one in the game who needs to invest to use its available strategies. The average attacker is poorly inclined to invest a huge effort in preparing an attack of the system, thus he judges that the cost $-k_2 C_\beta$ of developing the Trojan horse strategy to be -40. This

---

[4] This scenario is based around an Internet Bank known to the author.

is a relatively high price, but then the attacker is somewhat lazy and judges the effort to be more than what his time is worth. The value is thus subtracted from the final column. Similarly, he judges the cost of the hijack strategy to be -10, on the pay-off scale. This is less expensive to him than the Trojan horse, because there are tools already available on the Internet to help him. The resulting table is now like this:

Security	Steal credentials	Hijack session	Spoof site	Trojan in browser
Smart card	(90,10)	(90,0)	(40,60)	(30,30)
Certificate	(50,30)	(60,10)	(50,30)	(-10,50)
Password	(50,50)	(70,20)	(10,90)	(10,50)

After these alterations, there is no pure strategy equilibrium. Instead, there is now a mixed-strategy equilibrium, $(\frac{1}{2}\alpha_1 + \frac{1}{2}\alpha_2, \frac{4}{5}\beta_3 + \frac{1}{5}\beta_4)$, with the defender mixing 50–50 between smart cards and certificates, and the attacker mixing $\frac{4}{5}\beta_3$ (site-spoofing) and $\frac{1}{5}\beta_4$ (Trojan horse). The analysis mixes these as probabilities, but suppose we decide that a one in five chance of using a Trojan makes it worth disregarding this attack strategy altogether, then the game is solved by a pure strategy equilibrium of certificates versus site-spoofing.

Conclusions

This example, although somewhat contrived, tells us the relative stability of the conclusions drawn from placing value on different aspects of the system solution. It is not necessary to have precise information about the different pay-offs in order to make a reasonably informed decision about the optimal strategies. Why? In this case, the reason is that the pay-offs are quite stable to small perturbations. If we use our best guesses in order to find a suitable stable equilibrium, we can then test each assumption by saying: what if I perturb the value by a small amount? Is the equilibrium robust under this change? If it is, we have a good idea what conclusion the model predicts. On the other hand, if a small change leads to a quite different solution, then the onus is on the system analyst to find an accurate pay-off model, or to find additional strategies that can result in a more stable conclusion.

19.10   The garbage collection game

The difficult aspect of game theoretical modelling is turning the high-level concepts and aims listed above into precise numerical values. This is particularly true when the values that govern a game change over time.

To illustrate a possible solution to this problem, we consider an example of some importance, namely the clearing of garbage files from user disks (see fig. 19.4). The need for user garbage collection (called *tidying*) has been argued by several authors (see Burgess (1995); Burgess and Ralston (1997); Zwicky (1989)), but users do not like having even the most useless of files deleted from their home areas.

We shall model this game as a zero-sum game. The currency of this game must first be agreed upon. What value will be transferred from one player to the other in play?

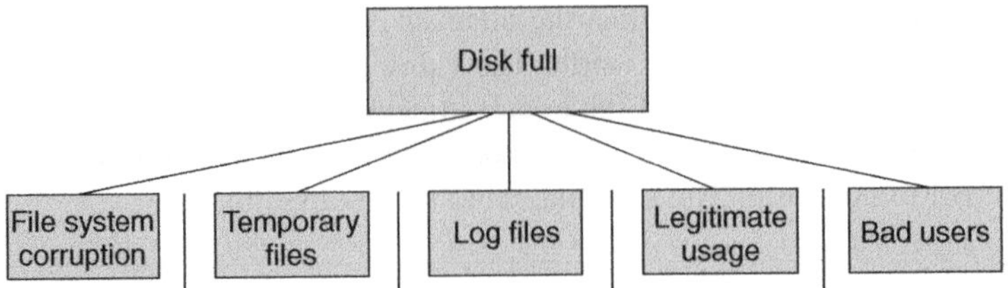

Counter-Strategies	File system corruption	Temporary files	Log files	Legitimate usage	Bad users
Force tidy	?	?	?	?	?
Ask users	?	?	—	?	?
Rotate logs	?	?	?	?	?
Check fs	?	?	?	?	?
Disk quotas	?	—	—	?	?

Disk full

Figure 19.4: Pay-off matrix and a fault tree showing how the fault tree feeds into the game as probabilities, and vice versa. The values in the matrix are probabilistic expressions expressing the likelihood of achieving each strategic goal, weighted by a currency scale for its relative importance. See Burgess (2000c) for details of this game.

There are three relevant measurements to take into account: (i) the amount of resources consumed by the attacker (or freed by the defender); sociological rewards: (ii) 'goodwill' or (iii) 'privilege' that are conferred as a result of sticking to the policy rules. These latter rewards can most easily be combined into an effective variable, 'satisfaction'. A 'satisfaction' measure is needed in order to set limits on individuals' rewards for cheating, or balance the situation in which the system administrator prevents users from using any resources at all. This is clearly not a defensible use of the system, thus the system defences should be penalized for restricting users too much. The characteristic matrix now has two contributions,

$$\pi = \pi_r(\text{resources}) + \pi_s(\text{satisfaction}). \tag{19.48}$$

It is convenient to define

$$\pi_r \equiv \pi(\text{resources}) = \frac{1}{2}\left(\frac{\text{Resources won}}{\text{Total resources}}\right). \tag{19.49}$$

Satisfaction π_s is assigned arbitrarily on a scale from plus to minus one half, such that,

$$-\frac{1}{2} \leq \pi_r \leq +\frac{1}{2}$$
$$-\frac{1}{2} \leq \pi_s \leq +\frac{1}{2}$$
$$-1 \leq \pi \leq +1. \tag{19.50}$$

The pay-off is related to the movements made through the lattice $\vec{d}$. The different strategies can now be regarded as duels, or games of timing.

Users/System	Ask to tidy	Tidy by date	Tidy above Threshold	Quotas
Tidy when asked	$\pi(1, 1)$	$\pi(1, 2)$	$\pi(1, 3)$	$\pi(1, 4)$
Never tidy	$\pi(2, 1)$	$\pi(2, 2)$	$\pi(2, 3)$	$\pi(2, 4)$
Conceal files	$\pi(3, 1)$	$\pi(3, 2)$	$\pi(3, 3)$	$\pi(3, 4)$
Change timestamps	$\pi(4, 1)$	$\pi(4, 2)$	$\pi(4, 3)$	$\pi(4, 4)$

These elements of the characteristic matrix must now be filled, using a model and a policy. A general expression for the rate at which users produce files is approximated by

$$r_u = \frac{n_b r_b + n_g r_g}{n_b + n_g},\tag{19.51}$$

where r_b is the rate at which bad users (i.e. problem users) produce files, and r_g is the rate for good users. The total number of users $n_u = n_b + n_g$. From experience, the ratio n_b/n_g is about 1%. The rate can be expressed as a scaled number between zero and one, for convenience, so that $r_b = 1 - r_g$.

The pay-off in terms of the consumption of resources by users, to the users themselves, can then be modelled as a gradual accumulation of files, in daily waves, which are a maximum around midday:

$$\pi_u = \frac{1}{2} \int_0^T dt \, \frac{r_u \left(\sin(2\pi t/24) + 1\right)}{R_{\text{tot}}},\tag{19.52}$$

where the factor of 24 is the human daily rhythm, measured in hours, and R_{tot} is the total amount of resources to be consumed. Note that by considering only good users or bad users, one has a corresponding expression for π_g and π_b, with r_u replaced by r_g or r_b respectively. An automatic garbage collection system (cfengine) results in a negative pay-off to users, that is, a pay-off to the system administrator. This may be written

$$\pi_a = -\frac{1}{2} \int_0^T dt \, \frac{r_a \left(\sin(2\pi t/T_p) + 1\right)}{R_{\text{tot}}},\tag{19.53}$$

where T_p is the period of execution for the automatic system. This is typically hourly or more often, so the frequency of the automatic cycle is some 20 times greater than that of the human cycle. The rate of resource-freeing r_a is also greater than r_u, since file deletion takes little time compared to file creation, and also an automated system will be faster than a human. The quota pay-off yields a fixed allocation of resources, which are assumed to be distributed equally amongst users and thus each quota slice assumed to be unavailable to other users. The users are nonchalant, so $\pi_s = 0$ here, but the quota yields

$$\pi_q = +\frac{1}{2} \left(\frac{1}{n_b + n_g} \right).\tag{19.54}$$

The matrix elements are expressed in terms of these.

$\pi(1, 1)$: Here $\pi_s = -\frac{1}{2}$, since the system administrator is as satisfied as possible by the users' behaviour. π_r is the rate of file creation by good users π_g, that is, only legal files are produced. Comparing the strategies, it is clear that $\pi(1, 1) = \pi(1, 2) = \pi(1, 3)$.

$\pi(1, 4)$: Here $\pi_s = 0$, reflecting the users' dissatisfaction with the quotas, but the system administrator is penalized for restricting the freedom of the users. With fixed quotas, users cannot generate large temporary files. π_q is the fixed-quota pay-off, a fair slice of the resources. Clearly $\pi(4, 1) = \pi(4, 2) = \pi(4, 3) = \pi(4, 4)$. The game has a fixed value if this strategy is adopted by system administrators. However, it does not mean that this is the best strategy, according to the rules of the game, since the system administrator loses points for restrictive practices, which are not in the best interest of the organization. This is yet to be determined.

$\pi(2, 1)$: Here $\pi_s = \frac{1}{2}$, since the system administrator is maximally dissatisfied with users' refusal to tidy their files. The pay-off for users is also maximal in taking control of resources, since the system administrator does nothing to prevent this, thus $\pi_r = \pi_u$. Examining the strategies, one finds that $\pi(2, 1) = \pi(3, 1) = \pi(3, 2) = \pi(3, 3) = \pi(4, 1) = \pi(4, 2)$.

$\pi(2, 2)$: Here $\pi_s = \frac{1}{2}$, since the system administrator is maximally dissatisfied with users' refusal to tidy their files. The pay-off for users is now mitigated by the action of the automatic system that works in competition, thus $\pi_r = \pi_u - \pi_a$. The automatic system is invalidated by user bluffing (file concealment).

$\pi(2, 3)$: Here $\pi_s = \frac{1}{2}$, since the system administrator is maximally dissatisfied with users' refusal to tidy their files. The pay-off for users is mitigated by the automatic system, but this does not activate until some threshold time is reached, that is, until $t > t_0$. Since changing the date cannot conceal files from the automatic system, when they are tidied above threshold, we have $\pi(2, 3) = \pi(4, 3)$.

Thus, in summary, the characteristic matrix is given by

$$
\pi(u, s) = \begin{pmatrix}
-\frac{1}{2} + \pi_g(t) & -\frac{1}{2} + \pi_g(t) & -\frac{1}{2} + \pi_g(t) & \pi_q \\
\frac{1}{2} + \pi_u(t) & \frac{1}{2} + \pi_u(t) + \pi_a(t) & \frac{1}{2} + \pi_u(t) + \pi_a(t)\theta(t_0 - t) & \pi_q \\
\frac{1}{2} + \pi_u(t) & \frac{1}{2} + \pi_u(t) & \frac{1}{2} + \pi_u(t) & \pi_q \\
\frac{1}{2} + \pi_u(t) & \frac{1}{2} + \pi_u(t) & \frac{1}{2} + \pi_u(t) + \pi_a(t)\theta(t_0 - t) & \pi_q
\end{pmatrix},
$$

(19.55)

where the step function is defined by,

$$
\theta(t_0 - t) = \begin{cases} 1 \ (t \geq t_0) \\ 0 \ (t < t_0) \end{cases},
$$

(19.56)

and represents the time delay in starting the automatic tidying system in the case of tidy-above-threshold. This was explained in more detail in (Burgess (2003)).

It is possible to say several things about the relative sizes of these contributions. The automatic system works at least as fast as any human; so, by design, in this simple model we have

$$\frac{1}{2} \geq |\pi_a| \geq |\pi_u| \geq |\pi_g| \geq 0, \tag{19.57}$$

for all times. For short times, $\pi_q > \pi_u$, but users can quickly fill their quota and overtake this. In a zero-sum game, the automatic system can never tidy garbage faster than users can create it, so the first inequality is always saturated. From the nature of the cumulative pay-offs, we can also say that

$$\left(\frac{1}{2} + \pi_u \right) \geq \left(\frac{1}{2} + \pi_u + \pi_a \theta(t_0 - t) \right) \geq \left(\frac{1}{2} + \pi_u + \pi_a \right), \tag{19.58}$$

and

$$\left| \frac{1}{2} + \pi_u \right| \geq \left| \pi_g - \frac{1}{2} \right|. \tag{19.59}$$

Applying these results to a modest strategy of automatic tidying of garbage, referring to fig. 19.5, one sees that the automatic system can always match users' moves. As drawn, the daily ripples of the automatic system are in phase with the users' activity. This is not realistic, since tidying would normally be done at night when user activity is low; however, such details need not concern us in this illustrative example.

The policy created in setting up the rules of play for the game penalizes the system administrator for employing strict quotas, which restrict their activities. Even so, users do

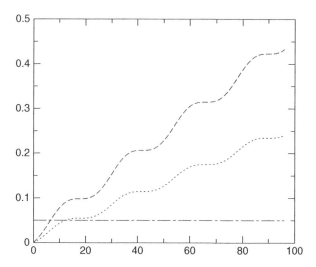

Figure 19.5: The absolute values of pay-off contributions as a function of time (in hours), for daily tidying $T_p = 24$. User numbers are set in the ratio $(n_g, n_b) = (99, 1)$, on the basis of rough ratios from the author's college environment, that is, 1% of users are considered mischievous. The filling rates are in the same ratio: $r_b/R_{tot} = 0.99$, $r_g/R_{tot} = 0.01$, $r_a/R_{tot} = 0.1$. The flat dot-slashed line is $|\pi_q|$, the quota pay-off. The lower wavy line is the cumulative pay-off resulting from good users, while the upper line represents the pay-off from bad users. The upper line doubles as the magnitude of the pay-off $|\pi_a| \geq |\pi_u|$, if we apply the restriction that an automatic system can never win back more than users have already taken. Without this restriction, $|\pi_a|$ would be steeper.

not gain much from this, because quotas are constant for all time. A quota is a severe handicap to users in the game, except for very short times before users reach their quota limits. Quotas could be considered cheating by the system administrator, since they determine the final outcome even before play commences. There is no longer an adaptive allocation of resources. Users cannot create temporary files that exceed these hard and fast quotas. An immunity-type model that allows fluctuations is a more resource-efficient strategy in this respect, since it allows users to span all the available resources for short periods of time, without consuming them forever.

According to the *minimax* theorem, if we have

$$\max_{\downarrow} \min_{\rightarrow} \pi_{rc} = \min_{\rightarrow} \max_{\downarrow} \pi_{rc}, \tag{19.60}$$

it implies the existence of a pair of single, pure strategies (r^*, c^*) that are optimal for both players, regardless of what the other does. If the equality is not satisfied, then the minimax theorem tells us that there exist optimal mixtures of strategies, where each player selects at random from a number of pure strategies with a certain probability weight.

The situation for our time-dependent example matrix is different for small t and for large t. The distinction depends on whether users have had time to exceed fixed quotas or not; thus 'small t' refers to times when users are not impeded by the imposition of quotas. For small t, one has

$$\max_{\downarrow} \min_{\rightarrow} \pi_{rc} = \max_{\downarrow} \begin{pmatrix} \pi_g - \frac{1}{2} \\ \frac{1}{2} + \pi_u + \pi_a \\ \frac{1}{2} + \pi_u \\ \frac{1}{2} + \pi_u + \pi_a\,\theta(t_0 - t) \end{pmatrix}$$

$$= \frac{1}{2} + \pi_u. \tag{19.61}$$

The ordering of sizes in the above minimum vector is

$$\frac{1}{2} + \pi_u \geq \frac{1}{2} + \pi_u + \pi_a\theta(t_0 - t) \geq \pi_u + \pi_a\theta(t_0 - t) \geq \pi_g - \frac{1}{2}. \tag{19.62}$$

For the opponent's endeavours, one has

$$\min_{\rightarrow} \max_{\downarrow} \pi_{rc} = \min_{\rightarrow} \left(\frac{1}{2} + \pi_u, \frac{1}{2} + \pi_u, \frac{1}{2} + \pi_u, \pi_q \right)$$

$$= \frac{1}{2} + \pi_u. \tag{19.63}$$

This indicates that the equality in eqn. (19.60) is satisfied and there exists at least one pair of pure strategies that is optimal for both players. In this case, the pair is for users to conceal files, regardless of how the system administrator tidies files (the system administrator's strategies all contribute the same weight in eqn. (19.63)). Thus, for small times, the users are always winning the game if one assumes that they are allowed to bluff by concealment. If the possibility of concealment or bluffing is removed (perhaps through an improved technology), then the next best strategy is for users to bluff by changing the date, assuming that the tidying looks at the date. In that case, the best system administrator strategy is to tidy indiscriminately at threshold.

For large times (when system resources are becoming or have become scarce), then, the situation looks different. In this case, one finds that

$$\max_{\downarrow} \min_{\rightarrow} \pi_{rc} = \min_{\rightarrow} \max_{\downarrow} \pi_{rc} = \pi_q. \qquad (19.64)$$

In other words, the quota solution determines the outcome of the game for any user strategy. As already commented, this might be considered cheating or poor use of resources, at the very least. If one eliminates quotas from the game, then the results for small times hold also at large times.

19.11   A social engineering game

The extensive form of a game is the form in which all possible moves are documented. The extensive form is more often associated with N-person game theory (where $N > 2$) than with simple two-person games, since the extensive game tree of a two-person game is often trivial, unless the moves are repeatable. A general introduction to N-person game theory and the extensive form is well beyond the scope of this book. However, we can examine some of the ideas through examples, as they contain some important insights.

So far, in decision-making, we have ignored the causality of strategies and actions taken by the agents within a human–computer system. The extensive form brings us back to this issue and rounds off the topic of decision-making by bringing together the concepts of information, causality and utility into a unified framework. Let us consider a simple example of causal decision-making, with three participants.

The following example is a special case of a general decision game. A generic three-person game tree for binary decision-making is shown in fig. 19.6. The yes–no decision can be interpreted in a number of ways to apply this to difference scenarios.

Consider the following scenario: A company or other enterprise offers a training programme to its employees so as to instruct them in policy. One of the aims of this training is to prevent attacks of the company by social engineers. Whether or not employees are trained, some of them will choose to obey company policy, while others will not. This decision may or may not be based on the information from the training process. A potential attacker of the company can observe the employees and have knowledge of the decisions made by management and by the individual employees (he might be an insider). How shall the employer, employees and attacker make their decisions?

In order to make rational decisions, there has to be a pay-off to each player. We shall consider first a game of perfect information, in which each player can see all of the moves and decisions throughout the game, in the order in which they occur. Note that the primary obstacle to understanding games in extensive form is in finding a suitable notation for describing the possible strategy combinations. On the one hand, a more complex notation can provide greater clarity, on the other hand, it can also overwhelm.

Player	Yes move	No move
P1 = Employer	Train personnel	Don't train personnel
P2 = Employees	Obey policy	Don't obey policy
P3 = Attacker	Attack enterprise	Don't attack enterprise

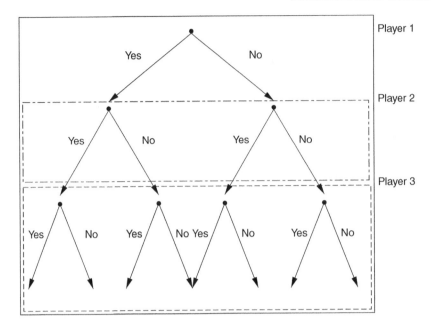

Figure 19.6: The extensive form of a three-person game with binary decision-making.

An alternative interpretation might be to imagine that the attacker is simply a bad case of 'gremlins', that is, chance, and to consider the worst-case scenario for the enterprise's training policy given that chance error plays its 'worst' hand.

A first step in solving this game is to change it into strategic form. One might suppose that the game tree as shown in fig. 19.6 catalogues all possible strategies, but this is not the case. The tree of moves is not the same as the tree of decisions about playing of an entire game. A game has two trees: a tree of moves (the game tree) and a tree of strategies that is not normally drawn. The strategy tree determines a complete set of contingencies for every player in every situation, and is often drawn as a table or pay-off matrix; however, the tree form preserves some of the causal structure in the game.

Any player's pure strategy must specify a course of action in each and every contingency that can occur throughout the game, given the information that is available to each player when he or she commences play. After all, it is not known at the outset of the game what will actually transpire between the players. A pure strategy must therefore correspond to a complete play of the game, by the player concerned, with all reasoning implicit. This means that, even if the response of a player is just one move, a strategy must specify alternative moves concomitant with the actual state.

The states of the binary decision game are described by the information strings using the binary symbols Y and N (for yes and no), or equivalently operations $\hat{O}_1$ and $\hat{O}_2$. Each player $\alpha = 1, 2, 3$ can send only one symbol or perform one operation, thus all game plays consist of possible sequences of these two symbols: YYY, YNY, YNN, YYN, NYY, NNY, NYN and NNN. However, this is not the same as the number of decisions, given that each player makes a move based on past information. Each decision has one of two forms: either the player decides on a move (operation), given that the game has arrived in a particular state,

$$S_\alpha(\hat{O}_1 \text{else} \hat{O}_2 \mid \text{state} = Q), \tag{19.65}$$

or the player decides to make a move, ignoring the information about past history:

$$S_\alpha(\hat{O}_1), \tag{19.66}$$

$$S_\alpha(\hat{O}_2). \tag{19.67}$$

Thus, as the state space grows exponentially, so the number of decisions grows exponentially. For Player 1 (the employer) there is only one state—the starting state, so the conditional moves make no sense, or one can say that they are not independent of the unconditional moves. Using Y, N notation for simplicity, these are

$$S_1(Y), S_1(N) \tag{19.68}$$

Player 2 inherits two possible states from Player 1 and thus can choose between

$$S_2(Y, N \mid Q = Y),$$
$$S_2(N, Y \mid Q = Y),$$
$$S_2(Y, N \mid Q = N),$$
$$S_2(N, Y \mid Q = N),$$
$$S_2(Y, Y \mid Q = Y),$$
$$S_2(N, N \mid Q = N). \tag{19.69}$$

The first four strategies can be summarized as two, that is, do the same as Player 1 or do the opposite of Player 1 (Employer). The latter two strategies are: do Y or N regardless of what Player 1 does, for example, do Y else do Y implies 'do anyway'.

Player 3 inherits four possible states from Player 2 and thus can choose all of the four choices at each node.

$$S_3(Y, N \mid Q = YY),$$
$$S_3(N, Y \mid Q = YY),$$
$$S_3(Y, Y \mid Q = YY),$$
$$S_3(N, N \mid Q = YY)$$
$$S_3(Y, N \mid Q = YN),$$
$$S_3(N, Y \mid Q = YN),$$
$$S_3(Y, Y \mid Q = YN),$$
$$S_3(N, N \mid Q = YN)$$
$$S_3(Y, N \mid Q = NY),$$
$$S_3(N, Y \mid Q = NY),$$
$$S_3(Y, Y \mid Q = NY),$$
$$S_3(N, N \mid Q = NY)$$
$$S_3(Y, N \mid Q = NN),$$
$$S_3(N, Y \mid Q = NN),$$

$$S_3(Y, Y \mid Q = NN),$$
$$S_3(N, N \mid Q = NN). \tag{19.70}$$

Notice that there must be an 'else' alternative at each branch: that is, each strategy is of the form, 'if the state of the system is Q, do X, else do Y'. If, for whatever reason, a player formally chooses a strategy on the basis of an irrelevant state, we must be able to evaluate the game, nevertheless[5]. What is the point of this? An intelligent player would hardly choose a strategy corresponding to a given state unless the system were in that state—but what if the player is somehow prevented from knowing the state, or suffers a lapse of judgement? This is where one strays into the realms of imperfect information: we shall not go down that path, but mention only the possibility in passing. Completeness requires us to catalogue all possible pathways in a game.

If we imagine that each path through the tree is a sequence of operations $\hat{O}_3 \hat{O}_2 \hat{O}_1$, then paths through the tree are represented, with dependencies explicit as

$$(3|21)(2|1)(1), \tag{19.71}$$

where each parenthesis is an operator, and we borrow the notation of conditional probability to make dependencies explicit. We must now transfer the pay-offs from the tree of moves to the table of pay-offs (now a three-dimensional table). Let us suppose hypothetically that an employer can earn 100 credits from a business in total. It costs 10 credits to train the staff in the company policy and procedures. Moreover, staff wages amount to a constant 10 credits, regardless of profits. If staffs are found to be not complying with policy, they are docked half their earnings, and thus they receive only 5 credits. An attacker is likely to corrupt an employee into cooperating with an attack if the employee does not follow policy, and might earn 20 credits from this, but must pay 10 to the corrupt employee who was the 'insider' to compensate for losses in wages. Clearly, this is rather simplistic, but serves to illustrate a point.

Path	Employer	Employee	Attacker	Pay-off vector
NNN	$100 - 5$	5	0	(95,5,0)
NNY	$100 - 5 - 20$	$5 + 10$	$20 - 10$	(75,15,10)
NYN	$100 - 10$	10	0	(90,10,0)
NYY	$100 - 10 - 20$	10	20	(70,10,20)
YNN	$90 - 5$	5	0	(85,5,0)
YNY	$90 - 5 - 20$	$5 + 10$	$20 - 10$	(65,15,10)
YYN	$90 - 10$	10	0	(80,10,0)
YYY	$90 - 10 - 20$	10	20	(60,10,20)

[5] Note that the simple form 'If (Q)... else...' is possible here because of the binary nature of the decisions. The tree grows very complex if there are more than two choices. In a game with three or more decisions at each node, there must be sufficient state information in the decision tree to distinguish a unique strategy path, in every decision, otherwise choices must be made *ad hoc*, at random.

These are the pay-offs to the various users. It is not clear from the table exactly what strategy is best for any of the players pursuing selfish interest. Thus, a method that can tell us the best rational choice in terms of the pay-off currency is of great interest, especially if it tells us the effect of changing the relative pay-offs between users. Could an employer maximize likely profits by paying employees a little more, faced with possible attack?

The game presented is an almost constant sum game, except that the constant is different in the two choices of Player 1 (the employer). If the employer trains the employees, then total profit is only 90, rather than 100, since this costs him the outlay of training[6].

The pay-off matrix is three dimensional, so we must be split it into two slices for $S_1 = Y$:

$$\Pi_{(S_1=Y,S_2,S_3)} = \begin{pmatrix} S_3/S_2 & (Y,Y|Y) & (Y,N|Y) & (N,Y|Y) & (N,N|Y) \\ \hline (Y,Y|YY) & YYY & YYY & YNY & YNY \\ (Y,N|YY) & YYY & YYY & YNN & YNN \\ (N,Y|YY) & YYN & YYN & YNY & YNY \\ (N,N|YY) & YYN & YYN & YNN & YNN \\ (Y,Y|YN) & YYY & YYY & YNY & YNY \\ (Y,N|YN) & YYN & YYN & YNY & YNY \\ (N,Y|YN) & YYY & YYY & YNN & YNN \\ (N,N|YN) & YYN & YYN & YNN & YNN \\ (Y,Y|NY) & YYY & YYY & YNY & YNY \\ (Y,N|NY) & YYN & YYN & YNN & YNN \\ (N,Y|NY) & YYY & YYY & YNY & YNY \\ (N,N|NY) & YYN & YYN & YNN & YNN \\ (Y,Y|NN) & YYY & YYY & YNY & YNY \\ (Y,N|NN) & YYN & YYN & YNN & YNN \\ (N,Y|NN) & YYY & YYY & YNY & YNY \\ (N,N|NN) & YYN & YYN & YNN & YNN \end{pmatrix}$$

(19.72)

---

[6] This could be handled by introducing a fourth player to whom we pay this value, but that would only serve to complicate matters here, since the fourth player plays no strategic role in the security situation being analysed here.

and $S_1 = N$:

$$\Pi_{(S_1=N,S_2,S_3)} = \begin{pmatrix}
\begin{array}{c|cccc}
S_3/S_2 & (Y,Y|N) & (Y,N|N) & (N,Y|N) & (N,N|N) \\
\hline
(Y,Y|YY) & NYN & NYN & NNN & NNN \\
(Y,N|YY) & NYN & NYN & NNN & NNN \\
(N,Y|YY) & NYY & NYY & NNY & NNY \\
(N,N|YY) & NYN & NYN & NNN & NNN \\
(Y,Y|YN) & NYY & NYY & NNY & NNY \\
(Y,N|YN) & NYN & NYN & NNN & NNN \\
(N,Y|YN) & NYY & NYY & NNY & NNY \\
(N,N|YN) & NYN & NYN & NNN & NNN \\
(Y,Y|NY) & NYY & NYY & NNY & NNY \\
(Y,N|NY) & NYY & NYY & NNN & NNN \\
(N,Y|NY) & NYN & NYN & NNY & NNY \\
(N,N|NY) & NYN & NYN & NNN & NNN \\
(Y,Y|NN) & NYY & NYY & NNY & NNY \\
(Y,N|NN) & NYN & NYN & NNY & NNY \\
(N,Y|NN) & NYY & NYY & NNN & NNN \\
(N,N|NN) & NYN & NYN & NNN & NNN \\
\end{array}
\end{pmatrix} \tag{19.73}$$

The outcomes of this game can be calculated (see Gambit (n.d.)), with the result that there are many possible equilibrium strategies. The pay-offs to the players are all pure strategy equilibria and are all quite similar, since there the pure strategies of the extensive game are not independent, in spite of the great number of distinguishable combinations. The one that is best for the employer assigns pay-offs (75,15,10) to the players. This comes from no staff training, no policy conformance and attack by an attacker. This is not a bad outcome perhaps, but the employer is losing ten units of profit to the attacker and has disloyal staff. This is not desirable.

Perhaps of greater concern, the employer is powerless to decide the outcome of the game in the equilibria. All of the solutions assume that no staff training is performed; thus, it is the actions of the other players who determine the employer's pay-off. There is no benefit to the employer in training staff, because they can get more by corrupt means. This is not a desirable situation, so we consider how to modify the rules of the game to achieve a more desirable result, and restore some of the control the employer has over destiny. Suppose, for instance, the employer doubles the wages of the employees and maintains the policy of halving of wages for failure to comply with company training.

Path	Employer	Employee	Attacker	Pay-off vector
NNN	$100 - 10$	10	0	(90,10,0)
NNY	$100 - 10 - 20$	$10 + 10$	$20 - 10$	(70,20,10)
NYN	$100 - 20$	20	0	(80,20,0)
NYY	$100 - 20 - 20$	20	20	(60,20,20)
YNN	$90 - 10$	10	0	(80,10,0)
YNY	$90 - 10 - 20$	$10 + 10$	$20 - 10$	(60,20,10)
YYN	$90 - 20$	20	0	(70,20,0)
YYY	$90 - 20 - 20$	20	20	(50,20,20)

We re-compute the Nash equilibria with these values. The best new equilibrium is in fact worse than before for the employer. Again, there are several equilibria; the outcomes for the players are all close to (60,20,20) with equilibrium strategies: no training, break policy and attack respectively for the players. Clearly, bribing the staff does not necessarily help. What else might we do? What we have not accounted for is the possible effect of training on deflecting the attacker. Suppose, instead of paying employees more, the employer seeks more effective training that halves the gain of the attacker. This need not cost any more—the employer simply finds a competent staff trainer. Now we have

Path	Employer	Employee	Attacker	Pay-off vector
NNN	$100 - 5$	5	0	(95,5,0)
NNY	$100 - 5 - 20$	$5 + 10$	$20 - 10$	(75,15,10)
NYN	$100 - 10$	10	0	(90,10,0)
NYY	$100 - 10 - 20$	10	20	(70,10,20)
YNN	$90 - 5$	5	0	(85,5,0)
YNY	$90 - 5 - 5$	5	5	(80,5,5)
YYN	$90 - 10$	10	0	(80,10,0)
YYY	$90 - 10 - 10$	10	10	(70,10,10)

With these pay-offs we have, again, many equilibria; however, amongst the equilibria is now the possibility of the equilibrium pay-offs (70,10,10) to the players using moves Train staff, Staff obey policy, Attacker attack. Thus, while the employer does not increase total profit, he can reduce losses to attackers.

We can continue re-evaluating how much resource to assign to each part of the system and indeed to our own value systems (what is truly important to us), to see how far we must go before the balance of the game tips over to a cost-effective result. The main point to be derived from this toy model is that by changing value allocations in a game, one can explore the rational consequences, even when they are somewhat convoluted and counter-intuitive.

19.12 Human elements of policy decision

Policy compliance by humans is a thorny issue, since humans do not always behave rationally. Game theory suggests that humans require a pay-off in order to tip their judgement in favour of compliance, on the scales of rational judgement. While game theory always assumes that humans will pursue purely selfish interest, this is clearly not necessarily true (though more true than many of us would care to imagine), so the game once again gives us mainly an insight into the worst-case scenario.

Humans are also multifaceted souls. The pay-off to a human might include a number of things. Sometimes thanks are enough, at other times monetary reward is craved. Choosing policy over humans is fraught with the irrationality of the possible response: humans require training or conditioning to accept policy. Whether one calls this training or brainwashing is a subject for an ethical debate, very interesting, but rather outside the scope of this chapter. Choosing a policy that is not understood by humans can lead to actions that are 'criminal' according to policy. Peer review of policy prevents policies that are overtly contrary to group wishes, but a group's wishes are not always rational or representative either. For humans to accept policy, they must often have the feeling of freedom to decide their own destiny, even if this is through a democratic process in which they actually have little influence. Responsibility within a system confers privilege; this can be used as a pay-off to motivate humans; it could also be withdrawn as a punishment, though disgruntled humans tend to sabotage systems, or become criminals within its policies.

19.13 Coda: extensive versus strategic configuration management

Let us now put together many of the themes of this book to consider the entire development of a human–computer system in terms of a game for attaining an ideal state (see Burgess (1998b)). If it is a game of perfect information, it must be deterministic and the size of the information sets must grow exponentially with each move. There is thus, in the limiting case, an infinite number of pure strategies required to attain the desired configuration in every possible situation. We must ask, however, whether or not the state we acquire is desirable and by what criteria. The view taken in this book is that stability is a key component of this criterion.

The path-dependent and path-independent approaches can be described using the game theoretical language of this chapter. They also have their analogues in thermodynamics of equilibrium and non-equilibrium systems. Asking whether or not complete path is required is the same as asking whether or not configuration must be a game of perfect information in order to 'win'. We have not properly discussed the matter of games of imperfect information here; this is a large and subtle topic that would require inappropriately long discussion (see Rasmusen (2001)), so we shall mention only the briefest sketch.

In a game of imperfect information, every decision in the game tree is made without knowing the whole history of the tree. All a player sees is the information in a coarse-grained form. The nodes of the game tree are grouped into sets called *information sets* that conceal the details of past history to the player. Information sets blur the distinction between individual nodes, making them indistinguishable to the player, by grouping together nodes in the game. Rasmusen calls these information sets 'clouds', since a player can tell that the game has reached an approximate location covered by a cloud, but cannot tell the exact node within the cloud.

Example 174 *In the configuration management approach used by the software cfengine, only the current state of the system is known, not the path by which this was achieved. The distinct routes by which the state occurred are hidden, and thus cfengine operates a game of imperfect information.*

Can a player win only on the basis of imperfect information? This depends on whether there is a path to a node within the winning game equilibrium from the current location. The key to this is to disallow deep trees. If every final state is attainable from every intermediate state, then this condition is satisfied. If we use only commuting operators (see section 15) this will be true, since there is no impediment to implementing a string of operations that makes a move in the game. If we use path-dependent operators, then the property of 'can get there from here' is spoiled unless we can steer the exact path throughout the game. If the something from the system environment should alter the state of the system, so that it is no longer in the equilibrium solution, it must be corrected by maintenance (see chapter 16), but now we are no longer guaranteed to be able to find a path back to where we were unless the operations are commutative. We can summarize this as follows.

Principle 10 (Regulation with imperfect information) *Only a system with finite state information (that functions with imperfect information) like a Markov process can be regulated indefinitely with a finite policy. Open systems must forget their history in order to satisfy this criterion.*

We can also turn this argument around to make it sound less like a limitation and more like an advantage. A system that achieves equilibrium cannot depend on the route it took to reach equilibrium. A steady state, that is, one that persists, as the end point of a process, (see section 9.10) is one in which the system has reached a stable dynamical equilibrium. That means that the system desires to change itself by amount zero in every time step once it has reached that state. Since zero does not depend on anything, this cannot depend on the path. Such a state is a property of a location, not a path. Consider the following analogy.

Example 175 *Imagine a ball rolling into a valley: once it reaches the bottom of the valley, it stays there regardless of the path it took—assuming only that once such path exists. The bottom of the valley is the most stable state and we would normally choose this as policy. Now, suppose we decide to place the ball on a ledge and make this our policy. Now the ball might not be stable to perturbations and if, by some perturbation, it falls out of that state, there is only a limited number of paths that will take it back there. If there is no single point of 'bottomness' in the valley (perhaps the whole base of the valley is at the same height), then these final states are an equivalence class and there is no need to distinguish them.*

The example above tells us that maintaining a truly stable state requires no particular information about the path. However, if we are asked to maintain an unstable state, precise information is required.

The reason why the extensive, path-dependent approach sometimes seems desirable in configuration management is that we seldom bother to construct the entire tree of actions or moves explicitly, or analyse the dependencies to determine their final importance. Moreover, if an extensive approach is to be coded as a policy, then we must find the pre-determined strategies that describe the tree (as in eqns. 19.68, 19.69 and 19.70); these are even more numerous.

Traugott has argued that complete path information is vital to maintain the correct 'congruent' state of configuration of a host (Traugott (2002)). In Traugott's proposal, one specifies a path always starting from a known base state so that the configuration follows the same path. This agrees with the view above, but is only distinct or special if the system is to be configured in a non-stable state, that is, a state that does not have maximum stability. The viewpoint we must take here is that a sufficient description of policy must guarantee a best stable state and that any path to this state is good enough. Once the system arrives at the state, the path it took becomes irrelevant.

With the expert operator approach (see section 15.2), we provide a set of primitives that, while not providing the actual strategy set, is guaranteed to cover the possibility space for any policy (as orthogonal vectors span a vector space), thus providing a path to any final state. Choosing a policy then constrains the space of all possible policies along each orthogonal operations axis and reductions can be made rationally. It might seem like a lot of work to build up a configuration in this way, but once it is done, it is guaranteed to be stable to decision fluctuations.

Applications and Further Study 19

- *Creating decision models that relate dependencies to their pay-off in a rational way.*

- *Understanding the mechanics of decision-making.*

- *Investigating how changes in assumptions can lead to changes in the rational outcome of the decision procedure (hence provide a formal procedure for testing one's assumptions).*

- *Evaluating the stability of policy conclusions to errors of assumption, by changing utility estimates until the rational outcome becomes significantly altered.*

20

Conclusions

...to see the general in the particular and the eternal in the transitory
—Alfred North Whitbread

System administration has been presented in this book as a rational endeavour, whose aim is to provide checks and balances to the design, construction and running of human–computer systems. These checks and balances can be investigated and described by traditional scientific inquiry. Some specific tools and techniques have been presented here to guide and inspire their usage in future work.

In this book, we have assumed that no system can be isolated from its environment—that opening a system for input means opening it to unpredictability. The task of the theoretician is then to look for a formal language by which to describe the observable phenomena of interacting humans and computers, including all of the uncertainties and fluctuations. For this, we need sets, graphs and functions of time and space (addresses) as well as statistical methods and the control of empiricism.

The approach taken here has been to allow the predictable and the unpredictable meet, by mapping the ideal of maintenance onto the idea of error correction of strings of digital operations (as in the communications theory of Shannon). The challenge lies in defining complex, multi-dependent processes as discrete operators that work predictably in noisy environments; if one can do that, classical error-correction methods automatically apply.

Key notions of stability, predictability, resource management, connectivity and flow then allow one to define the concept of a set of reasonable *policies*, or stable regions of behaviour for the system. Policies are alternative configurations that reflect system goals. By defining policy to be a point of stable operation for a system, one ensures that the system will not immediately come undone as soon as it is allowed to run. However, there might be several policies that can be sustained, so there is a freedom that remains.

The remaining freedom can be partially eliminated by rational means; it requires a value judgement, and here the language of games and decision theory can be used. Once this has been allowed to play its role, we are left with a far smaller number of alternatives that is sharpened by the decision constraints. This might still not select a unique policy; the remaining choices for policy are thus not rationally distinguishable. Human preference,

Analytical Network and System Administration. Managing Human–Computer Networks   Mark Burgess
© 2004 John Wiley & Sons, Ltd   ISBN 0-470-86100-2

or random choice, thus enters for the final selection of one or more of these policies; we refer to the remaining choice as a *strategy*.

Bringing the scientific tradition to a new field of research is not a trivial matter. The thickness of this brief, introductory volume is a testament to the effort required to make a precise, rational statements that can be challenged by experiment. Even so, this book is only a beginning: it suggests a platform or springboard in the form of a number of theorems and paradigms. For example,

- The maintenance theorem tells us what kinds of systems can have a probably stable policy.

- Convergence, closure and orthogonality tell us how systems can reach a stable fixed point of operation through a stochastic process, hence in a noisy environment.

- Shannon's theorem shows us that error correction is possible in maintainable systems.

- Queueing and fault network theorems and centrality measures indicate where problems and faults are likely to occur as well as how resources should be deployed most efficiently.

Some readers will find the liberal mixture of technology and sociology in this book disconcerting. These traditionally incompatible domains are treated here as one, with common methods. Subjective concerns are made into rational choices by formulating the *utility* of the choices on a measurable scale. Ultimately this allows us to employ algebraic reasoning. Utility theory, as originated by Von Neumann and Morgenstern, extends far beyond the tiny introduction offered here and should provide research theses for decades to come.

There are plenty of reasons not to be judgemental about the inadequacies of treating the human part of systems with approximate characterizations. We have survived such inadequacies in other sciences with no great injury. Moreover, it is always good to remember that all scientific models are descriptive and have underlying uncertainties.

The analogies between system administration and economics are not accidental; nor are the clear parallels between system administration and physics. All these fields share the goal of describing systems in which accountable resources are distributed, flow and interact. Physics is a game in which different forces compete for their share of the energy available in a system; we can rightfully describe the contents of this book as a physics of human–computer systems[1]. The main difference is that physics operates according to only one set of rules (one policy). In human–computer systems, each region can have its own laws; the question then becomes not only what happens within such regions but also what happens where these regions meet. This is a fascinating challenge that should be of interest to computer scientists and physicists alike.

Security, as an explicit viewpoint, has been avoided in this book, with good reason. Once mentioned, it tends to dominate discussions, and deflect from core issues. However, security is, in fact, addressed implicitly in a number of the topics: accessibility, vulnerability of nodes, percolation and decision theory. For more specific discussions of security, readers

---

[1] This viewpoint has often caused distress when I have expressed it in the past. 'This is not physics, it's computer science' say critics, but this artificial segregation misses the point completely. Science rarely advances by splintering off sub-cultures; it usually benefits when apparently dissimilar fields are fused.

are directed to (Bishop (2002)). One of the key problems in security is not technical but sociological: the sabotage of systems by disgruntled users. The Romans are reputed to have claimed that civilization is never more than three meals away from anarchy. In other words, for humans, systematic cooperation depends on the subtle bribery of the people: give them what they want and they will play by the rules of the system. Should the system fail to give them what they want, it will degenerate into a free-for-all.

Finally, it is worth directing a critical eye on the substance of this book. What does it actually achieve? Does it advance our ability to describe human–computer systems, or is it simply an indulgence in applying familiar forms to an insubstantial field of research? I believe that it does represent an advance. In particular,

- It invalidates certain *ad hoc* discussions that recur in the literature on subjective issues and suggests a rational replacement.

- It provides a number of conclusions about what makes one system better than another, with measurable certainty.

- It documents a conceptual basis so that discussions can be made uniform.

- It identifies current weaknesses and new goals for technological development.

I have always been a theorist by inclination. When I practice or experiment, it is not for the love of it, but in deference to its importance. Practice brings humility and experience, but only theory brings understanding. The trouble I have found is that few academics really believe in theory, when it comes down to it. They believe that it is acceptable in books, or in the classroom, but not in the infamous real world. I know colleagues who lecture of algorithms but will not apply the principles and knowledge to organize their surroundings. I know of economists who teach theory but resort to guesswork when it comes to application in 'the real world'. There are mathematicians who do not believe that any non-elementary mathematics should be applied to mundane problems. When it comes to 'real life', too many academics are sceptical of the validity of the knowledge they teach! As Faust exclaims in David Luke's superior translation:

> *'And I fear...*
> *Hard studies that have cost me dear.*
> *And now I sit, poor silly man,*
> *No wiser than when I began.*
> *They call me Professor and Doctor, forsooth,*
> *For misleading many an innocent youth...'*

Some have said of Network and System Administration: 'It is too early in the field to be talking of theory!' For others, it is too late: 'If you'd told me that before it might have changed my ways....' Never was more effort expended in justifying a viewpoint than in the defence of the irrational.

Why is there such disbelief in theory or resistance to formal rigour? Perhaps it is an insecurity in speaking the language of mathematics, perhaps a poverty of imagination caused by spoon-feeding television culture or the point and click passivity of modern society. If you 'don't get it', then the power of abstraction is lost on you. To the reader who has delved into this book, I am grateful. I implore you to stick it out and embrace the wealth of ideas that can be used to describe systems. The field of Network and System Administration has advanced in recent years, as much a result of theory as of practice. I am convinced that many of the answers to human–computer management lie in the fruits of theoretical investigation. I wish readers challenge and enjoyment in the perseverance in that pursuit.

Appendix A

Some Boolean formulae

The probability of that event B follows event A is

$$P(BA) = P(B|A)P(A), \tag{A.1}$$

that is, the probability that A happens, multiplied by the probability that B happens, given that A has already happened. There is an implicit causal sequence in these probabilities, because we are assuming that B depends on A somehow.

If the events are independent, that is, if the probability of A and B occurring is independent of the order of measurement, then it makes sense to refer to the concept of 'AND', that is, a symmetrical operator, with no memory of what came before. In this case, the probability of both events happening is merely coincidental, and is given by the overlap product of the probabilities:

$$P(A \textbf{ AND } B) = P(A \cap B) = P(A)P(B). \tag{A.2}$$

For independent events, the inclusive or exclusive ORs are the same:

$$P(A \textbf{ OR } A) = P(A \cup B) = P(A) + P(B) - P(A)P(B). \tag{A.3}$$

These expressions can be iterated, using the symmetry of their arguments for greater numbers of inputs. If the events are not independent, then the exclusive OR is given by

$$P(A \textbf{ XOR } A) = P(A \oplus B) = P(A) + P(B) - 2P(A)P(B). \tag{A.4}$$

A.1 Conditional probability

Let the set A consist of subsets $A = \{a_1, a_2, \ldots\}$, some of which might overlap. If the sets do not overlap, then

$$P(a_1) = \frac{N(a_1)}{\sum_i N(a_i)} \tag{A.5}$$

Analytical Network and System Administration. Managing Human–Computer Networks   Mark Burgess
© 2004 John Wiley & Sons, Ltd   ISBN 0-470-86100-2

In general, we can write the set that is complementary to a_1 as $\bar{a}_1$, that is, all of the elements that are not in a_1. Then

$$P(a_1) = \frac{N(a_1)}{N(a_1) + N(\bar{a}_1)}. \tag{A.6}$$

The conditional probability of two overlapping events is

$$\begin{aligned} P(a_1|a_2) &= \frac{N(a_1 \cap a_2)}{N(a_2)} = \frac{N(a_1 \cap a_2)/N}{N(a_2)/N} \\ &= \frac{P(a_1 \cap a_2)}{P(a_2)}. \end{aligned} \tag{A.7}$$

that is, knowledge that the search space is within a_2 increases the likelihood of finding the result, so the conditional probability is greater.

Now, by symmetry

$$\begin{aligned} P(a_2|a_1) &= \frac{N(a_1 \cap a_2)}{N(a_1)} = \frac{N(a_1 \cap a_2)/N}{N(a_1)/N} \\ &= \frac{P(a_1 \cap a_2)}{P(a_1)}. \end{aligned} \tag{A.8}$$

thus,

$$P(a_1 \cap a_2) = P(a_2|a_1)P(a_1) = P(a_1|a_2)P(a_2), \tag{A.9}$$

thus, one has Bayes' formula

$$P(a_2|a_1) = \frac{P(a_1|a_2)P(a_2)}{P(a_1)}. \tag{A.10}$$

This is really a definition of conditional probability.

A.2 Boolean algebra and logic

Thanks to Von Neumann et al, our present-day idea of computers is that of binary digital devices that perform Boolean logical operations. Such a device can simulate any computational process in principle. What remains in order to create systems that compute the results of mathematical or logical problems is the ability to combine information streams into functions that are things we want to evaluate.

Modern computers are based on the use of binary data and Boolean algebra or logic. It is straightforward to show that a simple set of linearly independent operations on bits can be used to perform simple binary arithmetic, and thus more complex calculations in combination. The commonly referred to operations in Boolean algebra are the unary (1:1) operator

NOT ¬	
In	Out
1	0
0	1

and the binary (2:1) operators

AND ∩				OR ∪				XOR ⊕		
In1	In2	Out		In1	In2	Out		In1	In2	Out
0	0	0		0	0	0		0	0	0
0	1	0		0	1	1		0	1	1
1	0	0		1	0	1		1	0	1
1	1	1		1	1	1		1	1	0

In digital electronics, these are simulated using multi-transistor circuit blocks.

It is easy to show that any Boolean logic operation can be constructed from the two operations ∩ (**AND**) and ¬ (**NOT**). This may be seen from the following identities:

$$P \cup Q = \neg(\neg P \cap \neg Q)$$
$$P \rightarrow Q = \neg P \cap Q$$
$$P \leftrightarrow Q = (P \rightarrow Q) \cap (Q \rightarrow P)$$
$$P \oplus Q = \neg(P \leftrightarrow Q). \tag{A.11}$$

or, in modern programming notation

```
P | Q = ! (!P & !Q)
P -> Q = !P & Q
P == Q = (P -> Q) & (Q -> P)
P ^ Q = ! (P == Q).
```

or, again, in more common notation

$$P \text{ **OR** } Q = \text{ **NOT** } (\text{ **NOT** } P \text{ **AND** } \text{ **NOT** } Q)$$
$$P \rightarrow Q = \text{ **NOT** } P \text{ **AND** } Q$$
$$P \text{ **EQUALS** } Q = (P \rightarrow Q) \text{ **AND** } (Q \rightarrow P)$$
$$P \text{ **XOR** } Q = \text{ **NOT** } (P \text{ **EQUALS** } Q). \tag{A.12}$$

The 'implication' symbol is defined by the truth table

$$0 \rightarrow 0 = 1$$
$$0 \rightarrow 1 = 1$$
$$1 \rightarrow 0 = 0$$
$$1 \rightarrow 1 = 1. \tag{A.13}$$

Appendix B

Statistical and scaling properties of time-series data

Consider a stochastic dynamical variable $q(t)$, whose complete behaviour is unknown.

In the following sections, we refer to a sample of data, measured by a sensor, or represented as an abstract function of time. The time span of the same is take to be from $t = 0$ to $t = T$. If the sample was measured at discrete regular time intervals, then t is a subset of discrete values labelled $t = [i]$.

The average of a measured or represented function or sample, taken over all points, is denoted simply by the expectation value brackets; this is also denoted by $E()$ in statistics literature:

$$\langle q(t) \rangle = E[q(t)]. \tag{B.1}$$

These brackets have no subscript. Subscripts are used to denote the average taken over a limited subset of the points.

Similarly, the variance over the entire sample is denoted by σ^2 with no subscript. Subscripts are used to denote the variance of a limited subset of the full sample, as defined below.

B.1   Local averaging procedure

Let us define a local averaging procedure, or method of *coarse-graining* (see fig 3.10).

The local averaging procedure re-averages data, moving from a detailed view to a less detailed view, by grouping neighbouring data together. In practice, one always deals with data that are sampled at discrete time intervals. We shall consider this case first, and then return to a continuous function approach, which is a useful approximation to the discrete case.

Analytical Network and System Administration. Managing Human–Computer Networks   Mark Burgess
© 2004 John Wiley & Sons, Ltd   ISBN 0-470-86100-2

Discrete time data

Consider the function $q(t)$ shown in fig. 3.10. Let the small ticks on the horizontal axis represent the true sampling of the data, and label these by $i = 0, 1, 2, 3, \ldots, I$. These have unit spacing. Now let the large ticks, which are more coarsely spread out, be labelled by $k = 1, 2, 3, \ldots, K$. These have spacing $\Delta t = m$, where m is some fixed number of the smaller ticks. The relationship between the small and the larger ticks is thus

$$i = (k - 1)\Delta t = (k - 1)m. \tag{B.2}$$

In other words, there are $\Delta t = m$ small ticks for each large one. To perform a coarse-graining, we replace the function $q(t)$ over the whole kth cell with an average value, for each non-overlapping interval Δt. We define this average by

$$\langle q(k) \rangle_{\mathrm{m}} \equiv \frac{1}{\Delta t} \sum_{i=(k-1)\Delta t+1}^{k\Delta t} q(i). \tag{B.3}$$

We have started with an abstract function $q(t)$, sampled it at discrete intervals, giving $q(i)$, and then coarse-grained the data into larger contiguous samples $\langle q(k) \rangle_{\mathrm{m}}$:

$$q(t) \rightarrow q(i) \rightarrow \langle q(k) \rangle_{\mathrm{m}}. \tag{B.4}$$

The variance of data $q(i)$ over the kth cell is thus

$$\sigma^2(k) = \frac{1}{\Delta t} \sum_{i=(k-1)\Delta t+1}^{k\Delta t} (q(i) - \langle q(k) \rangle_{\mathrm{m}})^2 \tag{B.5a}$$

$$= \langle q^2(k) \rangle_{\mathrm{m}} - \langle q(k) \rangle_{\mathrm{m}}. \tag{B.5b}$$

The mean of the entire set of samples (summed over either i or k variables) is the same:

$$\langle q \rangle = \frac{1}{I} \sum_{i=0}^{I} q(i) = \frac{1}{K} \sum_{k=0}^{K} \langle q(k) \rangle_{\mathrm{m}}, \tag{B.6}$$

This follows from the linearity of the sums. The same is not true of the variances, however.

i-coordinates (small ticks)

$$\langle q \rangle = \frac{1}{I} \sum_{i=0}^{I} q(i) \tag{B.7}$$

The variance:

$$\sigma^2 = \frac{1}{I} \sum_{i=0}^{I} (q(i) - \langle q \rangle)^2$$

$$= \langle q^2 \rangle - \langle q \rangle^2. \tag{B.8}$$

Recall this expression for comparison below.

k-coordinates (long ticks)

The average is the same as for the small ticks:

$$\langle q \rangle = \frac{1}{K} \sum_{k=0}^{K} \langle q(k) \rangle_{\mathrm{m}},$$

$$= \frac{1}{K} \sum_{k=0}^{K} \left(\frac{1}{\Delta t} \sum_{i=(k-1)\Delta t+1}^{k\Delta t} q(i) \right)$$

$$= \frac{1}{K\Delta t} \left(\frac{1}{K} \sum_{k=0}^{K} \sum_{i=(k-1)\Delta t+1}^{k\Delta t} \right) q(i)$$

$$= \frac{1}{I} \sum_{i=1}^{I} q(i)$$

$$= \langle q \rangle. \tag{B.9}$$

However, the variance is not the same:

$$\sigma_K^2 = \frac{1}{K} \sum_{k=0}^{K} (\langle q(k) \rangle_{\mathrm{m}} - \langle q \rangle)^2$$

$$\equiv \langle ((\langle q(k) \rangle_{\mathrm{m}} - \langle q \rangle)^2 \rangle_K$$

$$= \langle \langle q(k) \rangle_{\mathrm{m}}^2 \rangle_K - 2 \langle \langle q(k) \rangle_{\mathrm{m}} \langle q \rangle \rangle_K + \langle q \rangle^2$$

$$= \langle \langle q(k) \rangle_{\mathrm{m}}^2 \rangle_K - 2 \langle \langle q(k) \rangle_{\mathrm{m}} \rangle_K \langle q \rangle + \langle q \rangle^2 \tag{B.10}$$

Now

$$\langle \cdots \rangle = \langle \langle \cdots \rangle_{\mathrm{m}} \rangle_K, \tag{B.11}$$

thus

$$\sigma_K^2 = \langle \langle q(k) \rangle_{\mathrm{m}}^2 \rangle_K - \langle q \rangle^2 \tag{B.12a}$$

$$\neq \sigma_I^2. \tag{B.12b}$$

The two expressions thus differ by

$$\sigma^2 - \sigma_K^2 = \langle q^2 \rangle - \langle \langle q(k) \rangle_{\mathrm{m}}^2 \rangle_K$$

$$= \langle \langle q^2 \rangle_{\mathrm{m}} - \langle q(k) \rangle_{\mathrm{m}}^2 \rangle_K$$

$$= \langle \sigma_{\mathrm{m}}^2(k) \rangle_K, \tag{B.13}$$

which is the average variance of the coarse-grained cells.

Continuous time data

We can now perform the same procedure using continuous time. This idealization will allow us to make models using continuous functions and functional methods, such as functional integrals. Referring once again to the figure, we define a local averaging procedure by

$$\langle q(\bar{t}) \rangle_{\Delta t} = \frac{1}{\Delta t} \int_{\bar{t}-\Delta t/2}^{\bar{t}+\Delta t/2} q(\tilde{t}') \, d\tilde{t}'.$$

(B.14)

The coarse-grained variable $\bar{t}$ is now the more slowly varying one. It is convenient to define the parameterization

$$\tilde{t} = (t - t')$$

(B.15a)

$$\bar{t} = \frac{1}{2}(t + t'),$$

(B.15b)

on any interval between points t and t'. The latter is the mid-point of such a cell, and the former is the offset from that origin. The variance of the fundamental variable, over such a grain is

$$\sigma^2(\bar{t}) = \frac{1}{\Delta t} \int dt' (q(t') - \langle q(\bar{t}) \rangle_{\Delta t})^2$$
$$= \langle q^2(\bar{t}) \rangle_{\Delta t} - \langle q(\bar{t}) \rangle_{\Delta t}^2.$$

(B.16)

t-coordinates (infinitesimal ticks)

Over a total sample, running from 0 to T

$$\sigma^2 = \frac{1}{T} \int_0^T dt' (q(t') - \langle q(t') \rangle)^2$$
$$= \langle q^2 \rangle - \langle q \rangle^2.$$

(B.17)

$\bar{t}$-coordinates (Δt ticks)

Define the average over the τ cells of width Δt by

$$\langle \langle q(\bar{t}) \rangle_{\Delta t} \rangle_{\bar{t}} = \frac{1}{\tau} \int_0^\tau \langle q(\bar{t}) \rangle_{\Delta t}.$$

(B.18)

Noting that $\tau = T/\Delta t$ and $d\tau = d\bar{t}/\Delta t$, one confirms that

$$\langle \langle q(\bar{t}) \rangle_{\Delta t} \rangle_{\bar{t}} = \langle q \rangle$$

(B.19)

that is,

$$\langle \langle \cdots \rangle_{\Delta t} \rangle_{\bar{t}} \equiv \langle \cdots \rangle.$$

(B.20)

Over a total sample, running from 0 to $T = \tau \Delta t$, the directly calculated variance is

$$\sigma^2 = \frac{1}{T} \int_0^T dt' (q(t') - \langle q(t') \rangle)^2$$
$$= \langle q^2 \rangle - \langle q \rangle^2.$$

(B.21)

The variance of the coarse-grained variables differs once again,

$$\sigma^2 = \frac{1}{\tau} \int_0^\tau (\langle q(\bar{t}) \rangle_{\Delta t} - \langle q \rangle)^2$$

$$= \langle \langle q(\bar{t}) \rangle_{\Delta t}^2 \rangle_{\bar{t}} - \langle q(\bar{t}) \rangle^2. \tag{B.22}$$

The difference

$$\sigma^2 - \sigma_{\bar{t}}^2 = \langle q(\bar{t}) \langle q(\bar{t})^2 \rangle_{\Delta t} \langle q(\bar{t}) \rangle_{\Delta t}^2$$

$$= \langle \sigma^2(\bar{t}) \rangle_{\bar{t}}, \tag{B.23}$$

which, again, is the average of the local variances.

B.2   Scaling and self-similarity

The scaling hypothesis, for a function $q(t)$, under a dilatation by an arbitrary constant α, is expressed by

$$q(\alpha t) = \Omega(\alpha) \, q(t). \tag{B.24}$$

In other words, the assumption is that stretching the parameterization of time $t \to \alpha t$, leads to a uniform stretching of the function $q(t)$, by a factorizable magnification $\Omega(\alpha)$. The function retains its same 'shape', or functional form; it is just magnified by a constant scale.

This property is clearly not true of an arbitrary function. For example, $q(t) = \sin(\omega t)$ does not satisfy the property. Our interest in such functions is connected with dynamical systems that exist and operate over a wide range of scales. Physical systems are always limited by some constraints, so this kind of scaling law is very unlikely to be true over more than a limited range of α values. Nevertheless, it is possible to discuss functions that, indeed, scale in this fashion, for all values of α, as an idealization. Such functions are said to be *scale invariant*, *dilatation invariant*, or self-similar.

In addition to perfect self-similarity, or dilatation invariance of a function, physical systems sometimes exhibit other forms of self-similarity.

Dynamical invariance tells us that the equations that describe how the function $q(t)$ behaves, or is constrained, are invariant under the change of scale. This is a weaker condition, which means that the behaviour of a complete system is invariant, but that $q(t)$ itself need not be.

$$S[\Omega^{-1}(\alpha) q(\alpha t)] \to S[q(t)]. \tag{B.25}$$

Statistical invariance tells us that the average properties of a stochastic variable or a physical system are invariant; that is, the function need only satisfy the scaling law on average.

B.3   Scaling of continuous functions

From eqn. (B.24), the symmetry between $q(t)$ and $\Omega(s)$, tells us that

$$q(x) \sim \Omega(x), \tag{B.26}$$

that is, they must possess similar scaling properties. In fact, $q(t)$ and $\Omega(s)$ must be homogeneous functions, in order to satisfy this relationship:

$$q(t) = t^H$$
$$\Omega(s) = s^H, \tag{B.27}$$

for some power H. In other words, one has

$$s^{-H} q(st) = q(t). \tag{B.28}$$

Consider a stochastic process $q(t)$, whose average properties show invariance over a wide range of scales, compared to the limiting resolution of the data. Consider what happens if we scale the basic definition of the local averaging procedure; the procedure, starting from the basic function, is as follows:

(1) The coarse-graining parameterization is $t = \Delta t \cdot \bar{t} + \tilde{t}$, that is, $q(t) \rightarrow q(\bar{t}, \tilde{t})$, where $\bar{t}$ is the slowly varying parameter, and $\tilde{t}$ is the more rapidly varying parameter.

(2) Average over the intervals of size Δt, by integrating or summing over $\tilde{t}$.

(3) Rescale the coarse-graining interval from $\Delta t \rightarrow s\Delta t$.

Beginning with the definition of the averaging procedure:

$$\langle q(\bar{t}) \rangle_{\Delta t} = \frac{1}{s\Delta t} \int_{\bar{t}-\Delta t/2}^{\bar{t}+\Delta t/2} q(\tilde{t}')\,(s\mathrm{d}\tilde{t}'). \tag{B.29}$$

one may scale the variable of integration, so that the left-hand side of the equation is the same. From the assumption of the scale invariance of $q(t)$, in eqn. (B.28), one may write

$$\langle q(\bar{t}) \rangle_{\Delta t} = \frac{1}{s\Delta t} \int_{\bar{t}-\Delta t/2}^{\bar{t}+\Delta t/2} \left[\frac{q(s\tilde{t}')}{s^H} \right] \mathrm{d}(s\tilde{t}'). \tag{B.30}$$

We now extend the limits of the integral, without amplifying the function itself, in order to perform a further coarse-graining, incorporating s times more points:

$$\langle q(\bar{t}) \rangle_{\Delta t} = \frac{1}{s\Delta t} \int_{\bar{t}-s\Delta t/2}^{\bar{t}+s\Delta t/2} \left[\frac{q(s\tilde{t}')}{s^H} \right] \mathrm{d}(s\tilde{t}'). \tag{B.31}$$

By rewriting slightly, we now observe that the function has the form of an average of a new quantity, over the larger interval $s\Delta t$:

$$\langle q(\bar{t})\rangle_{\Delta t} = \frac{1}{s\Delta t}\int_{\bar{t}-s\Delta t/2}^{\bar{t}+s\Delta t/2}\left[\frac{q(s\tilde{t}')}{s^H}s\right]d\tilde{t}'. \tag{B.32}$$

$$\langle q(\bar{t})\rangle_{\Delta t} = \frac{\langle q(st)\rangle_{s\Delta t}}{s^{H-1}} \tag{B.33}$$

The same procedure can be applied to the variance that behaves simply as the square of the average:

$$\sigma_{\Delta t}^2 = \int dt\, dt' \langle q(t)q(t')\rangle$$

$$= \int (s\,dt\, s\,dt')\frac{\langle q(t)q(t')\rangle}{s^{2H}} \tag{B.34}$$

$$\sigma_{\Delta t}^2 = \frac{\sigma_{2\Delta t}^2}{s^{2H-2}}. \tag{B.35}$$

Appendix C

Percolation conditions

C.1 Random graph condition

We reproduce here the argument of Newman et al. (2001) to derive the condition for the probable existence of a giant cluster for a uni-partite random graph with degree distribution p_k, and correct it for smaller graphs.

The method of generating functions is a powerful way of encapsulating the properties of a whole graph in a single analytical expression. Let k represent the degree of each node, and p_k be the probability distribution for the occurrence of nodes of degree k within the graph. We have,

$$\sum_k p_k = \sum_k \frac{n_k}{N} = 1,$$
(C.1)

where n_k is the number of nodes of degree k, and N is the total number of nodes. The generating function for this distribution is simply the polynomial, in a dummy source variable J, whose kth power coefficient is p_k, that is,

$$G(J) = \sum_{k=0}^{k_{\max}} p_k J^k,$$
(C.2)

so that the probability distribution is recovered by the derivatives:

$$p_k = \frac{1}{k!} \frac{\mathrm{d}^k G(J)}{\mathrm{d} J^k}\bigg|_{J=0},$$
(C.3)

and the average degree of nodes in the graph is

$$z \equiv \langle k \rangle = J \frac{\mathrm{d}}{\mathrm{d} J} G(J)\bigg|_{J=0}.$$
(C.4)

Note that $k_{\max}$ is normally taken to be infinite to approximate large graphs. We can use this generating function to evaluate average (probabilistic) properties of the graph.

Analytical Network and System Administration. Managing Human–Computer Networks   Mark Burgess
© 2004 John Wiley & Sons, Ltd   ISBN 0-470-86100-2

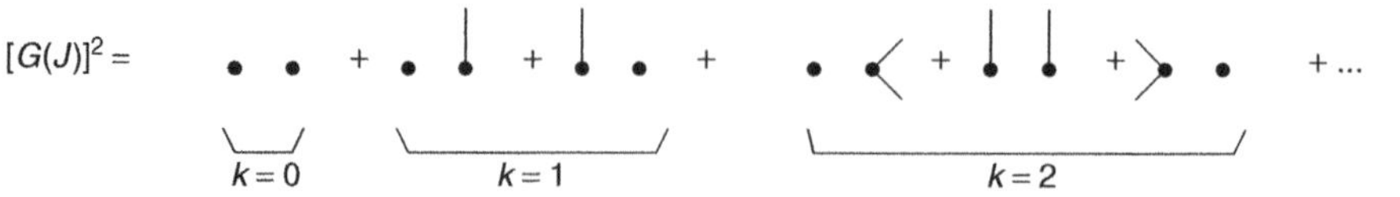

Figure C.1: Graphical form of the first three terms in the second power of the generating function. The nth power of $G(J)$ generates the probabilities of finding a total degree of k from a cluster of n nodes, that is, the probability that n nodes have k outgoing edges.

If we pick an arbitrary node and follow one of the edges (links) of the graph to another node, the probability of arriving at a node of degree k' is proportional to k', since a highly connected node is proportionally more likely to be arrived at than a poorly connected node (there are more ways for it to occur). Thus, in our average picture, the probability of getting to a node of degree k is

$$P_k = \frac{k\, p_k}{\displaystyle\sum_k k\, p_k} = \frac{k}{\langle k \rangle} p_k. \tag{C.5}$$

This distribution is generated by the normalized derivative of $G(J)$, like this:

$$G_1(J) \equiv \frac{\displaystyle\sum_k k\, p_k\, J^{k-1}}{\displaystyle\sum_k k\, p_k} = \frac{1}{\langle k \rangle} \frac{\mathrm{d}}{\mathrm{d}J} G(J). \tag{C.6}$$

Following Newman et al. (2001) we note that, if a distribution p_k is generated by $G(J)$, then a number of related generating functions are obtained by taking powers of $G(J)$. Suppose that there are m independent ways of obtaining the probability $P_{k'}$, from different, but equivalent, graph configurations p_k, then the function that generates the right combinatorics for p_k is the mth power of $G(J)$.

$$\gamma_m(J) \equiv [G(J)]^m = \sum_\kappa \pi_\kappa J^\kappa. \tag{C.7}$$

This is easy to see when $m = 2$ (see fig. C.1):

$$\gamma_2(J) = [G(J)]^2 = \left[\sum_k p_k J^k \right]^2$$
$$= \sum_{i,j} p_i p_j J^{i+j}$$
$$= p_0 p_0 J^0 + (p_0 p_1 + p_1 p^0) J^1$$
$$+ (p_0 p_2 + p_1 p_1 + p_2 p_0) J^2 + \cdots$$

If we compare the coefficients of J^k in eqns. (C.7) and (C.8), the we see that

$$\pi_0 = p_0 p_0$$
$$\pi_1 = p_0 p_1 + p_1 p_0$$
$$\pi_2 = p_0 p_2 + p_1 p_1 + p_2 p_0$$
$$\pi_k = C(i, j, k), \tag{C.8}$$

where $C(p_m)$ is the sum of all combinations such that $i + j = k$.

Thus, suppose now that we wish to calculate the average number of nodes within a connected cluster, that is, the *size* of the cluster. We can obtain this result by summing the nodes that themselves have connected neighbours. This can be achieved by using an *effective* generating function, of the form:

$$W_c[J] = J \sum_k p_k [\chi_c(J)]^k. \tag{C.9}$$

Here, we postulate the existence of a constrained function $\chi_c(J)$, that pertains to a given cluster c, within the graph, and generates the distribution of degrees recursively at all connected sub-nodes of a cluster, starting from some arbitrary point. An additional power of J is added here, by convention, so that the counting starts from 1. The constraint is derived using a recursive definition that sums over clusters of connected nodes. Suppose we define the normalized distribution

$$\chi_c(J) = \sum_k c_k J^k \tag{C.10}$$

$$= \frac{\sum_k k p_k [\chi_c(J)]^k}{\langle k \rangle} \tag{C.11}$$

$$= J G_1(\chi_c(J)). \tag{C.12}$$

Equation (C.11) is a constraint equation; its right-hand side is interpreted as a sum of probabilities for arriving at a node of degree k, from some arbitrary starting point that has a number of k nearest neighbours each with degree distributions generated by $\chi_c(J)^k$. That is,

$$\chi(J) \propto \sum_k \quad \text{Probability of picking a} \quad \times \quad \text{Probable ways of connecting}$$
$$\text{node of degree } k \qquad \qquad \text{to } k \text{ nodes from a random node}$$

The recursive definition indicates that the same average probabilities exist at each node of the graph; only the limit of total nodes in the cluster stops the iteration. Substituting in the generic form with coefficients c_k leads to an eigenvalue equation for the vector c_k, with a matrix of probabilities. The principal eigenvector gives the appropriate solution for the largest cluster. Remarkably, we do not need to know the solution of $\chi_c(J)$ in order to find out when the size of connected clusters becomes dangerously large for system security. Instead, the constraint can be eliminated.

Differentiation of $W_c(J)$ with respect to the source J gives a quantity that is analogous to the result in eqn. (C.4), but with a new kind of average that includes both nearest neighbour degrees, next-nearest neighbour degrees and so on:

$$\langle\langle k \rangle\rangle \equiv \frac{d}{dJ} W_c[J]$$

$$= 1 + J \frac{d}{dJ} G(\chi_c(J))$$

$$= 1 + J \frac{dG(\chi_c)}{d\chi_c} \cdot \frac{d\chi_c}{dJ}\bigg|_{J=1}. \tag{C.13}$$

The result is an average estimate of the size of a connected cluster. Using eqn. (C.12), we find that

$$\frac{d\chi_x}{dJ} = \left(1 - \frac{dG_1(J)}{dJ}\right)^{-1}, \tag{C.14}$$

Thus, the average size of a randomly picked cluster is

$$\langle\langle k \rangle\rangle = 1 + \frac{J \dfrac{dG(J)}{dJ}}{\left(1 - \dfrac{dG_1}{dJ}\right)}\bigg|_{J=1}. \tag{C.15}$$

Here we note that self-consistently $W[1] = 1$, as long as $W[J]$ has no singularities. In the general case, we must define $\Gamma_c = \ln W_c$ and $\langle\langle k \rangle\rangle = d\Gamma/dJ$ at $J = 1$. A giant component or cluster is defined to be a cluster that is of order N nodes. If such a cluster exists, then other smaller clusters of order $\log N$ might also exist (Molloy and Reed (1998)). The condition for a giant cluster is thus that the denominator in this fraction becomes small, or

$$\frac{dG_1(J)}{dJ}(1) = 1. \tag{C.16}$$

Using eqn. (C.6), we find the critical point for the emergence of a giant cluster. The large-graph condition for the existence of a giant cluster (of infinite size) is simply

$$\sum_k k(k-2) \, p_k \geq 0. \tag{C.17}$$

This provides a simple test that can be applied to a human–computer system, in order to estimate the possibility of complete failure via percolating damage. If we only determine the p_k, then we have an immediate machine-testable criterion for the possibility of a systemwide security breach. The condition is only slightly more complicated than the simple Cayley tree approximation; but (as we will see below) it tends to give more realistic answers.

C.2 Bi-partite form

Random bi-partite graphs are also discussed in Newman et al. (2001) and a corresponding expression is derived for giant clusters. Here, we can let p_k be the fraction of users with degree k (i.e. having access to k files), and q_k be the fraction of files to which k users have access. Then, from Newman et al. (2001), the large-graph condition for the appearance of a giant bi-partite cluster is

$$\sum_{jk} jk(jk - j - k) \, p_j q_k > 0. \tag{C.18}$$

This result is still relatively simple, and provides a useful guideline for avoiding the possibility of systemwide infection—in those cases in which such is practical, one seeks to hold the whole system below the percolation threshold, by not satisfying the inequality in (C.18). The left-hand side of (C.18) can be viewed as a weighted scalar product of the two vectors of degree distributions:

$$q^T W p = p^T W q > 0, \tag{C.19}$$

with $W_{jk} = jk(jk - j - k)$ forming a symmetric, graph-independent weighting matrix.

C.3 Small-graph corrections

The problem with the above expressions is clearly that they are derived under the assumption of there being a smooth differentiable structure to the average properties of the graphs. For a small graph with N nodes (either uni-partite or bi-partite), the criterion for a giant cluster becomes inaccurate. Clusters do not grow to infinity, they can only grow to size N at the most, hence we must be more precise and use a dimensionful scale rather than infinity as a reference point. The correction is not hard to identify; we require

$$\left. \frac{J \dfrac{dG(J)}{dJ}}{\left(1 - \dfrac{dG_1}{dJ}\right)} \right|_{J=1} \gg 1, \tag{C.20}$$

for the uni-partite case. This more precise percolation criterion states that, at percolation, the average size of clusters is of the same order of magnitude as the number of nodes. However, for a small graph the size of a giant cluster and of below-threshold clusters [N and $\log(N)$, respectively] are not that different (Molloy and Reed (1998)). The above criterion translates into

$$\frac{\langle k \rangle^2}{-\sum_k k(k-2) \, p_k} \gg 1. \tag{C.21}$$

Thus, the threshold point can be taken to be as follows. The small-graph condition for widespread percolation in a uni-partite graph of order N is

$$\langle k \rangle^2 + \sum_k k(k-2) \, p_k > \log(N). \tag{C.22}$$

This can be understood as follows. If a graph contains a giant component, it is of order N and the size of the next largest component is typically $O(\log N)$; thus, according to the theory of random graphs, the margin for error in estimating a giant component is of order $\pm \log N$. In the criterion above, the criterion for a cluster that is much greater than unity is that the right-hand side is greater than zero. To this, we now add the magnitude of the uncertainty in order to reduce the likelihood of an incorrect conclusion.

Similarly, for the bi-partite graph, one has the small-graph condition for widespread percolation in a bi-partite graph of order N:

$$\langle k \rangle^2 \langle j \rangle^2 + \sum_{jk} jk(jk - j - k) \, p_j p_k > \log(N/2). \tag{C.23}$$

These expressions are not much more complex than the large-graph criteria. Moreover, they remain true in the limit of large N. Hence, we expect these small-graph criteria to be the most reliable choice for testing percolation in small systems. This expectation is borne out in the examples below. In particular, we find that since the average coordination number $\langle k \rangle$ enters into the small-graph percolation criteria, the earlier problem of ignoring isolated nodes in the uni-partite case is now largely remedied.

Bibliography

Abdu H, Lutfiya H and Bauer M 1999 A model for adaptive monitoring configurations. *Proceedings of the VI IFIP/IEEE IM Conference on Network Management* p. 371.

Ahmad I and Dhodhi MK 1996 Multiprocessor scheduling in a genetic paradigm. *Parallel Computing* **22**, 395–406.

Albert R and Barabási A 2002 Statistical mechanics of complex networks. *Reviews of Modern Physics* **74**, 47.

Albert R, Jeong H and Barabasi A 1999 Diameter of the world-wide web. *Nature* **401**, 130.

Anderson E, Burgess M and Couch A 2001 *Selected Papers in Network and System Administration*. John Wiley & Sons, Chichester.

Apthorpe R 2001 A probabilistic approach to estimating computer system reliability. *Proceedings of the Fifteenth Systems Administration Conference (LISA XV)* (USENIX Association: Berkeley, CA) p. 31.

Baase S and Gelder AV 1999 *Computer Algorithms* (3rd edition). Addison-Wesley, Reading, MA.

Balakrishnan V 1997 *Graph Theory*. Schaum's Outline Series, McGraw-Hill, New York.

Barabási A 2002 *Linked*. Perseus, Cambridge, MA.

Barabasi A and Albert R 1999 Emergence of scaling in random networks. *Science* **286**, 509.

Barabasi A, Albert R and Jeong H 2000 Scale-free characteristics of random networks: topology of the world-wide web. *Physica A* **281**, 69.

Beran J 1994 *Statistics for Long Memory Processes*. Chapman & Hall, Boca Raton, FL.

Berge C 2001 *The Theory of Graphs*. Dover Publications, New York.

Bishop M 2002 *Computer Security: Art and Science*. Addison-Wesley, New York.

Bonacich P 1987 Power and centrality: a family of measures. *American Journal of Sociology* **92**, 1170–1182.

Box G, Jenkins G and Reinsel G 1994 *Time Series Analysis*. Prentice Hall, NJ.

Breipohl A 1970 *Probabilistic systems analysis*. John Wiley & Sons, New York.

Buchanan M 2002 *Nexus: Small Worlds and the Groundbreaking Science of Networks*. W.W. Norton & Co., New York.

Bunke H and Csirik J 1995 Parametric string edit distance and its application to pattern recognition. *IEEE Transactions on Systems, Man and Cybernetics* **25**, 202.

Burgess M 1993 Cfengine www site. *http://www.iu.hio.no/cfengine*.

Burgess M 1995 A site configuration engine. *Computing Systems* (MIT Press, Cambridge, MA) **8**, 309.

Burgess M 1998a Automated system administration with feedback regulation. *Software Practice and Experience* **28**, 1519.

Analytical Network and System Administration. Managing Human–Computer Networks   Mark Burgess
© 2004 John Wiley & Sons, Ltd   ISBN 0-470-86100-2

Burgess M 1998b Computer immunology. *Proceedings of the Twelfth Systems Administration Conference (LISA XII)* (USENIX Association: Berkeley, CA) p. 283.

Burgess M 2000a The kinematics of distributed computer transactions. *International Journal of Modern Physics* **C12**, 759–789.

Burgess M 2000b *Principles of Network and System Administration.* John Wiley & Sons, Chichester.

Burgess M 2000c Theoretical system administration. *Proceedings of the Fourteenth Systems Administration Conference (LISA XIV)* (USENIX Association: Berkeley, CA) p. 1.

Burgess M 2002a *Classical Covariant Fields.* Cambridge University Press, Cambridge.

Burgess M 2002b Two dimensional time-series for anomaly detection and regulation in adaptive systems. *IFIP/IEEE 13th International Workshop on Distributed Systems: Operations and Management (DSOM 2002)* p. 169.

Burgess M 2003 On the theory of system administration. *Science of Computer Programming* **49**, 1.

Burgess M 2004 Cfengine's immunity model of evolving configuration management. *Science of Computer Programming*, in press.

Burgess M and Canright G 2003 Scalability of peer configuration management in partially reliable and ad hoc networks. *Proceedings of the VII IFIP/IEEE IM conference on Network Management* p. 293.

Burgess M and Ralston R 1997 Distributed resource administration using cfengine. *Software Practice and Experience* **27**, 1083.

Burgess M and Reitan T 2004 Theory of events and their maintenance. Paper in preparation.

Burgess M and Sandnes F 2001 Predictable configuration management in a randomized scheduling framework. *Proceedings of the 12th International Workshop on Distributed System Operation and Management (IFIP/IEEE).* INRIA Press, France, p. 293.

Burgess M, Canright G and EngøK 2004 A definition of clusters and roles in directed graphs. Paper in preparation.

Burgess M, Canright G and Engø K 2003a A graph theoretical model of computer security: from file access to social engineering. Submitted to *International Journal of Information Security.*

Burgess M, Canright G, Stang TH, Pourbayat F, Engo K and Weltzien A 2003b Archipelago: a network security analysis tool. *Proceedings of the Seventeenth Systems Administration Conference (LISA XVII)* (USENIX Association: Berkeley, CA) p. 153.

Burgess M, Hagen S and Sandnes F 2003c Voluntary rpc: how distributed automation can be accomplished in a pervasive computing environment. Paper in preparation.

Burgess M, Haugerud H, Reitan T and Straumsnes S 2001 Measuring host normality. *ACM Transactions on Computing Systems* **20**, 125–160.

Buzen J 1973 Computational algorithms for closed queueing networks with exponential servers. *Communications of the ACM* **16**, 527.

Buzen J 1976 Fundamental laws of computer system performance. *Proceedings of SIGMETRICS'76* p. 200.

Canright G and Engø K 2004 A natural definition of clusters and roles in undirected graphs. To appear in *Science of Computer Programming*, Special edition on network and system administration.

Cheung S and Kramer J 1996 Checking subsystem safety properties in compositional reachability analysis. *18th International Conference on Software Engineering (ICSE'18)*, Berlin, Germany.

Chung F 1997 Spectral graph theory. *Regional Conference Series in Mathematics, American Mathematical Society* **92**, 1–212.

Cormen T, Leiserson C, Rivest R and Stein C 2001 *Introduction to Algorithms.* MIT Press, Cambridge, MA.

Couch A 2000 An expectant chat about script maturity. *Proceedings of the Fourteenth Systems Administration Conference (LISA XIV)* (USENIX Association: Berkeley, CA) p. 15.

Couch A and Daniels N 2001 The maelstrom: Network service debugging via "ineffective procedures". *Proceedings of the Fifteenth Systems Administration Conference (LISA XV)* (USENIX Association: Berkeley, CA) p. 63.

Couch A and Gilfix M 1999 It's elementary, Dear Watson: applying logic programming to convergent system management processes. *Proceedings of the Thirteenth Systems Administration Conference (LISA XIII)* (USENIX Association: Berkeley, CA) p. 123.

Couch A and Sun Y 2003 On the algebraic structure of convergence. Submitted to *DSOM 2003*.

Couch A, Hart J, Idhaw E and Kallas D 2003 Seeking closure in an open world: a behavioural agent approach to configuration management. *Proceedings of the Seventeenth Systems Administration Conference (LISA XVII)* (USENIX Association: Berkeley, CA) p. 129.

Cover T and Thomas J 1991 *Elements of Information Theory*. John Wiley & Sons, New York.

Cowan C n.d. StackGuard Project. *http://www.cse.ogi.edu/DISC/projects/immunix/StackGuard/*.

Damianou N, Dulay N, Lupu E and Sloman M 2000 *Ponder: A Language for Specifying Security and Management Policies for Distributed Systems*. Imperial College Research Report DoC 2000/1.

Date C 1999 *Introduction to Database Systems* (7th edition). Addison-Wesley, Reading, MA.

David R and Alla H 1994 Petri nets for modelling of dynamic systems—a survey. *Automatica* **30**, 175–202.

D'haeseleer P, Forrest S and Helman. P n.d. 1996 An immunological approach to change detection: algorithms, analysis, and implications. *Proceedings of the 1996 IEEE Symposium on Computer Security and Privacy*.

Diao Y, Hellerstein J and Parekh S 2002 Optimizing quality of service using fuzzy control. *IFIP/IEEE 13th International Workshop on Distributed Systems: Operations and Management (DSOM 2002)* p. 42.

Dresher M 1961 *The Mathematics of Games of Strategy*. Dover Publications, New York.

Duda R, Hart P and Stork D 2001 *Pattern Classification*. Wiley-Interscience, New York.

Dunbar R 1996 *Grooming, Gossip and the Evolution of Language*. Faber and Faber, London.

Durbin R, Krigh SE and Mitcheson G 1998 *Biological Sequence Analysis*. Cambridge University Press, Cambridge.

Endsley M 1995 Towards a theory of situation awareness in dynamic systems. *Human Factors* **37**, 32.

Gambit n.d. Game theory analyser. *http://econweb.tamu.edu/gambit*.

Gilfix M and Couch A 2000 Peep (the network aualizer): monitoring your network with sound. *Proceedings of the Fourteenth Systems Administration Conference (LISA XIV)* (USENIX Association: Berkeley, CA) p. 109.

Glance N, Hogg T and Huberman B 1991 Computational ecosystems in a changing environment. *International Journal of Modern Physics* **C2**, 735.

Goudarzi K and Kramer J 1996 Maintaining node consistency in the face of dynamic change. *Proc. of 3rd International Conference on Configurable Distributed Systems (CDS '96)*, Annapolis, MD, USA, IEEE Computer Society Press, p. 62.

Griffin T and Wilfong G 2002 On the correctness of ibgp configuration. *ACM SIGCOMM'02)*.

Grimmett G and Stirzaker D 2001 *Probability and Random Processes* (3rd edition). Oxford Scientific Publications, Oxford.

Gupta P and Kumar P 2000 The capacity of wireless networks. *IEEE Transactions on Information Theory* **46**(2), 388–404.

Hellerstein J, Zhang F and Shahabuddin P 1999 An approach to predictive detection for service management. *Proceedings of IFIP/IEEE IM VI* p. 309.

Herzog U 1994 Network planning and performance engineering. In *Network and Dsitributed Systems Management*. M. Sloman (Ed.), Addison-Wesley, Wokingham, UK, p. 349.

Hofstadter D 1979/1981 *Gö del, Escher, Bach: An Eternal Golden Braid*. Penguin Books, Middlesex, UK.

Holgate M and Partain W 2001 The arushra project: a framework for collaborative unix system administration. *Proceedings of the Fifteenth Systems Administration Conference (LISA XV)* (USENIX Association: Berkeley, CA) p. 187.

Hoogenboom P and Lepreau J 1993 Computer system performance problem detection using time series models. *Proceedings of the USENIX Technical Conference, (USENIX Association: Berkeley, CA)* p. 15.

Horgan J 1996 *The End of Science*. Addison-Wesley, New York.

Hughes B 1995 *Random Walks and Random Environments (Volume 1: Random Walks)*. Oxford Science Publications, Oxford.

IEEE n.d. 1992 *A Standard Classification for Software Anomalies*. IEEE Computer Society Press.

Jackson M 1975 *Principles of Program Design*. Academic Press, New York.

Jain R 1991 *The Art of Computer Systems Performance Analysis*. Wiley-Interscience, New York.

Jerne N 1964 The generative grammar of the immune system. Nobel lecture.

Kasahara H and Narita S 1984 Practical multiprocessor scheduling algorithms for efficient parallel processing. *IEEE Transactions on Computers* **C-33**(11), 1023–1029.

Kleinberg J 1999 Authoritative sources in a hyperlinked environment. *Journal of the ACM* **46**, 604.

Leland W, Taqqu M, Willinger W and Wilson D 1994 On the self-similar nature of ethernet traffic. *IEEE/ACM Transactions on Networking* 1–15.

Lewis H and Papadimitriou C 1997 *Elements of the Theory of Computation (2nd edition)*. Prentice Hall, New York.

Li J, Blake C, DeCouto D, Lee H and Morris R 2001 Capacity of ad hoc wireless networks. *Proc. 7th ACM Intl. Conf. on Mobile Computing and Networking* pp. 61–69.

Libes D 1990 Using *expect* to automate system administration tasks. *Proceedings of the Fourth Large Installation System Administrator's Conference (LISA IV)* (USENIX Association: Berkeley, CA, 1990) p. 107.

Logic n.d. Mathematical logic around the world. *http:///www.uni-bonn.de/logic/world.html*.

McRuer D 1980 Human dynamics in man-machine systems. *Automata* **16**, 237.

Meyer J, Movaghar A and Sanders W 1985 Stochastic activity networks: structure, behavior and application. *Proceedings of the International Conference on Timed Petri Nets* p. 106.

Sun Microsystems n.d. Java programming language. *http://java.sun.com/aboutJava/*.

Molloy M and Reed B 1998 The size of the giant component of a random graph with a given degree sequence. *Combinatorics, Probability and Computing* **7**, 295.

Myerson R 1991 *Game theory: Analysis of Conflict*. Harvard University Press, Cambridge, MA.

Nash J 1996 *Essays on Game Theory*. Edward Elgar, Cheltenham.

Natvig B 1998 *Pålitelighetsanalyse med teknologiske anvendelser*. University of Oslo Compendium, Oslo, Norway.

Neumann J and Morgenstern O 1944 *Theory of Games and Economic Behaviour*. Princeton University Press, Princeton.

Newman MEJ, Strogatz S and Watts D 2001 Random graphs with arbitrary degree distributions and their applications. *Physical Review E* **64**, 026118.

NRC UNRC 1981 *Fault Tree Handbook*. NUREG-0492, Springfield.

Omari S, Boutaba R and Cherakaoui O 1999 Policies in snmpv3-based management. *Proceedings of the VI IFIP/IEEE IM Conference on Network Management* p. 797.

Oommen B and Kashyap R 1998 A formal theory for optimal and information theoretic syntactic pattern recognition. *Patter Recognition* **31**, 1159.

Page L, Brin S, Motwani R and Winograd T 1998 The pagerank citation ranking: bringing order to the web. Technical report, Stanford University, Stanford, CA.

Paxson V and Floyd S 1995 Wide area traffic: the failure of Poisson modelling. *IEEE/ACM Transactions on networking* **3**(3), 226.

Pearl J 1988 *Probabilistic Reasoning in Intelligent Systems: Networks of Plausible Inference*. Morgen Kaufmann, San Francisco.

Pearl J 2000 *Causality*. Cambridge University Press, Cambridge.

Perelson A and Weisbuch G 1997 Immunology for physicists. *Reviews of Modern Physics* **69**, 1219.

Qie X and Narain S 2003 Using service grammar to diagnose bgp configuration errors. *Proceedings of the Seventeenth Systems Administration Conference (LISA XVII)* (USENIX Association: Berkeley, CA) p. 243.

Rapoport A 1970 *N-Person Game Theory: Concepts and Applications*. Dover Publications, New York.

Rasmusen E 2001 *Games and Information* (3rd edition). Blackwell Publishing, Oxford.

Rasmussen J 1983 Skills, rules, and knowledge; signals, signs and symbols, and other distinctions in humans performance models. *IEEE Transactions on Systems, Man and Cybernetics* **13**, 257.

Rouse W 1989 On capturing human skills and knowledge: Algorithmic approaches to model identification. *IEEE Transactions on Systems, Man and Cybernetics* **19**, 558.

Sandnes F 2001 Scheduling partially ordered events in a randomized framework - empirical results and implications for automatic configuration management. *Proceedings of the Fifteenth Systems Administration Conference (LISA XV)* (USENIX Association: Berkeley, CA) p. 47.

Sato K 1999 *Levy Processes and Infinitely Divisible Distributions*. Cambridge studies in advanced mathematics, Cambridge University Press, Cambridge.

Scherr A 1967 *An Analysis of Timeshared Computer Systems*. MIT Press, Cambridge, MA.

Seltzer M and Small C 1997 Self-monitoring and self-adapting operating systems. *Proceedings of the Sixth workshop on Hot Topics in Operating Systems*, Cape Cod, MA, USA, IEEE Computer Society Press.

Shannon C and Weaver W 1949 *The Mathematical Theory of Communication*. University of Illinois Press, Urbana.

Sheridan T 1996 Allocating functions among humans and machines. In D. Beevis, P. Essens and H. Schuffel (Eds.), *Improving Function Allocation for Integrated Systems Design*. Wright-Patterson Airforce Base, CSERIAC State-of-the-Art Report, pp. 179–198.

Somayaji A and Forrest S 2000 Automated response using system-call delays. *Proceedings of the 9th USENIX Security Symposium* (USENIX Association: Berkeley, CA) p. 185.

Sommerville I 2000 *Software Engineering* (6th edition). Addison-Wesley, New York.

Steinder M and Sethi A 2002 Distributed fault localization in hierarchically routed networks. *IFIP/IEEE 13th International Workshop on Distributed Systems: Operations and Management (DSOM 2002)*.

Steinder M and Sethi A 2003 A survey of fault localization techniques in computer networks. *Science of Computer Programming* (To appear).

Tapscott D and McQueen R 1995 *Digital Economy: Competing in the New Networked Economy*. McGraw-Hill Education, Boston.

Traugott S 2002 Why order matters: turing equivalence in automated systems administration. *Proceedings of the Sixteenth Systems Administration Conference (LISA XVI)* (USENIX Association: Berkeley, CA) p. 99.

Tridgell A and Mackerras P 1996 The rsync algorithm. Technical report of the Australiean National University.

Watt D 1991 *Programming Language Syntax and Semantics*. Prentice Hall, New York.

Whittaker E and Robinson G 1929 *Calculus of Observations*. Blackie and Son Ltd., London.

Willinger W and Paxson V 1998 Where mathematics meets the internet. *Notices of the American Mathematical Society* **45**(8), 961.

Willinger W, Paxson V and Taqqu M 1996 Self-similarity and heavy tails: structural modelling of network traffic. *in* A practical guide to heavy tails: statistical techniques and applications pp. 27–53.

XML-RPC n.d. Internet remote procedure call. *http://www.xmlrpc.com/spec*.

Høyland A and Rausand M 1994 *System Reliability Theory: Models and Statistical Methods*. John Wiley & Sons, New York.

Zadeh L 1973 Outline of a new approach to the analysis of complex systems and decision process. *IEEE Transactions on Systems, Man and Cybernetics* **3**, 28–44.

Zwicky E 1989 Disk space management without quotas. *Proceedings of the Workshop on Large Installation Systems Administration III* (USENIX Association: Berkeley, CA, 1989) p. 41.

Index

A postiori belief, 41
A priori belief, 41
A. Turing, 109
Ad hoc network, 15
Ad hoc networks
    Percolation, 167
Ad hoc systems, 194
Addresses, 61, 92
Adjacency matrix, 15, 75, 149, 170
Administration, 2
Algorithm, 52
Alphabet, 4, 114
Approximation
    Continuum, 65
    Gaussian continuum, 14
    Stirling, 126
Architecture, 192
Assets, 159
Associations, 93
Asynchronous Transfer Mode, 107
ATM, 107
Authority, 83, 193
Automata, 66
Average constancy, 136
Average time before failure, 11

Back up of systems, 29
Bayes' formula, 40, 76
Bayesian statistics, 38
Belief and science, 42
Bi-partite graph percolation, 89
Birnbaum measure, 9
Birth–death process, 177
Boolean algebra, 76
Boolean logic, 29, 75

Bottom up, 196
Bridges, 170
Brouwer fixed point, 152
Burst, 38

C. Shannon, 109
Causality, 1, 16, 24, 28, 34, 55
Cause trees, 24
Central limit theorem, 29
Centrality, 9, 24, 77, 142, 169
Cfengine, 6
Chains, 62, 137
Checksums, 205
Chomsky hierarchy, 69
Clark–Wilson security model, 199
Classification, 28
Classification as knowledge, 38
Closed system, 56
Clouds of information, 68
Cluster, 38, 166
Clusters in graphs, 166
Coarse graining, 79
Coarse-graining, 111
Coherent systems, 5
Commutation of operations, 7
Competition or Cooperation, 45
Compression of data, 127
Configuration, 18, 64, 160
    Generation, 7
Configuration error, 29
Configuration management, 9, 16
Configuration space, 160
Congruence, 69, 118
Congruent state, 69
Connectivity, 14, 77

Analytical Network and System Administration. Managing Human–Computer Networks   Mark Burgess
© 2004 John Wiley & Sons, Ltd   ISBN 0-470-86100-2

Conservation law, 162
Constancy, 136
Constraint, 50
Constraints, 196
Contention, 180
Continuous time, 81
Continuum approximation, 65, 102, 103
Control
    Requirements for, 111
Convergence, 3, 8, 24
    Coherence, 5
    Geometric series, 47
    Monotonicity, 5
Convergence of learning, 48
Cooperation or competition, 45
Copy on write, 30
Cost of ownership, 53
Covert channels, 165
Critical dependency, 32
Currency, 37, 161
Cut sets, 7
Cutset, fault tree, 32

Damped oscillation, 138
Data compression, 127
Data plots, 26
Data structure, 46
Database models, 2
De-multiplexing, 11
Decision theory, 34
Decisions, 35, 43, 203
    As games, 35
    Classifiers, 35
Definition
    Secure system, 6
Degeneracy, 127
Degree of a node, 75
Degree of freedom, 50
Delta distribution, 34
Dependence, 1
    Critical, 195
    Strong, 195
    Weak, 195
Dependencies
    In games, 63

Dependency, 25, 194
    Component faults, 10
    Critical, 32
    Normalization, 1
    Principle of minimum, 203
    Serial, 4
Dependency diagrams, 86
Dependency problems, 27
Derivative, 102
Design
    Functional, 197
Deterministic, 98
Deterministic system, 56
DFS, 30
Diagnostic trees, 24
Diagram
    Dependency, 86
    Entity relation, 86
    Flow, 85
    Functional structure, 85
    Mind map, 84
    Transition, 85
Diagrams, 73
Digitization, 111
Dijkstra shortest path, 165
Discrete time, 80
Distance function, 95
Distributed Computing Environment, 30
Distribution, measurements, 32
DNA, 11, 127
Domination of strategy, 44
Downtime, 11
Duality of description, 14

Eigencentre, 79, 169
Eigenstates, 148
Eigenvalues, 148
Eigenvector centrality, 77
Enterprise, 197
Entities, 3
Entity-relation diagram, 86
Entropy, 117
    Common, 122
    Conditional, 121
    Distribution of maximum, 148

Geometrical interpretation, 123
Joint, 120
Principle of maximum, 129, 188
Properties, 118
Relative, 122
State, 128
Environment
Constant, 27
Equilibrium, 141
Dynamical, 142
Nash, 44
Error correction, 12
Error law, 35
Errors
Experimental, 14
Human, 2
Of observation, 30
Random, 16, 30
Systematic, 31
Ethernet traffic, 105
Event handling, 174
Event tree analysis, 25
Event-driven systems, 201
Expectation value, 29
Experiment
Design, 28
Experimental uncertainty, 14
Expert operators, 5
Experts, 37
Extensive form of game, 41
Extensive instruction, 4
External stability, 139

Fail-over, 11
Failure, 28
Fault
Probability, 13
Fault tree analysis, 28
Fault trees, 28
Faults
Birnbaum measure, 9
Cut sets, 7
Dependency, 10
Deterministic model, 3
Emergent, 1
Random, 1

Stochastic model, 7
Systemic, 1
FCFS, 178
Fidelity, 120
FIFO, 178
Finite state machine, 66
First Come First Served, 178
Fixed point, 24, 138, 151
Fixed points, 98
Flow
Algorithmic, 193
Resource, 193
Flow diagrams, 85
Floyd shortest path, 165
Fluctuation
Gaussian, 14
Maximum entropy, 188
Oscillation, 136
Spectra, 133
Fluctuations, 99
Arrival process, 29
Distribution, 29
Huge, 22
Reason for, 21
Scaling, 83
Fourier analysis, 39, 40
Freedom, 50
Frequency histogram, 27
Functional design, 197
Functional design and events, 201
Functional structure diagram, 85

Game
Extensive form, 41
Strategic form, 40
Games
As decisions, 35
Zero sum, 42
Garbage collection, 29, 54, 181
Gaussian continuum approximation, 14
Gaussian distribution, 29, 35, 99
Gaussian power, 32
Generating functions, 86
Giant cluster, 166
Giant clusters, 86
Gnutella, 80

Google, 81
Grammar of machine behaviour,
        69
Graph
        Cluster, 166
        Definition, 74
        Directed, 75
        Giant component, 166
        Kernel, 140
        scheduling, 175
        Stability, 139
        Undirected, 75
Graphs, 137
        Bridges, 170
Grid computing, 189

Hamming distance, 95, 125
Hash function, 205
Hawk-Dove game, 47
Heavy-tailed distribution, 38
Help desk, 201
Hidden Markov model, 68
Hidden variables, 29, 141
Histograms, 26
Homeostasis, 24
Hopelessness, 125
Human error, 2
Human group sizes, 202
Human–computer interaction, 4
Hurst exponent, 144, 148
Hypotheses, 21

IDS, 70
Immune system, 39
Imperfect information, 38
Importance ranking, 78
Imputations, 156
Infinite variance, 22
Information, 109
        Algorithmic, 47
        Geometrical interpretation, 123
        Interpretation, 119
        Mutual, 122
        Partial, 165
        Perfect, 165
        Representation, 91, 109
        Resource, 47

State, 128
Theory of, 1
Information clouds, 68
Information sets, 68
Integrity, 1
Intermediate value theorem, 152
Internal stability, 139
Interpretation of patterns, 69
Interpretation of theory, 14
Interruption, 189
Intrusion Detection System, 70

J. Von Neumann, 109
Jitter, 35
Joint probability, 119
Junk mail, 45
Junk mail filter, 44

Kakutani fixed point, 154
Kendall notation, 178
Kernel, 140
Keys, 92
Knowledge, 37
Knowledge as classification, 38

Lévy distribution, 143, 189
Labels, 92
Lagrange's method, 130
Language interpretation, 69
Law of errors, 35
Laws of queueing, 181
Learning, 37
Limit cycle, 137
Little's law, 184
Local averaging procedure, 18, 41,
        79
Local maximum, 139
Logic, Boolean, 29, 75

Macrostate, 64
Macrostates, 64
Maintenance, 109
        Convergent, 24
        Defined, 18
        Fluctuation, 21
        Theorem, 27
Makespan, 174

Mapping
    Fuzzy, 24
Mappings, 61
Markov chain, 63
Martingale, 10
Maximum, 102, 138
    Local, 139
Maximum entropy distribution, 12, 131, 148
Maximum entropy fluctuation, 188
Maximum entropy principle, 129
Mean downtime, 11
Mean time before failure, 11
Mean time before failure (MTBF), 12
Mean time to repair (MMTR), 12
Mean value, 31
Memoryless, 63, 178
Mesh topology, 20
Metric, 95
Mind maps, 84
Minimal dependence, 203
Minimax theorem, 43
Minimum, 102, 138
MMTR, 12
Mobile nodes, 21
Model
    Mesh with central policy, 20
    Star, 17
Money, 54
Morse code, 109
MTBF, 12
Mutual information, 122

Nash equilibrium, 44
Network
    Scale free, 142
Network intrusion, 70
Network management
    Defined, 2
Networks
    As graphs, 76
Node degree, 75
Node removal, 141
Nodes
    Mobile, 21

Noise, 22
Noisy channel, 117
Non-cooperative game, 52
Non-deterministic, 29
Non-zero sum game, 52
Normal distribution, 35
Normal error law, 35
Normal form
    First, 4
    Second, 6
    Third, 7
Normalization, 1, 5
NP complexity, 70
Nyquist theorem, 15

Observation, 14
Open system, 21, 56
    Conservation, 162
Operand, 68
Operator, 68
    Expert, 5
    Ordering, 7
Optimization, 37
Organization, 193
Orthogonality, 7
Oscillation, 99

Parallel utilization, 182
Parallelism, 4
Parameterization, 14
Parameters, 27
Parametrization, 92
Pareto distribution, 38
Parsing and automata, 69
Pattern detection, 9
Pattern recognition, 39
Patterns
    Interpretation, 69
Pay off, 37
Peer to peer, 198
Percolation, 86, 166
    Transition, 165
Perfect information, 38
Periodic functions, 40
Periodicity, 188
Persistent state, 26, 63

Petri net, 87
Philosophy of science, 17
Planck distributions, 37
Point of failure, 203
Policy, 13, 16, 18, 71, 154, 174
    Alternatives, 156
    Autonomy, 21
    Backup, 30
    Centralized, 20
    Coalition, 22
    Configuration and, 7
    Convergent, 25
    Defined formally, 23
    Definition, 5
    Definition rigorous, 24
    Expressing in variables, 94
    Fixed point, 154
    Heuristic definition, 4
    High level, 193
    Low level, 193
    Scheduling, 178
    Security, 6
    Time limit, 47
Policy current, 16
Power law behaviour, 38
Power law distribution, 31
Primary key, 3
Principal eigenvector, 169
Prisoner's dilemma, 46
Probability
    Joint, 10, 119
    Tree, 28
Probability distribution, 14
Probability distributions, 32
Process, 46
    Renewal, 29
Protocol, 52

Quality of service, 106, 176
Queue
    M/M/k, 186
    M/M/1, 179
    Multiple servers, 186
Queue notation, 178
Queues, 177
Quotas, 163

Random graph percolation, 86
Random process, 29
Random variables, 29
Rate of change, 102
Rational decisions, 34
Receiver, 110
Redundancy, 4, 11, 195
    Folk theorem, 6
Regions, 170
Regions of a system, 168
Relative entropy, 122
Reliability, 14, 17
Reliable system, 58
Renewal process, 29
Representations, 109
Resource management, 9
Resources, 159
    Access to, 165
    Allocation, 163, 168
    Consumable, 162
    Currency, 161
    Flow, 193
    Organization of human,
        202
    Reclamation, 163
    Representation, 160
    Reusable, 162
    Using centrality, 169
    Where to attach, 163
Response time law, 185
Responsibility, 55
Root cause analysis, 25

Saddle point, 43, 102
Scale free network, 142
Scale invariant, 145
Scaling behaviour, 83
Scatter, 35
Scheduling, 173
Scientific approach, 13
Search algorithm, 81
Search problem
    Immunology, 40
Security, 5
Security policy, 6
Self-similarity, 83, 145

Semi-group, 10
    Stochastic, 10
Series utilization, 182
Service, 29
    Client model, 200
    Peer model, 200
Services
    Flow, 12
Sets, 59
Shakespeare, 1
Shannon, 1
Shannon channel capacity theorem, 28
Shannon's theorem, 12
Shannon–Nyquist theorem, 15
Signal
    Analogue, 111
    Digital, 111
Simple Network Management Protocol,
    115
Single point of failure, 195
SNMP, 115, 179
Source, 110
Spam mail, 45
Spanning tree, 175
Spectrum of frequencies, 40
Stability, 5, 98, 135
    Defined, 136
    External, 139
    Graph, 139
    Internal, 139
    Kernel, 140
    Multi-, 155
    Scaling, 145
    Statistical, 22, 136, 143
Standard deviation, 31, 32
Standard error of the mean, 36, 37
Star models, 17
State
    Ideal, 71
    Locally averaged, 26
    Persistent, 26
    Policy defined, 71
    Steady, 129
State machine, 66
States, 62
Statistical equilibrium, 79

Steady state, 129
Stirling's approximation, 126
Stochastic network, 87
Stochastic variables, 29
Strategic form of a game, 40
Strategic instruction, 4
Strategy, 34
    Domination, 44
    Mixed, 38
    Pure, 38
    Sub-optimal, 37
Structural importance, 9
Structure function, 3
Sub-optimal strategies, 37
Symmetry, 23, 51
Symmetry breaking, 155
Syntax in machine behaviour, 69
System
    Backup, 29
    Defined, 2, 45
    Dynamical, 47
    Open, 21
    Regions, 168
    Static, 46
System administration
    Defined, 1
System verification, 205

Task
    Defined, 19
Task management, 173
Technology's value, 20
Theoretical studies, 6
Time management
    Human, 189
Timescales, 21, 104
Time series, 26, 97
Top-down, 196
Traffic analysis, 38
Transient, 63
Transition diagram, 85
Transition function, 138
Transition matrix, 63, 138
Transitions, 97
Translation invariance, 65
Trend, 136

Truth tables, 76
Turing machine, 67
Turning point, 102
Two-dimensional time, 47

UML, 89
Uncertainty, 38, 107
    Communication, 119
    Conditional, 121
    Continuum approximation, 105
    Statistical, 114
Uncertainty principle, 31

Unified Modelling Language, 89
Unreliable system, 58
Utility, 37
Utilization law, 182

Value, 54
Value of a game, 39
Variables, 91
Verification of a system, 205

World Wide Web, 81
Worst-case scenario, 37

Printed and bound by CPI Group (UK) Ltd, Croydon, CR0 4YY

27/10/2024

14580294-0002